Illicium, Pimpinella and Foeniculum

Medicinal and Aromatic Plants — Industrial Profiles

Individual volumes in this series provide both industry and academia with in-depth coverage of one major genus of industrial importance.

Series Edited by Dr Roland Hardman

Volume 1
Valerian, edited by Peter J. Houghton
Volume 2
Perilla, edited by He-ci Yu, Kenichi Kosuna and Megumi Haga
Volume 3
Poppy, edited by Jenö Bernáth
Volume 4
Cannabis, edited by David T. Brown
Volume 5
Neem, edited by H.S. Puri
Volume 6
Ergot, edited by Vladimír Křen and Ladislav Cvak
Volume 7
Caraway, edited by Éva Németh
Volume 8
Saffron, edited by Moshe Negbi
Volume 9
Tea Tree, edited by Ian Southwell and Robert Lowe
Volume 10
Basil, edited by Raimo Hiltunen and Yvonne Holm
Volume 11
Fenugreek, edited by Georgios Petropoulos
Volume 12
Ginkgo biloba, edited by Teris A. Van Beek
Volume 13
Black Pepper, edited by P.N. Ravindran
Volume 14
Sage, edited by Spiridon E. Kintzios
Volume 15
Ginseng, edited by W.E. Court
Volume 16
Mistletoe, edited by Arndt Büssing
Volume 17
Tea, edited by Yong-su Zhen
Volume 18
Artemisia, edited by Colin W. Wright
Volume 19
Stevia, edited by A. Douglas Kinghorn
Volume 20
Vetiveria, edited by Massimo Maffei
Volume 21
Narcissus and Daffodil, edited by Gordon R. Hanks
Volume 22
Eucalyptus, edited by John J.W. Coppen
Volume 23
Pueraria, edited by Wing Ming Keung
Volume 24
Thyme, edited by E. Stahl-Biskup and F. Sáez
Volume 25
Oregano, edited by Spiridon E. Kintzios
Volume 26
Citrus, edited by Giovanni Dugo and Angelo Di Giacomo
Volume 27
Geranium and Pelargonium, edited by Maria Lis-Balchin
Volume 28
Magnolia, edited by Satyajit D. Sarker and Yuji Maruyama
Volume 29
Lavender, edited by Maria Lis-Balchin
Volume 30
Cardamom, edited by P.N. Ravindran and K.J. Madhusoodanan
Volume 31
Hypericum, edited by Edzard Ernst
Volume 32
Taxus, edited by H. Itokawa and K.H. Lee
Volume 33
Capsicum, edited by Amit Krish De
Volume 34
Flax, edited by Alister Muir and Niel Westcott
Volume 35
Urtica, edited by Gulsel Kavalali
Volume 36
Cinnamon and Cassia, edited by P.N. Ravindran, K. Nirmal Babu and M. Shylaja
Volume 37
Kava, edited by Yadhu N. Singh
Volume 38
Aloes, edited by Tom Reynolds
Volume 39
Echinacea, edited by Sandra Carol Miller
Assistant Editor: He-ci Yu
Volume 40
Illicium, Pimpinella and Foeniculum, edited by Manuel Miró Jodral

Illicium, Pimpinella and Foeniculum

Edited by Manuel Miró Jodral

Medicinal and Aromatic Plants — Industrial Profiles

CRC PRESS

Boca Raton London New York Washington, D.C.

Library of Congress Cataloging-in-Publication Data

Catalog record is available from the Libray of Congress

Direct all inquiries to CRC Press LLC, 2000 N.W. Corporate Blvd., Boca Raton, Florida 33431.

Visit the CRC Press Web site at www.crcpress.com

International Standard Book Number 0-415-32246-4
Printed in the United Kingdom by Biddles/IBT Global 1 2 3 4 5 6 7 8 9 0
Printed on acid-free paper

Contents

Preface to the Series

There is increasing interest in industry, academia, and the health sciences in medicinal and aromatic plants. In passing from plant production to the eventual product used by the public, many sciences are involved. This series brings together information which is currently scattered through an ever-increasing number of journals. Each volume gives an in-depth look at one plant genus, about which an area specialist has assembled information ranging from the production of the plant to market trends and quality control.

Many industries are involved, such as forestry, agriculture, chemicals, food, flavor, beverage, pharmaceutical, cosmetics, and fragrance. The plant raw materials are roots, rhizomes, bulbs, leaves, stems, barks, wood, flowers, fruits, and seeds. These yield gums, resins, essential (volatile) oils, fixed oils, waxes, juices, extracts, and spices for medicinal and aromatic purposes. All these commodities are traded worldwide. A dealer's market report for an item may say "drought in the country of origin has forced up prices."

Natural products do not mean safe products, and account of this has to be taken by the above industries, which are subject to regulation. For example, a number of plants which are approved for use in medicine must not be used in cosmetic products.

The assessment of "safe to use" starts with the harvested plant material, which has to comply with an official monograph. This may require absence of, or prescribed limits of, radioactive material, heavy metals, aflatoxin, pesticide residue, as well as the required level of active principle. This analytical control is costly and tends to exclude small batches of plant material. Large-scale, contracted, mechanized cultivation with designated seed or plantlets is now preferable.

Today, plant selection is not only for the yield of active principle, but for the plant's ability to overcome disease, climatic stress and the hazards caused by mankind. Such methods as *in vitro* fertilization, meristem cultures and somatic embryogenesis are used. The transfer of sections of DNA is giving rise to controversy in the case of some end uses of the plant material.

Some suppliers of plant raw material are now able to certify that they are supplying organically farmed medicinal plants, herbs and spices. The Economic Union directive CVO/EU No. 2092/91 details the specifications for the *obligatory* quality controls to be carried out at all stages of production and processing of organic products.

Fascinating plant folklore and ethnopharmacology leads to medicinal potential. Examples are the muscle relaxants based on the arrow poison, curare, from species of *Chondrodendron*, and the antimalarials derived from species of *Cinchona* and *Artemisia*. The methods of detection of pharmacological activity have become increasingly reliable and specific, frequently involving enzymes in bioassays and avoiding the use of laboratory animals. By using bioassay-linked fractionation of crude plant juices or extracts, compounds can be specifically targeted which, for example, inhibit blood platelet aggregation, or have antitumor, or antiviral, or any other required activity. With the assistance of robotic devices, all the members of a genus may be readily screened. However, the plant material must be fully authenticated by a specialist.

The medicinal traditions of ancient civilizations such as those of China and India have a large armamentarium of plants in their pharmacopoeias which are used throughout Southeast Asia. A similar situation exists in Africa and South America. Thus, a very high percentage of the world's population relies on medicinal and aromatic plants for their medicine. Western medicine is also responding. Already in Germany all medical practitioners have to pass an examination in phytotherapy before being allowed to practice. It is noticeable that medical, pharmacy, and health-related schools throughout Europe and the United States are increasingly offering training in phytotherapy.

Multinational pharmaceutical companies have become less enamored of the single compound, magic-bullet cure. The high costs of such ventures and the endless competition from "me-too" compounds from rival companies often discourage the attempt. Independent phytomedicine companies have been very strong in Germany. However, by the end of 1995, 11 (almost all) had been acquired by the multinational pharmaceutical firms, acknowledging the lay public's growing demand for phytomedicines in the Western world.

The business of dietary supplements in the Western world has expanded from the health store to the pharmacy. Alternative medicine includes plant-based products. Appropriate measures to ensure their quality, safety, and efficacy either already exist or are being answered by greater legislative control by such bodies as the U.S. Food and Drug Administration and the recently created European Agency for the Evaluation of Medicinal Products based in London.

In the United States, the Dietary Supplement and Health Education Act of 1994 recognized the class of phytotherapeutic agents derived from medicinal and aromatic plants. Furthermore, under public pressure, the U.S. Congress set up an Office of Alternative Medicine, which in 1994 assisted the filing of several Investigational New Drug (IND) applications, required for clinical trials of some Chinese herbal preparations. The significance of these applications was that each Chinese preparation involved several plants and yet was handled as a *single* IND. A demonstration of the contribution to efficacy of *each* ingredient of *each* plant was not required. This was a major step toward more sensible regulations in regard to phytomedicines.

My thanks are due to the staff of CRC Press who have made this series possible and especially to the volume editors and their chapter contributors for the authoritative information.

Dr. Roland Hardman

Foreword

At the beginning of the third millennium, phytotherapy is experiencing one of its most exciting moments, not only due to the increase in the consumption of medicinal plants, but also to the important phytochemical, pharmacological, and clinical investigations that are being conducted to develop products with sufficient guarantees of quality, efficacy, and security. Species with essences rich in anethole, such as star anise (*Illicium verum*), aniseed (*Pimpinella anisum*), and fennel (*Foeniculum vulgare*), are commercially important as a result of their uses in medicine and diet. These species are studied in depth in the book. The simultaneous study of anise fruits, commercially known as "seeds," is habitual in the pharmacopoeias and phytotherapy texts because their therapeutic applications (bronchial expectorant, gastrointestinal spasmolytic, carminative, etc.) are fundamentally caused by anethole, although other phytochemical components are different. Other species of economic interest belonging to the genera cited are also discussed in the text.

Chinese star anise is widely used as a spice for flavoring food and beverages. Its adulteration with the fruits of a species of similar morphology, Japanese star anise (*I. anisatum*), of which the sesquiterpenlactone anisatin is the main cause of toxicity, is relatively frequent. In recent years, a sanitary warning has been issued in Spain as a result of this adulteration, causing a fall in the market for *I. verum* until all imported versions were analyzed. The phytochemistry of *I. verum,* as well as that of other species of *Illicium*, is also studied in the text, which provides the book with an additional discussion of current affairs.

The text includes a revision of the botanical characterization of the species, including the methods of cultivation, harvesting, and drying/storage; chemical constituents; therapeutic and pharmacological properties; and the monographs, formulations, economic importance, and market trends, which will be useful for university professors and students, sanitation professionals, investigators, and persons working with aromatic and medicinal plants.

I wish to finish this brief prologue with words of gratitude. First, I would like to thank Dr. Roland Hardman, who proposed the coordination of this book to me, for his constant aid in the updating of the content. In addition, I want to express my gratitude to all the contributors, whose dedication and infinite patience have allowed the book to finally see the light after multiple vicissitudes. Finally, of course, I would like to thank the editorial staff of CRC Press, which continues to publish the texts of the *Medicinal and Aromatic Plants — Industrial Profiles* series.

Manuel Miró Jodral
Granada, Spain

Contributors

Ulrich Bomme
Bavarian State Research Center
for Agriculture
Freising, Germany

María Dolores Contreras
University of Granada
Granada, Spain

Anna De Pasquale
University of Messina
Messina, Italy

Amit Krishna De
Indian Science Congress Association
Calcutta, India

Anthony C. Dweck
Dweck Data
Wiltshire, United Kingdom

Yoshiyasu Fukuyama
Tokushima Bunri University
Tokushima, Japan

Enza Maria Galati
University of Messina
Messina, Italy

María Reyes González-Tejero
University of Granada
Granada, Spain

Jian-mei Huang
Tokushima Bunri University
Tokushima, Japan

Sheng-lin Li
Beijing University of Chinese Medicine and
Pharmacology
Beijing, China

Emi Okuyama
Josai International University
Tōgane, Japan

Juan Carlos Orte
University of Granada
Granada, Spain

Antonio Rapisarda
University of Messina
Messina, Italy

Jürgen Reichling
University of Heidelberg
Heidelberg, Germany

Richard M.K. Saunders
The University of Hong Kong
Hong Kong, China

Chun-shu Yang
Beijing University of Chinese Medicine and
Pharmacology
Beijing, China

1 General Introduction to *Pimpinella* and *Illicium*

Anthony C. Dweck

CONTENTS

0-415-32246-4/04/$0.00+$1.50

1.1 ANISEED SPECIES

The *Pimpinella* and *Illicium* species are two genera that give the material commonly known as anise and star anise. Burnet saxifrage is also a well-known species of the *Pimpinella* genus, though it is neither burnet nor saxifrage. The genus *Pimpinella* (family: Umbelliferae) includes

- *P. alba*
- *P. alpinum*
- *P. angelicaefolium* Lam. [= *Aegopodium podagraria* L.], goatweed
- *P. anisum* L. [Syn. *Anisum vulgare* Gaertn., *A. officinarum* Moench., *P. aromaticum* Bieb.], anise, aniseed, sweet cumin, anis, anason, anasur, anise seed, anison, anisu, anisum, anisoon, kleiner anis, susser kummel, anis vert, fruit d'anis, maha-duru, anisi fructus, sweet fennel, anysun, saunf, yan kok, and jintan manis (see Figure 1.1)
- *P. anisum* Willd.
- *P. aromatica* Bieb.
- *P. caffra*
- *P. diversifolia*
- *P. domingensis* Willd. [= *Ciclospermum leptophylum* var. *leptophylum*]
- *P. involucrata*
- *P. leptophylla* Pers. [= *Ciclospermum leptophylum* var. *leptophylum*]
- *P. maggiore* [= *Sanguisorba officinalis*]
- *P. magna*, Greater burnet
- *P. major* L. (Hudson), [Syn. *P. magna*?], great burnet, greater burnet saxifrage
- *P. peregrina* L.
- *P. saxifraga*, small burnet, saxifrage, burnet saxifrage, lesser burnet, petit boucage, boucage saxifrage, kleine bibernelle, small pimpernel; and in Shropshire, England, known as bennet and old man's plaything (Grigson, 1996); the old medieval name was dipinella

FIGURE 1.1 *P. anisum*

P. schweinfurthii
P. thellugiana Wolff
P. villosa
P. wallichiana

The genus *Illicium* (family Magnoliaceae) includes

I. anisatum Lour.
I. difengpi K.I.B. et K.I.M.
I. dunnianum Tutch.
I. floridanum
I. griffithii
I. henryi Diels.
I. lanceolatum A.C. Smith [= *I. religiosum* Sieb. et Zuce., *I. anisatum*]
I. majus
I. modestum
I. parviflorum
I. religiosum Sieb. et Zuce [Syn. *I. anisatum*], Japanese anise, Japanese star anise, bastard anise, shikimi
I. san-ki Perr.
I. simonsii Maxim.
I. tashiroi
I. verum Hook. [Syn. *I. anisatum*, *Anisi stellatum*], anice stellato, aniseed stars, badiana, stermanis, anise étoile, Chinese star anise, Chinese anise, eight-horned anise, eight horns, star anise, pak kok, bunga lawang, poy kak bua

1.2 PLANTS BEARING SIMILAR NAMES

Osmorbiza longistylis is known as anise root, and *O. occidentalis* is known as sweet anise or western anise root (Coffey, 1993). *Myrrhis odorata* (L.) Scop. or sweet cicely is also known locally by the name anise in Durham (England), and wild anise or annaseed in Cumbria (England) (Grigson, 1996). *Carum kelloggii* Gray is known as wild anise (Saunders, 1976).

1.3 *P. ANISUM* L.

1.3.1 Biblical References

> Woe unto you, scribes and Pharisees, hypocrites!, for ye pay the tithe of mint and anise and cumin, and have omitted the weightier matters of the law, judgment, mercy, and faith: these ought ye to have done, and not to leave the other undone.
>
> **—Matthew 23:23**

1.3.2 History

Anise was cultivated and well appreciated by the ancients. The Arabs called it by the name *anysun.* The Egyptians called it *ìnst,* and the hieroglyphic name can be traced to Pharaonic texts as a component of refreshing drinks for stomach ailments, bladder problems, and other gastric illness (Manniche, 1989). They certainly appreciated its aromatic qualities. About the same time, the Chinese were also using anise as a carminative and expectorant (*Lawrence Review of Natural Products,* 1991).

Anison was known to the ancient Greeks, and the Greek physician Pedanius Dioscorides mentioned the use of aniseed in medicine and that he preferred aniseed from Crete, with the Egyptian material as his second preference. Pythagorus, the Greek geometrician and mathematician, declared *anisonas* bread a great delicacy in 550 B.C., and praised it cooked or raw (Keville, 1991; Sturtevant, 1972). Theophrastus (372–278 B.C.) mentioned aniseed, as did Pliny (23–77 A.D.), who said "*anesum,* green or dry, is desirable in all seasonings or sauces" (Sturtevant, 1972).

Anise was also used in the ancient and Ayurvedic (Indian) medicinal system (Heinerman, 1988). The Indian name for anise is the same as that for dill, and it is considered similar in properties and uses (Attygale, 1994). In Sri Lanka, the seed has traditionally been used as a carminative and aromatic (Jayaweera, 1982).

Anise flavored a spicy cake, called *mustaceus* by the Romans, which was made with an unfermented wine and included powdered aniseed mixed with honey (Back, 1987; Gordon, 1980); it also contained cumin and other digestive herbs (Bremness, 1991). This cake, considered a great delicacy, was served at banquets and weddings to aid digestion and, possibly more important, was considered to have aphrodisiac properties (Gordon, 1980). It is probably from this rich cake that the traditional European wedding cake was derived.

The Romans also used anise medicinally (Back, 1987).

> Anise warms the abdomen, dispels gas (especially after eating beans) and is helpful for belching, vomiting, chronic diarrhoea, abdominal pains, sluggish digestion and hernia. The tribal people of the Amazon find it especially good for children with stomach-aches. Anise is used as a sedative, especially for nervousness and to induce sleep. It is thought to prevent fainting. In the Amazon, this herb is thought to function as a female tonic during the menses by eliminating sad thoughts.*

* Raintree Group, Inc., Austin, TX, 1997.

Palladius (at the beginning of the third century) gave directions for its sowing (Sturtevant, 1972), and Charlemagne (in the ninth century) instructed that anise should be grown on the imperial farms in Germany (Sturtevant, 1972), as were all of the herbs and spices that he found growing in St. Gall's Monastery (Bremness, 1991).

In the thirteenth century, Albertus Magnus referred to the plant as *Roman fennel.*

In 1305, Edward I granted a patent giving the right to levy a toll on it at the Bridge of London for the purpose of raising money to repair the bridge (Wood and Bache, 1883).

In 1480, King Edward IV had his personal linen scented with "lytil bagges of fustian stuffed with ireos and anneys."

In 1536, Ruellius recorded the use of anise in France and gave it the common name *Roman fennel.*

In 1542, Boore, in his *Dyetary of Helth,* said of aniseed and fennel, "These herbes be seldom used, but theyr seedes be greatly occupyde." Before this date, the plant seems to have been grown as a pot herb in England (Sturtevant, 1972).

In 1551, William Turner (1520–1568), in *A New Herball,* used anise to "maketh the breth sweter." It was taken in the form of comfits (seeds coated with sugar) (Norman, 1991).

In the seventeenth century, Quintyne wrote about the use of the leaves in salads. John Josselyn, traveling in New England, made a list of the plants taken by the settlers and noted whether they had prospered or failed; of coriander, dill, and anise, he wrote: "they thrive exceedingly." It was taken to North America by the Quakers as a medicinal herb crop, and in 1806, McMahon mentioned anise as a culinary herb (McMahon, 1806).

In Victorian times, there existed a sweet known as a *dragati,* an aniseed-flavored "ball" with a hard sugar coating, which became the much-loved *dragée* in France. It is interesting to note that this term has been adopted by the pharmacy as any sweet sugar-coated pill.

1.3.3 Traditional Uses

1.3.3.1 Flavoring

Mouthwashes and toothpastes or dentifrices were common uses for anise (Harry, 1963). It was also used to mask the flavor of unpleasant-tasting medicines, as a flavor for some teas (Launert, 1989), and in the preparation of various liqueurs. One of the oldest traditional uses is as a flavoring in sweets, the best illustration of this use being aniseed balls.

It is a flavoring in pernod (Graves, 1990), anisette (Leung and Foster, 1996; Wichtl, 1994), ouzo (Greek aniseed spirit), and pastis, and it is an ingredient of Benedictine, Boonekamp, Danziger Goldwasser, etc. (Wichtl, 1994).

1.3.3.2 Perfumery

Anise is the component of various perfumes (Harry, 1963), and it is also used as a component of potpourri (Back, 1987), in which the crushed seeds can be used for their fragrance (Bremness, 1991) or simply for their appearance.

The fragrance of anise is described as penetrating, the taste warm, aromatic, and sweetish. It imparts its virtues wholly to alcohol, but only partially to water (Phelps-Brown, 1993) and sparingly to boiling water (Wood and Bache, 1883). Because of the traditional use of anise oils with licorice in licorice sweets, the flavor of anise is often confused with that of licorice (Leung, 1980).

1.3.3.3 Cosmetic

The tea will reduce skin oiliness (Heinerman, 1988), and the seeds can be ground and added to a face pack (Bremness, 1991).

1.3.3.4 Medicinal

The tea is used for children's flatulence, upper respiratory tract problems, and bronchial asthmatic attacks (Buchman, 1987). The tisane is also used as an expectorant (British Herbal Manufacturing Association, 1996), as a cough suppressant (Fluck, 1988), and for pertussis (Newall et al., 1996), and it is good for sore throats and bronchial infections (Ody, 1996). It is a pectoral (relieves infections of the chest and lungs) and is used, not only in cough medicines, but also in lozenges (Potter, 1985). It has been cited for use in whooping cough (Hoffman, 1991) because of its antispasmodic action.

Sweetened with a little honey, anise is a soothing carminative for babies (for colic) and is useful for a hacking cough (Ceres, 1984). It is also used as a sedative for children (Launert, 1989) and may help reduce nausea (Phelps-Brown, 1993). It is an aid to digestion, nervous indigestion in particular. The dried seeds may also be chewed as a digestive aid (Gordon, 1980).

Anise is a strong galactogogue, and so helps nursing mothers to produce more milk (Schauenberg and Paris, 1990). In addition, it has been traditionally used to facilitate childbirth (Leung and Foster, 1996; Ody, 1996).

Anise oil is useful in destroying body lice (Spoerke, 1990), head lice, and itching insects (Buchman, 1987), and the oil can be used by itself (Hoffman, 1991), which makes it useful for pediculosis, the skin condition caused by lice (Newall et al., 1996).

It can be used for scabies (Ody, 1996), where it may be applied externally in an ointment base (Hoffman, 1991). It can also be used in oil or in an ointment base as a stimulating liniment and against vermin (Wichtl, 1994). It is mildly laxative (Fluck, 1988), but it is probably more often used to counter the griping pains that can occur with constipation (Hoffman, 1991).

The following therapeutic benefits have been noted:

- Promotes menstruation (Leung and Foster, 1996) or emmenagogue (stimulates the menstrual flow) (Wichtl, 1994)
- Has a mild tonic effect on the liver (Ody, 1996)
- Helps insomnia when taken as a few seeds in a cup of hot milk at bedtime (Gordon, 1980)
- Acts as an antispasmodic and antiseptic (Wichtl, 1994)
- Acts as a diuretic and diaphoretic (Spoerke, 1990)

1.3.3.5 Aromatherapy

Anise is used in aromatherapy to help ease difficulty in breathing (Price, 1987). The oil is also thought to be an aphrodisiac (Wichtl, 1994), though the action is unclear from any source. In some references it is said to be specifically a female aphrodisiac (Ody, 1996), whereas in others it is said to increase libido and alleviate symptoms of male climacteric (Leung and Foster, 1996).

It has been described as very mildly narcotic, similar to fennel (*Foeniculum vulgare*) (Launert, 1989), and is reported as capable of causing delirium (in large doses) (Graves, 1990).

1.3.3.6 Food

The leaves of anise can be used in salads. The seeds are used in Italy to flavor diverse pastries; in Germany, they are put in breads; and in England, they are used in special breads, in rye bread, and even in cheese (Sturtevant, 1972).

Anise, star anise (to a lesser extent), anise oil, and star anise oil are widely used as flavoring ingredients in all major categories of foods, including frozen dairy desserts, sweets (e.g., licorice confections), baked goods, gelatins, and puddings, as well as in meat and meat products. The highest average maximum use levels for anise oil are about 0.06% (570 ppm) in alcoholic beverages and 0.07% (681 ppm) in sweets (Leung and Foster, 1996). Anise and star anise are widely used as domestic spices; the former is mainly used by Westerners, whereas the latter is used primarily by Asians, especially in Chinese foods.

1.3.3.7 Animals

> The powdered seed is largely employed in condition and other condiments for horses.
>
> **—Potter, 1985**

The scent of aniseed is fascinating to dogs and is often used to decoy them away from a scent (Graves, 1990), and the seed can be used as bait in mouse traps (Bremness, 1991).

1.3.3.8 Miscellaneous

Anise oil contains antifungal substances (Shukla and Tripathi, 1987), and one of the components, anisic acid, is sold under the name SL-688 as a specific treatment for molds and is used for that purpose at a concentration of 0.25% (Straetman, 1993). The oil may also demonstrate parasiticide activity (Hoffman, 1991), and the tea is reputed to improve memory (Heinerman, 1988).

1.3.4 Preparations

Following is a list of different ways of taking anise:

- Dried fruits: Dose is 0.5 to 1 g or by infusion.
- Oil: Dose is 0.05 to 0.2 mL.
- Spirit BPC (British Pharmaceutical Codex) (1949): Dose is 0.3 to 1 mL.
- Distilled anise water BPC (1934): Dose is 15 to 30 mL (British Herbal Manufacturing Association, 1983).

To take anise via infusion, the seeds should be gently crushed just before use to release the volatile oils. Pour 1 cup boiling water over 1 to 2 teaspoons of the seeds and let stand covered for 5 to 10 minutes; take 1 cup three times daily. To treat flatulence, the tea should be drunk slowly before meals (Hoffman, 1991).

One drop of the oil may be taken internally by mixing it into half a teaspoonful of honey (Hoffman, 1991). Another method of taking anise is to bring 2 pints of water to the boil, add 7 teaspoons aniseed, reduce heat to a lower setting, and simmer contents down to 1.5 pints. While still warm, strain and add 4 teaspoons honey and glycerin. Take 2 teaspoons of this syrup every few hours to relieve hacking coughs, or three times daily to strengthen the memory. If using as a tea, omit the honey and drink 2 cups once or twice daily for skin problems, milk needs, or to relieve stomach problems (Heinerman, 1988).

To make tea, cover 1 to 5 grams of the seeds (pounded or coarsely powdered immediately before use) with boiling water and allow to draw in a closed vessel for 10 to 15 minutes (1 teaspoon = ca. 3.5 g) (Wichtl, 1994).

Take 1 to 2 mL tincture three times a day or dilute 10 drops of essential oil in 25 mL carrier oil as a chest rub (Ody, 1993).

1.3.5 Combinations with Anise

Anise combines well with *Mentha piperita* in flatulent colic, with *marrubium, tussilago, symplocarpus,* and *lobelia* in bronchitis, and with *prunus* in tracheitis. The oil (1%) may be combined with oil of sassafras (1%) in an ointment base for scabies (British Herbal Manufacturing Association, 1983).

For flatulent colic and indigestion, mix aniseed with equal amounts of fennel and caraway. For bronchitis, it combines well with coltsfoot, white horehound, and lobelia (Hoffman, 1991). Combine it with 1 to 2 mL wild lettuce for irritant coughs, or with 2 to 3 mL thyme or hyssop tincture for infections. Add 10 drops eucalyptus oil to form a chest rub (Ody, 1993).

1.3.6 Pharmacopoeial Monographs

Pharmacopoeial monographs mentioning anise include the *British Herbal Pharmacopoeia*, 1983 (British Herbal Manufacturing Association, 1993), the *British Pharmacopoeia*, Vol. 1 (1980), and *Martindale, The Complete Drug Reference,* 30th and 27th eds. Other pharmacopoeias include those from Austria, Brazil, Czechoslovakia, Egypt, France, Germany, Greece, Hungary, Italy, The Netherlands, Romania, Russia, and Switzerland.

Both crude and essential oils are listed in the *National Formulary* (1985) and the *Food Chemicals Codex* (1972). Preparations containing 5 to 10% essential oil are used as respiratory inhalants in Germany (German Commission E Monograph [B Anz. no. 122, dated 6 June, 1988]).

1.3.7 Legal Recognition

The regulatory status for anise is generally recognized as safe (#182.10 and #182.20). Anise seed and star anise seed are subjects of German official monographs; 3.0 g of seed or 0.3 g essential oil (mean daily dose) are allowed as a bronchial expectorant for upper respiratory tract congestion and as a gastrointestinal spasmolytic (Leung and Foster, 1996).

Aniseed is used extensively as a spice and is listed by the Council of Europe as a natural source of food flavoring (category N2). This category allows small quantities of aniseed to be added to foodstuffs, with a possible limitation of an active principal (as yet unspecified) in the final product (Newall et al., 1996)

1.3.8 Legal Category (Licensed Products)

The legal category of anise oil is General Sales List, Schedule 1, Table A.

1.3.9 Remedies and Supplements

The use of anise is widespread in modern medicines, as shown by this list of preparations (Ody, 1996):

BioCare Artemisia Complex
Bioforce Bronchosan
Bumbles Propolis, Aniseed, and Liquorice Lozenges
Culpeper Cough Relief Mixture
Culpeper Herbal Mixture for Acidity, Indigestion, and Flatulence
Frank Roberts Acidosis Tablets
Hactos Cough Mixture
Herbcraft Peppermint Formula
Lane's Cut-a-Cough
Lane's Honey and Molasses Cough Mixture
Neal's Yard Echinacea and Mallow Linctus
Neal's Yard Horehound and Aniseed Linctus
Neal's Yard Horehound and Honey Lozenges
Potter's Acidosis
Potter's Lightning Cough Remedy
Potter's Malted Kelp Tablets

Potter's Vegetable Cough Remover
Revitonil
Weleda Carminative Tea
Weleda Clairo Tea
Weleda Cough Elixir
Weleda Fragador Tablets
Weleda Herb and Honey Cough Elixir
Weleda Laxadoron Tablets

In addition, a celebrated empirical preparation sold under the name *Dalby's Carminative* is supposed to owe its virtues to these oils, united with water by means of magnesia (Graves, 1990).

1.3.10 Simples

Neal's Yard remedies use anise oil.

1.3.11 Safety

No significant toxic activity has been ascribed to anise oil, which has generally been recognized as safe and is approved for food use (*Lawrence Review of Natural Products*, 1991). When applied to human skin in 2% concentrations in petrolatum base, anise oil produced no topical reactions, and the oil has not been considered a primary irritant. However, anethole has been associated with sensitization and skin irritation and may cause erythema, scaling, and vesiculation. Contact dermatitis reactions to aniseed and aniseed oil have been attributed to anethole (Newall et al., 1996).

Bergapten is known to cause photosensitivity reactions, and concern has been expressed over its possible carcinogenic risk (Newall et al., 1996). Reactions have been reported with products such as creams and toothpastes flavored with aniseed oil, and the oil has been reported to cause contact sensitivity, cheilitis, and stomatitis.

Ingestion of the oil may result in pulmonary edema, vomiting, and seizures in doses as small as 1 to 5 mL (*Lawrence Review of Natural Products*, 1991). Although anise itself is low in toxicity, the oil distilled from the herb may cause some skin irritation, nausea, vomiting, and seizures (Spoerke, 1990). According to Tisserand (1988), the oil is safe to use, whereas sensitization is doubtful or borderline. High doses can produce intoxication (Fluck, 1988).

1.3.11.1 Contraindications

Anise is contraindicated where there is an allergy to aniseed or anethole, and the use of aniseed oil should be avoided in dermatitis or any inflammatory or allergic skin condition. Side effects include occasional allergic reactions of the skin, respiratory tract, and gastrointestinal tract. There are no known interactions with other remedies. Anise requires the warning that external use of aniseed preparations must be restricted to the inhalation of the essential oil (German Commission E, Monograph, 1988).

> Bergapten may cause photosensitivity in sensitive individuals. Excessive doses may interfere with anticoagulant and MAOI therapy. The documented estrogenic activity of anethole and its dimers may affect existing hormone therapy, including the oral contraceptive pill and hormone replacement therapy, if excessive doses are ingested. In view of the structural similarity reported between anethole and myristicin, consumption of large amounts of aniseed may cause neurological effects similar to those documented for nutmeg.
>
> **—Newall et al., 1996**

1.3.11.2 Toxicology

The acute oral LD/50 (median lethal dose) in rats is 2.25g/kg. No percutaneous absorption of the oil occurred through mouse skin within 2 hours. The oral LD/50 of anethole in rats is 2090 mg/kg (*Lawrence Review of Natural Products*, 1991) and is thus rated as a moderate acute toxin. The *cis* isomer of anethole is from 15 to 38 times more toxic to animals than the *trans* isomer (*Lawrence Review of Natural Products*, 1991).

Mild liver lesions were observed in rats fed repeated anethole doses (695 mg/kg) for an unspecified duration. Hepatic changes have been described in rats fed anethole in their daily diet (1%) for 15 weeks, although at a level of 0.25% there were no changes after 1 year. In therapeutic doses, anethole is reported to cause minimal hepatotoxicity (Newall et al., 1996).

1.3.11.3 Pregnancy and Lactation

The safety of aniseed taken during pregnancy and lactation has not been established; however, there are no known problems, provided that doses taken do not greatly exceed the amounts used in foods (Newall et al., 1996).

1.3.11.4 Adulterants

A case of poisoning is on record from the accidental admixture of the fruits of *Conium maculatum*, which bear some resemblance to those of anise but may be distinguished by their crenate or notched ridges and the absence of oil tubes. Moreover, they are broader in proportion to their length and are generally separated into half fruits, whereas those of anise are whole (Wood and Bache, 1883).

Occasionally (in former times more often and now very rarely), the highly toxic coniine-containing fruit of *C. maculatum* L., hemlock, is encountered in individual lots of aniseed. At present, almost all lots of aniseed are also contaminated with up to 1% coriander fruit (Wichtl, 1994).

1.3.12 Deodorizing Properties

> Dr. Ruschenberger, U.S.N., has shown that oil of anise has a remarkable power of deodorising sulphide of potassium; a drop of the oil having entirely deprived of offensive odor a drachm of lard with which five grains of the sulphide had been incorporated.
>
> **—Wood and Bache, 1883**

1.4 *P. SAXIFRAGA* AND *P. MAJOR*

1.4.1 Origin of the Name

The specific name of this plant derives from an observation that its root can penetrate into the cracks in stones, which led to the belief that it was able to break down carbonates (Latin: *saxafrago*, "I break stones"). As a result, it has been used in the past as a lithontriptic (Chiej, 1988).

P. saxifraga is known as burnet saxifrage, which is confusing as it is neither a burnet (family Rosaceae) nor a saxifrage (family Saxifragaceae). *P. major* is the greater burnet saxifrage and is so closely related that the two plants are considered together in this chapter.

1.4.2 History

The Romans employed *P. major* in the treatment of heart disease and male infertility (Lanska, 1992). William Turner (1520–1568) described it in *A New Herball* in 1551, "for the propertye that it hath in breaking of the stone in a mannis bodye." Culpeper (1616–1654) discussed the burnet saxifrage in the *Complete Herbal* of 1653 and mentioned the lesser and the greater varieties of English burnet saxifrage:

> They are both of them herbs of the moon. The saxifrages are hot as pepper; and Tragus saith, by his experience, that they are wholesome. They have the same properties the parsleys have, but in provoking urine, and causing the pains thereof, and of the wind and colic, are much more effectual, the roots or seed being used either in powder, or in decoctions, or any other way; and likewise helps the windy pains of the mother, and to precure their courses, and to break and void the stone in the kidneys, to digest cold, viscous, and tough phlegm in the stomach, and is an especial remedy against all kinds of venom. Castoreum being boiled in the distilled water thereof, is singularly good to be giveth to those that are troubled with cramps and convulsions. Some do use to make the seeds into comfits (as they do caraway seeds) which is effectual to all the purposes aforesaid. The juice of the herb dropped into the most grievous wounds of the head, dries up their moisture, and heals them quickly. Some women use the distilled water to take away freckles or spots in the skin or face; and to drink the same sweetened with sugar for all the purposes aforesaid.

During the Middle Ages, *P. major* (the greater burnet saxifrage) was considered to be effective for the treatment of plague and cholera, and it was also used as a spice in beer (Lanska, 1992).

1.4.3 Traditional Uses

1.4.3.1 Flavor

There are a number of descriptions that relate to the taste of this herb, which is described by one author as sweet and astringent (Tsarong, 1994), but by another as cucumber-like, though odorless (Potter, 1994). Yet another describes the taste of the leaves as having a slightly sweet, hot, and mildly astringent taste with a hint of cucumber flavor (Lanska, 1992). The root is very hot and acrid to one author, burning the mouth like pepper (Grieve, 1998); another says that it is only the fresh root that has this property (Culpeper, 1983). On drying or on being kept for a long time, the pungency of the root is considerably diminished (Grieve, 1998).

Small bunches of the leaves and shoots are tied together and suspended in a cask of beer to impart an agreeable aromatic flavor to the drink (Grieve, 1998); Lanska (1992) said that *P. major* was used to spice the beer. In addition, it is believed that the herbs will correct sharp-tasting or spoiled wine (Grieve, 1998). The oil is used as a commercial flavoring for liqueurs (Ody, 1996).

1.4.3.2 Medicinal

In Tibet, the entire plant is used as a tonic to promote vigor and to treat loss of clarity of senses caused by the lack of essential blood and the imbalance of the three primary processes (Tsarong, 1994).

The fresh root can be chewed for toothache and paralysis of the tongue (Grieve, 1998) and is used to prepare a tincture which is administered for sore throats, pharyngitis, laryngitis, and bronchitis (Schauenberg and Paris, 1990). Infusions or tinctures of the drug are employed as a gargle for inflammation of the mouth and throat (Wichtl, 1994). The rootstock (used fresh or dried) has antiinflammatory and mildly astringent properties (Launert, 1989).

P. major is said to dissolve mucus, and on this account is used as a gargle in hoarseness and some cases of throat infection (Grieve, 1998); Fluck (1988) said that the drug stimulates the secretion of certain mucosa, especially in the bronchi. It is used as an antispasmodic (Schauenberg and Paris, 1990), an antitussive, and a mild expectorant in bronchitis (Wichtl, 1994), and it is sometimes included in catarrh remedies (Ody, 1996). The roots are also used (in folk medicine) to treat coughs as well as inflammations (Lanska, 1992) and infections of the upper respiratory tract, such as pharyngitis, tracheitis, as well as angina (Wichtl, 1994).

The root is an emmenagogue and a cholagogue (releases bile from the gall bladder). It is used in the forms of decoction, powder, tincture, fluid extract, and syrup and is still employed in cases of hepatic insufficiency (Chiej, 1988). The tincture, when taken consistently, helps to restore the appetite (Chiej, 1988) and improve the digestion and may be used to assist in the treatment of certain digestive disorders (Lanska, 1992). In folk medicine, the drug is also sometimes taken as a stomachic (Wichtl, 1994). The oil and resin are also useful to help relieve flatulent indigestion (Grieve, 1998).

The roots have mild diuretic effects (Lanska, 1992) and may be used in cases of cystitis (Ody, 1996). The root is also prescribed in asthma and dropsy (Grieve, 1998). The fresh root is antidiarrheic (Schauenberg and Paris, 1990). In homoeopathy, the tincture is prescribed for headaches and nosebleeds (Schauenberg and Paris, 1990).

1.4.3.3 Cosmetic

Externally, *P. major* is used in lotions to help regenerate the skin of older people, and the distilled water is used as an eye lotion (Chiej, 1988). Alcoholic extracts are a component of some mouthwashes (Wichtl, 1994). A decoction has been known to remove freckles (Grieve, 1984).

1.4.3.4 Food

One author states that the young leaves can be mixed into salads to add an aromatic, parsley-like taste (Chiej, 1988); another author says that the leaves give a cucumber taste (Ody, 1996). The fresh young leaves may be finely chopped and added to soups, sauces, mayonnaise, spreads, stuffings, herb butters, salads, cream, and yogurt, and can be used with grilled poultry, fish, and salads and to season vegetables. Its leaves are not usually dried for use, but are cut up finely (about 1 tablespoon of cut leaves per person is normally used) (Lanska, 1992). Burnet saxifrage is used in Italian, French, Spanish, German, and English cuisines.

1.4.3.5 Animals

Cows that feed on this plant have their flow of milk increased (Grieve, 1998).

1.4.4 Preparations

Boiling water is poured over 3 to 10 g of very finely cut drug; alternatively, cold water can be added and the mixture brought briefly to the boil. As a cough remedy, 1 cup is taken three or four times a day, sweetened with honey (1 teaspoon = 2.5 g) (Wichtl, 1994).

The plant may be taken internally as an infusion for expectorant for catarrh (of the respiratory tract). Externally, this infusion may be used as an application to aid wound healing. Half a liter (1 pint) of cold water is added to 1 tablespoon of root (as coarse powder) and raised to the boil (Fluck, 1988).

The drug can be used for soothing coughs or to counter the effects of bronchitis and laryngitis. An infusion is prepared by adding boiling water to the root (2 teaspoons per cup) and allowing it to stand for 15 minutes; this drink should be taken hot two to three times daily. It can also be used as a mouthwash; for gargling in cases of tonsillitis, ulcers, infected gums, etc. (Launert, 1989).

1.4.5 Combinations with Burnet Saxifrage

An infusion of the leaves is generally used in combination with other stomach remedies like meadowsweet and centaury (Culpeper, 1983).

1.4.6 Legal Category (Licensed Products)

The legal category for burnet saxifrage is General Sales List (Potter, 1994).

1.4.7 Remedies and Supplements

The use of burnet saxifrage in herbal medicine is not widespread, but a few examples are found (Ody, 1996), such as Bioforce Bronchosan and Ricola Swiss Herb Lozenges. Burnet saxifrage root or extracts made from it are components of bronchial remedies, such as Melrosum; arthritis remedies; and remedies such as Befelka Tinctur for the circulation (Wichtl, 1994).

1.4.8 Contraindications

Burnet saxifrage should not be eaten in large quantities, as it irritates the kidneys (Lanska, 1992).

1.4.9 Plants Confused with Burnet Saxifrage

The roots of burnet saxifrage are often mistaken for the roots of common parsnip (*Pastinaca sativa*), which are straight and have a parsnip-like smell (Lanska, 1992). The roots can also be mistaken for the roots of cow parsnip (*Heracleum sphondylium*), which have an unpleasant smell and a sharp burning taste (Lanska, 1992). It should not be confused with *Sanguisorba minor* Scop., or the greater burnet (Rosaceae), which is used as a spice in salads (Wichtl, 1994).

1.4.10 Adulteration

Adulteration is extremely frequent, and on some occasions the genuine drug is unobtainable. Most often it is adulterated with hogweed or cow parsnip root (from *Heracleum sphondylium* L.), which is known in commerce as Radix *P. franconiae*; with wild parsnip root (from *Pastinaca sativa* L.); and with the roots of other *Pimpinella* species. In recent years, *P. peregrina* L. has been increasingly cultivated in southern Germany and is being offered as Radix Pimpinellae DAB 6. The composition of the essential oil and coumarins in *P. peregrina*, *P. saxifraga*, and *P. major* is extremely similar, so they may be worth considering as alternatives to burnet saxifrage (Wichtl, 1994).

1.5 *I. VERUM* HOOK.

1.5.1 Origin of the Name

The generic name for *I. verum* Hook. comes from the Latin *illicio* or *illicere* (to attract), referring to the fragrance (Coombes, 1994; Simonetti, 1991).

1.5.2 History

The Japanese plant the trees of star anise in their temples and on tombs and burn the pounded bark as incense, to produce a perfumed smoke. *I. verum* Hook. is an extremely ancient species, known in China as far back as 100 B.C. The use of star anise has remained very much within its native region, though the Chinese have taken it to the countries in which they settled. Old recipes from the seventeenth century reveal that by that time, star anise was used in the West for fruit syrups and jams (Norman, 1991).

1.5.3 Traditional Uses

1.5.3.1 Flavor

The odor of *I. verum* Hook. is strongly aromatic, and the taste is similar to aniseed, but somewhat more bitter (Simonetti, 1991). Because of the traditional use of anise oils with licorice in licorice sweets, the flavor of anise is often confused with that of licorice and is described as licorice-like (Leung and Foster, 1996).

1.5.3.2 Food

The fruits of *I. verum* Hook. are used in liqueurs, and especially anisette, an ancient liqueur that is still very popular (Bianchini and Corbetta, 1975). The fruits are used in the liqueur industry as a tincture or distillate (Simonetti, 1991), and they are largely employed, especially on the continent of Europe, in flavoring for liqueurs (Wood and Bache, 1883).

The drug can be used as it is, or it can be ground in the kitchen (Simonetti, 1991). Uses of star anise are similar to those of aniseed, and it is an important spice in Chinese cookery (Potter, 1994), being used for seasoning dishes, especially sweets (Grieve, 1998). The Chinese also mix the fruit with coffee and tea to improve their flavor. The Muslims of India season some of their dishes with the capsules, and these capsules are also largely imported into Germany, France, and Italy for the flavoring of spirits (Sturtevant, 1972).

Both anise and star anise are widely used as domestic spices; the former is mainly used by Westerners, whereas the latter is used primarily by Asians, especially in Chinese foods (Leung and Foster, 1996).

Star anise oil is employed in pharmacy practice and in the food industry in the same way as (very expensive) genuine anise oil (which has not been available in sufficient quantities for many years): as a component of alcoholic drinks, liqueurs, toothpastes, sweets, pharmaceutical preparations, and occasionally as a perfume in soap. Star anise is a component of herbal mixtures that are used for making mulled wine (Wichtl, 1994).

1.5.3.3 Medicinal

The medicinal uses of star anise are very similar to those already mentioned for anise. The oil has a sedative effect on the nervous spasms caused, for example, by coughs and asthma (Bianchini, 1975).

Star anise is also used for catarrh of the respiratory tract and for dyspeptic complaints, as well as being used as a bronchosecretolytic and as a gastrointestinal spasmolytic (German Commission E Monograph, 1985). Preparations containing 5 to 10% essential oil are used as a respiratory inhalant in Germany.

It is often chewed in small quantities after each meal to promote digestion and to sweeten the breath (Grieve, 1998), and it has carminative properties (Simonetti, 1991) and helps to relieve flatulence (Lust, 1986). The fruit is used in Asia as a remedy for colic and rheumatism (Grieve, 1998).

Star anise oil is used as stimulant, mild spasmolytic, weak antibacterial, and expectorant in cough mixtures and lozenges, among other preparations. Internally it is used for dyspeptic complaints, and externally it is used as an inhalant for treatment of congestion of the respiratory tract (Leung and Foster, 1996). It is used as a component of prepared antitussives and in gastrointestinal remedies (Wichtl, 1994).

1.5.4 Preparations

- Rarely, 0.5 to 1.0 g of the drug is coarsely powdered immediately before use, covered with boiling water, and allowed to draw in a closed vessel for 10 minutes and strained (1 teaspoon = approximately 3.2 g) (Wichtl, 1994).
- To make an infusion, steep 1 teaspoon crushed seeds in 1 cup water and take 1 to 2 cups per day (Lust, 1986).
- For the tincture, a dose is from 1/4 to 1/2 teaspoon (Lust, 1986).
- A dose of powdered seeds is 10 to 20 g.
- Ol. Anis. BPC: Dose 1 to 3 mL.
- Aq. Anis. Conc. BPC: Dose 5 to 15 mL.
- Emuls. Anis. BPC: Dose 5 to 30 mL.
- Emuls. Anis. et Menth. pip. BPC (1949): Dose 5 to 30 mL.

- Sp. Anis. BPC (1949): Dose 5 to 20 mL (Potter, 1985).
- As for aniseed, anise oil BP: Dose 0.05 to 0.2 mL.
- Anise spirit BPC (1949): Dose 0.3 to 1.2 mL.
- Conc. anise water BPC: Dose 0.3 to 1.0 mL (Potter, 1994).
- The mode of administration for internal use or for inhalation is that the drug is crushed immediately before use, as well as being used via other galenical preparations. Unless otherwise prescribed, the average daily dose is 3.0 g drug or 0.3 g essential oil (German Commission E Monograph, 1988).

1.5.5 Safety

Toxicity, irritation, and sensitization all have been tested, and the oil is safe to use (Tisserand, 1988). There are no known contraindications, no known side effects and no known interactions with other remedies (German Commission E Monograph, 1988).

1.5.6 Legal Category (Licensed Products)

The legal category for anise oil is General Sales List, Schedule 1, Table A. Anise oil and star anise oil are used interchangeably in the United States, both being officially recognized as anise oil in the United States Pharmacopoeia and *Food Chemicals Codex* (1972).

The regulatory status for anise and star anise oil is generally recognized as safe (#182.10 and #182.20). Anise seed and star anise seed are subjects of German official monographs; 3.0 g of seed or 0.3 g essential oil (mean daily dose) is allowed as a bronchial expectorant for upper respiratory tract congestion and as a gastrointestinal spasmolytic.

1.5.7 Adulteration

Commercial star anise seeds may be adulterated with the poisonous, similar but slightly smaller and darker seed of another plant. These seeds have a sharp and bitter odor, resembling cardamom rather than anise (Lust, 1986). Cases of poisoning have occurred through the use of the fruit of *Illicium anisatum* (Syn. *Illicium religiosum*), Japanese anise, bastard anise, or shikimi, which closely resembles star anise. This false star anise is devoid of the anise odor but has an odor resembling laurel, clove, and nutmeg. The poisonous principle has been identified as sikimine by Ekyman (Wood and Bache, 1883), the sesquiterpene anisatin (Leung and Foster, 1996), or a lactone glycoside known as skimmin (Tyler et al., 1988). However, in practice, this adulteration is very rarely encountered (Leung and Foster, 1996).

1.5.8 Plants Confused with Star Anise

Japanese star anise should not be confused with true star anise (Chinese star anise). Japanese star anise is obtained from a related species, *I. lanceolatum* A.C. Smith, formerly believed to be *I. anisatum* L., or *I. religiosum* Sieb. et Zuce., which grows in southern China, Taiwan, and Japan. It looks like a smaller, deformed version of Chinese star anise and is highly poisonous. A 10 to 15% aqueous extract is used in China as agricultural insecticide (Leung and Foster, 1996). The fruits produced by *I. anisatum* do not smell of aniseed but of cardamom.

1.5.9 Other Species of Anise

I. floridanum, found in Florida and along the coast of the Gulf of Mexico, has its bark, leaves, and probably also seed vessels endowed with a spicy odor and taste, similar to those of anise.

I. parviflorum, a shrub found in Michaux and in the hilly regions of Georgia and Carolina, has a flavor closely resembling that of sassafras root.

References

Attygalle, J., *Sinhalese Materia Medica*, Navrang, New Dehli, India (in collaboration with Lake House Book Shop, Colombia, Sri Lanka), 1994.

Back, P., *The Illustrated Herbal*, Hamlyn, London (through Octopus Books, printed in Hong Kong by Mandarin), 1987.

Bianchini, F., and Corbetta, F., *The Fruits of the Earth*, Bloomsbury Books, London, 1975.

Bremness, L., *The Complete Book of Herbs*, Colour Library Books, © Dorling Kindersley, Godalming, Surrey, United Kingdom, (1988) 1991.

British Herbal Manufacturing Association, *The British Herbal Pharmacopoeia*, Bournemouth, Dorset, United Kingdom, 1983, p. 160.

British Herbal Manufacturing Association, *The British Herbal Pharmacopoeia*, 4th ed., Biddles, Guildford, Surrey, United Kingdom, 1996.

Buchman, D.D., *Herbal Medicine: The Natural Way to Get Well and Stay Well*, Century Hutchinson, London, 1987.

Ceres, *The Healing Power of Herbal Teas*, Thorsons, London, 1984.

Chiej, R., *The Macdonald Encyclopaedia of Medicinal Plants*, Macdonald Orbis, London, 1988.

Coffey, T., *The History and Folklore of North American Wildflowers*, Facts on File, New York, 1993.

Coombes, Allen J., *Dictionary of Plant Names*, Hamlyn, London, 1994.

Culpeper, N., *Culpeper's Colour Herbal*, W. Foulsham, Slough, Berkshire, United Kingdom, 1983.

Culpeper, N., *Culpeper's Complete Herbal: A Book of Natural Remedies for Ancient Ills*, Wordsworth, Ware, Hertfordshire, United Kingdom, 1995.

Fluck, H., *Medicinal Plants*, W. Foulsham, Slough, Berkshire, United Kingdom, 1988.

Food Chemicals Codex, 2nd ed., National Academy of Sciences, Washington, D.C., 1972.

German Commission E Monograph (B Anz. no. 122, dated 6 June, 1988), 1988.

Gordon, L., *A Country Herbal*, Webb and Bower, Exeter, Devon, United Kingdom, 1980.

Graves, G., *Medicinal Plants: An Illustrated Guide to More than 180 Plants that Cure Disease and Relieve Pain*, Bracken Books, London, 1990.

Grieve, M., *A Modern Herbal: The Medicinal, Culinary, Cosmetic and Economic Properties, Cultivation and Folklore of Herbs, Grasses, Fungi, Shrubs and Trees with All Their Modern Scientific Uses*, Tiger Books International, London, 1998.

Grigson, G., *The Englishman's Flora*, Helicon, Oxford, 1996.

Harry, R.G., *The Principles and Practice of Modern Cosmetics*, 2nd ed., Leonard Hill, London, 1963.

Heinerman, J., *Heinerman's Encyclopaedia of Fruits, Vegetables and Herbs*, Parker, New York, 1988.

Hoffmann, D., *Thorsons Guide to Medical Herbalism: A Comprehensive and Practical Introduction*, Thorsons, London, 1991.

Jayaweera, D.M.A., *Medicinal Plants used in Ceylon*, Part 5, National Science Council of Sri Lanka, Colombo, Sri Lanka, 1982.

Keville, K., *The Illustrated Herb Encyclopaedia: A Complete Culinary, Cosmetic, Medicinal and Ornamental Guide to Herbs*, Grange Books, London, 1991.

Lanska, D., *The Illustrated Guide to Edible Plants*, Chancellor, London, 1992.

Launert, E., *The Hamlyn Guide to Edible & Medicinal Plants of Europe and Northern Europe*, 4th ed., Hamlyn, London, (1981) 1989.

Lawrence Review of Natural Products, Pimpinella, Facts and Comparisons, St. Louis, 1991.

Leung, A.Y., *Encyclopaedia of Common Natural Ingredients Used in Food, Drugs and Cosmetics*, Wiley, New York, 1980.

Leung, A.Y. and Foster, S., *Encyclopaedia of Common Natural Ingredients Used in Food, Drugs and Cosmetics*, 2nd ed., Wiley, New York, 1996.

Lust, J., *The Herb Book*, 16th imprint, Bantam, Sun Valley, CA, 1986.

Manniche, L., *An Ancient Egyptian Herbal*, British Museum Publications, London, 1989, p. 135.

McMahon, B., *Amer. Gard.*, 199, 1806.

National Formulary, 16th ed., American Pharmaceutical Association, Washington, D.C., 1985.

Newall, C.A., Anderson, L.A., and Phillipson, J.D., *Herbal Medicines—A Guide for Health-Care Professionals*, The Pharmaceutical Press, London, 1996.

Norman, J., *The Complete Book of Spices*, Colour Library Books, Godalming, Surrey, United Kingdom, 1991.

Ody, P., *The Herb Society's Complete Medicinal Herbal—A Practical Guide to Medicinal Herbs, with Remedies for Common Ailments*, Dorling Kindersley, London, 1993.

Ody, P., *Handbook of Over-the-Counter Herbal Medicines*, Kyle Cathie, London, 1996.

Phelps-Brown, O., *The Complete Herbalist (A Reprint of the Long-Lost Classic)*, Newcastle Publishing, Van Nuys, CA, 1993.

Potter's New Cyclopaedia of Botanical Drugs and Preparations, Wren, R.C., Ed., C.W. Daniels, London, 1985.

Potter's New Cyclopaedia of Botanical Drugs and Preparations, Wren, R.C., Ed., C.W. Daniels, London, 1994.

Price, S., *Practical Aromatherapy*, Thorsons, London, 1987.

Saunders, C.F., *Edible and Useful Wild Plants of the United States and Canada*, Dover Books, New York, 1976.

Schauenberg, P. and Paris, F., *Guide to Medicinal Plants*, Lutterworth, Cambridge, 1990.

Shukla, H.S. and Tripathi, S.C., Antifungal substance in the essential oil of anise (*Pimpinella anisum* L), *Agric. Biol. Chem.* 51, 1991–1993, 1987.

Simonetti, G., *The MacDonald Encyclopaedia of Herbs and Spices*, MacDonald Illustrated, London, 1991.

Spoerke, D.G., *Herbal Medications*, Woodbridge Press, Santa Barbara, CA, 1990.

Straetman, Udo, private communication, 1993.

Sturtevant, E.L., in *Sturtevant's Edible Plants of the World*, Hedrick, U.P., Ed., Dover, New York, 1972.

Tisserand, R., *The Essential Oil Safety Data Manual*, The Tisserand Aromatherapy Institute, Brighton, Sussex, United Kingdom, 1988.

Tsarong, T.J., *Tibetan Medicinal Plants*, Tibetan Medical Publications, Kalimpong, WB, India, 1994.

Tyler, V.E., Brady L.R., and Robbers, J.E., *Pharmacognosy*, 9th ed., Lea and Febiger, Philadelphia, 1988.

Wichtl, M., in *Herbal Drugs and Phytopharmaceuticals: A Handbook for Practice on a Scientific Basis*, Bisset, N.G., Ed., Medpharm Scientific, Stuttgart, 1994.

Wood, G. and Bache, F., *The Dispensatory of the United States of America*, 15th ed., Lippincott, Philadelphia, 1883.

2 Plant Sources of the Genus *Illicium*

Richard M.K. Saunders

CONTENTS

2.1 INTRODUCTION

Illicium L. (Linnaeus 1759, 1050). Syn: *Skimmi* Kaempf. *ex* Adans., *Fam. fl. pl.*, **2**, 364 (1763); *Cymbostemon* Spach, *Hist. vég. phan.*, **7**, 444 (1839); *Badianifera* L. *ex* Kuntze, *Rev. gen. pl.*, **1**, 6 (1891).

The aromatic fruits of *Illicium* and *Pimpinella* have been used in similar medicinal ways because of their common possession of anethol-rich essences; as a consequence, they have traditionally been grouped together in pharmacopoeias and other phytotherapy texts. The two genera are phylogenetically very distinct, however, structural and molecular studies indicate that although *Illicium* forms one of the earliest evolutionary branches in the angiosperms, *Pimpinella* occupies a relatively advanced phylogenetic position in the Euasterid II clade (Angiosperm Phylogeny Group, 1998).

Illicium L. (Illiciaceae) is a genus of flowering trees and shrubs primarily growing in subtropical evergreen forests, although some species extend into north temperate deciduous forests and others grow in tropical montane forests. The primitive status of the genus is reflected in various aspects of the floral morphology, including the gradual transition between outer, green, sepal-like tepals and inner, pigmented, petal-like tepals; the large but indefinite number of free floral organs; the spiral development of floral organs (Robertson and Tucker, 1979; Erbar and Leins, 1983; Ronse Decraene and Smets, 1993); the elongated floral axis; and the developmentally conduplicate carpels (Robertson and Tucker, 1979). The fruits are highly distinctive, dry, star-shaped folliceta (Figure 2.1), which dehisce by splitting along the adaxial margin to expose the seeds (Figure 2.2).

2.2 TAXONOMIC HISTORY

The genus was formally established by Linnaeus in his *Systema Naturae* (1759) with the publication of the protologue of *I. anisatum* L. Numerous new species names were subsequently introduced, although no attempt was made to comprehensively revise the taxonomy of the genus until Smith (1947) undertook his monograph. Smith recognized 42 species, with a disjunct distribution in eastern North America, Mexico, and the West Indies (5 species) and eastern Asia (37 species). Several regional revisions have subsequently been published (Saunders, 1995, 1997; Zhang, 1996).

Smith (1947) classified the genus into two sections: sect. *Illicium* (under the synonym sect. *Badiana* Spach) and sect. *Cymbostemon* (Spach) A.C. Sm. The former section is characterized by thin, membranous inner tepals that are somewhat lax at anthesis and are narrowly oblong, ligulate,

0-415-32246-4/04/$0.00+$1.50

FIGURE 2.1 Fruit of *Illicium stapfii* Merr. from Borneo: The follicetum is dry and is dehiscing along the adaxial margin, exposing the seed inside. Photograph © Richard M.K. Saunders.

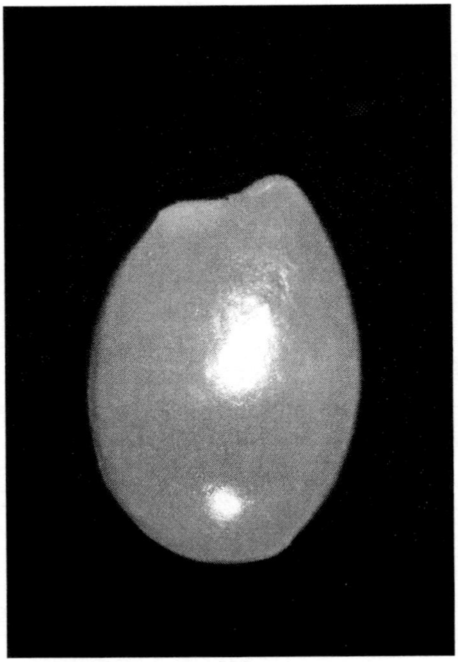

FIGURE 2.2 Seed of *Illicium stapfii* Merr. Photograph © Richard M.K. Saunders.

FIGURE 2.3 (see color insert following page 22) *Illicium anisatum* L., with elongated tepals (typical of sect. *Illicium*). Reproduced from Hooker (1842: pl. 3965). Photograph © Royal Botanic Gardens, Kew.

FIGURE 2.4 (see color insert following page 22) *Illicium verum* Hook. *f.*, with short tepals (typical of sect. *Cymbostemon*). Reproduced from Hooker (1888: pl. 7005). Photograph © Royal Botanic Gardens, Kew.

or lanceolate in shape (Figure 2.3). In contrast, sect. *Cymbostemon* possesses carnose to papyraceous inner tepals that are never lax and that are usually ovate to suborbicular in shape (Figure 2.4). Keng (1965) suggested that tepal shape was a significant taxonomic character because it reflected more fundamental differences in floral vasculature; the significance of the vascular distinctions are unclear, however, as Saunders (1995) has indicated that a variety of different vascular arrangements are found in the tepals of several species of sect. *Cymbostemon*. Species of sect. *Cymbostemon* invariably have trisyncolpate pollen (with three furrows that fuse over the distal pole), whereas species of sect. *Illicium* typically have trizonocolpate pollen (with three furrows that do not fuse over the distal pole) (Erdtman, 1952; Ikuse, 1956; Wodehouse, 1959; Hayashi, 1960; Walker, 1976; Lieux, 1980; Lin, 1989; Liu and Yang, 1989; Saunders, 1995). Although the palynological data therefore broadly corroborate the sectional classification, the correlation is not absolute, as *I. floridanum* Ellis possesses trisyncolpate pollen (Wodehouse, 1959; Lieux, 1980) despite being assigned to sect. *Illicium* on the basis of floral morphology. This incongruence is particularly significant because *Illicium* is otherwise extremely stenopalynous.

Smith's (1947) infrageneric classification of the genus has been widely accepted. The only publication to offer an alternative to this classification (Zhang, 1996) elevated sections *Illicium* and *Cymbostemon* to the subgeneric level. Zhang furthermore recognized two sections within subgenus *Illicium* based on foliar venation differences, namely, sect. *Illicium* and sect. *Impressicosta* Y.-W. Law and Q. Lin.

Hao et al. (2000) recently published a phylogenetic analysis of the genus based on sequences of internal transcribed spacers (ITS) of nuclear ribosomal DNA. They found that the two sections recognized by Smith (1947) were unnatural, and that the ITS phylogeny was congruent with the palynological data.

2.3 NOMENCLATURE

Illicium seeds were first imported into Europe by Cavendish in 1587 (Robinson, 1908) as a spice for flavoring food and liqueurs. Although the seeds originated in China, they were imported via the Philippines; this resulted in the mistaken belief that the plants were of Philippine origin, which led Clusius (1601) to describe it under the name *Anisum philippinarum insularum*. This spice is referable to *I. verum* Hook. *f.*

Linnaeus used the generic name *Badianifera* in his *Materia Medica* (1749) in reference to a plant of presumed medical value (again referable to *I. verum*). *Badianifera* was not mentioned in any of Linnaeus's subsequent publications, however, and the earliest nomenclaturally valid publication of a name for the genus was *Illicium*, published in Linnaeus's *Systema Naturae* (1759). The only species cited under this new generic name was *I. anisatum* L. The name *Illicium* is derived from the Latin word *illicere,* meaning an attractant, presumably in reference to the presence of aromatic oils.

The commercially important species *I. verum* has often been confused with *I. anisatum*, which is toxic and potentially fatal if consumed. This confusion is particularly problematic because *I. anisatum* has been used in traditional Chinese herbal medicine, and both species are therefore retailed commercially in Asia. The confusion is largely the result of similarities between the fruits, as the flowers are highly distinctive: *I. verum* possesses flowers with rounded pink, red or purple tepals (Figure 2.4), whereas *I. anisatum* possesses elongated white or yellow tepals (Figure 2.3). Small (1996) outlined the confused use of vernacular names for the two species: *I. verum* is variously called "star anise," "Chinese anise," and "China star anise" in the literature, whereas *I. anisatum* is called "star anise," "Chinese anise," "Japanese anise," and "Japanese star anise." Although the scientific nomenclature is clearly preferable because it is unambiguous, the use of the vernacular name "star anise" for fruits of *I. verum* appears to be well established. Small (1996) therefore recommended that fruits of *I. anisatum* (as the only species of *Illicium* native to Japan) be called "Japanese star anise" to distinguish it from true "star anise," although he noted that caution is still necessary because *I. verum* is cultivated in Japan.

COLOR FIGURE 2.3 *Illicium anisatum* L., with elongated tepals (typical of sect. *Illicium*). Reproduced from Hooker (1842: pl. 3965). Photograph © Royal Botanic Gardens, Kew.

COLOR FIGURE 2.4 *Illicium verum* Hook. *f.*, with short tepals (typical of sect. *Cymbostemon*). Reproduced from Hooker (1888: pl. 7005). Photograph © Royal Botanic Gardens, Kew.

2.4 PHYLOGENETIC RELATIONSHIPS

The systematic position of *Illicium* has received considerable attention. In most of the early taxonomic literature it was regarded as a member of the family Magnoliaceae (e.g., Bentham and Hooker 1862; Dalla Torre and Harms, 1900–1907), although this approach is rarely adopted today. The genus was subsequently excluded from the Magnoliaceae by Smith (1947) and Bailey and Nast (1948) on the basis of morphological and anatomical criteria, and Smith (1947) formally published the new familial name Illiciaceae solely for the genus *Illicium*. Both Smith (1947) and Bailey and Nast (1948) suggested a close relationship with the Schisandraceae, a family of scrambling and twining woody vines; this is reflected in all current angiosperm classification schemes (e.g., Dahlgren, 1980; Takhtajan, 1980; Cronquist, 1981; Thorne, 1992; Angiosperm Phylogeny Group, 1998), with most people using the ordinal taxon Illiciales collectively for the two families. Supportive data (generally interpreted using cladistic phylogenetic techniques) have been obtained from a wide variety of different sources, including general morphology (Donoghue and Doyle, 1989; Loconte and Stevenson, 1991), embryology (Hayashi, 1965), cytology (Ehrendorfer et al., 1968), phytochemistry (Sy et al., 1997), and the analysis of nucleic acid sequence data (Chase et al., 1993; Qiu et al., 1993; Soltis et al., 1997).

Analysis of morphological data has highlighted the Winteraceae as the closest relatives of the Illiciales (Donoghue and Doyle, 1989; Loconte and Stevenson, 1991). More recently, analysis of molecular data has contradicted this connection and revealed possibly closer relationships with the Austrobaileyaceae (Chase et al., 1993; Qiu et al., 1993; Soltis et al., 1997).

References

Angiosperm Phylogeny Group, An ordinal classification for the families of flowering plants, *Ann. Missouri Bot. Gard.*, 85, 531–553, 1998.

Bailey, I.W. and Nast, C.G., Morphology and relationships of *Illicium*, *Schisandra* and *Kadsura*. I. Stem and leaf, *J. Arnold Arbor.*, 29, 77–89, 1948.

Bentham, G. and Hooker, J.D., *Genera Plantarum*, Vol. 1, Part 1. A. Black, London. 1862.

Chase, M.W., Soltis, D.E., Olmstead, R.G., Morgan, D., Les, D.H., Mishler, B.D., Duvall, M.R., Price, R.A., Hills, H.G., Qiu, Y.-L., Kron, K.A., Rettig, J.H., Conti, E., Palmer, J.D., Manhart, J.R., Sytsma, K.J., Michaels, H.J., Kress, W.J., Karol, K.G., Clark, W.D., Hedrén, M., Gaut, B.S., Jansen, R.K., Kim, K.-J., Wimpee, C.F., Smith, J.F., Furnier, G.R., Strauss, S.H., Xiang, Q.-Y., Plunkett, G.M., Soltis, P.S., Swensen, S.M., Williams, S.E., Gadek, P.A., Quinn, C.J., Eguiarte, L.E., Golenberg, E., Learn Jr, G.H., Graham, S.W., Barrett, S.C.H., Dayanandan, S., and Albert, V.A., Phylogenetics of seed plants: An analysis of nucleotide sequences from the plastid gene *rbc*L, *Ann. Missouri Bot. Gard.*, 80, 528–580, 1993.

Clusius, C., *Rariorum Plantarum Historia,* Christophori Plantini, Antwerp, 1601.

Cronquist, A., *An Integrated System of Classification of Flowering Plants,* Columbia University Press, New York, 1981.

Dahlgren, R.M.T., A revised system of classification of the angiosperms, *Bot. J. Linn. Soc.*, 80, 91–124, 1980.

Dalla Torre, C.G. and Harms, H., *Genera Siphonogamarum ad Systema Englerianum Conscripta,* Wilhelm Engelmann, Leipzig, 1900–1907.

Donoghue, M.J. and Doyle, J.A., Phylogenetic analysis of angiosperms and the relationships of the Hamamelidae, in *Evolution, Systematics, and Fossil History of the Hamamelidae*, Vol. 1, Crane, P.R. and Blackmore, S., Eds., Clarendon Press, Oxford, 1989, p. 17–45.

Ehrendorfer, F., Krendl, F., Habeler, E., and Sauer W., Chromosome numbers and evolution in primitive angiosperms, *Taxon*, 17, 337–353, 1968.

Erbar, C. and Leins, P., Zur Sequenz von Blütenorganen bei einigen Magnoliiden, *Bot. Jahrb.*, 103, 433–449, 1983.

Erdtman, G., *Pollen Morphology and Plant Taxonomy,* Almqvist and Wiksell, Stockholm, 1952.

Hao, G., Saunders, R.M.K., and Chye, M.-L., A phylogenetic analysis of the Illiciaceae based on sequences of internal transcribed spacers (ITS) of nuclear ribosomal DNA, *Pl. Syst. Evol.*, 223, 81–90, 2000.

Hayashi, Y., On the microsporogenesis and pollen morphology in the family Magnoliaceae, *Sci. Rep. Tôhoku Imp. Univ. Ser. IV Biol.*, 26, 45–52, 1960.

Hayashi, Y., The comparative embryology of the Magnoliaceae (*s.l.*) in relation to the systematic consideration of the family, *Sci. Rep. Tôhoku Imp. Univ. Ser. IV Biol.*, 31, 29–44, 1965.

Hooker, J.D., *Illicium verum*, *Curtis's Bot. Mag.*, 114, pl. 7005, 1888.

Hooker, W.J., *Illicium religiosum*, *Curtis's Bot. Mag.*, 69, pl. 3965, 1842.

Ikuse, M., *Pollen Grains of Japan,* Hirokawa, Tokyo, 1956.

Keng, H., Observations on the flowers of *Illicium, Bot. Bull. Acad. Sin.*, 6, 61–73, 1965.

Lieux, M.H., An atlas of pollen of trees, shrubs, and woody vines of Louisiana and other southeastern States, part I. Ginkgoaceae to Lauraceae, *Pollen Spores*, 22, 17–57, 1980.

Lin, Q., A study of the pollen morphology of genus *Illicium* L, *Bull. Bot. Res.*, 9, 115–124, 1989.

Linnaeus, C., *Materia Medica*, Laurentii Salvii, Stockholm, 1749.

Linnaeus, C., *Genera Plantarum*, Laurentii Salvii, 5th ed., Leiden, 1754.

Linnaeus, C., *Systema Naturae*, Laurentii Salvii, 10th ed., Stockholm, 1759.

Liu, H. and Yang, C.-S., Pollen morphology of Illiciaceae and its significance in systematics, *Chin. J. Bot.*, 1, 104–115, 1989.

Loconte, H. and Stevenson, D.W., Cladistics of the Magnoliidae, *Cladistics*, 7, 267–296, 1991.

Qiu, Y.-L., Chase, M.W., Les, D.H., and Parks, C.R., Molecular phylogenetics of the Magnoliidae: cladistic analyses of nucleotide sequences of the plastid gene *rbc*L, *Ann. Missouri Bot. Gard.*, 80, 587–606, 1993.

Robertson, R.E. and Tucker, S.C., Floral ontogeny of *Illicium floridanum*, with emphasis on stamen and carpel development, *Am. J. Bot.*, 66, 605–617, 1979.

Robinson, C.B., Perrottet and the Philippines, *Philipp. J. Sci. Sect. C Bot.*, 3, 303–306, 1908.

Ronse Decraene, L.P. and Smets, E.F., The distribution and systematic relevance of the androecial character polymery, *Bot. J. Linn. Soc.*, 113, 285–350, 1993.

Saunders, R.M.K., Systematics of the genus *Illicium* L. (Illiciaceae) in Malesia. *Bot. J. Linn. Soc.*, 117, 333–352, 1995.

Saunders, R.M.K., Illiciaceae, *Fl. Males. Ser. I*, 13, 169–184, 1997.

Small, E., Confusion of common names for toxic and edible "star anise" (*Illicium*) species, *Econ. Bot.*, 50, 337–339, 1996.

Smith, A.C., The families Illiciaceae and Schisandraceae, *Sargentia*, 7, 1–224, 1947.

Soltis, D.E., Soltis, P.S., Nickrent, D.L., Johnson, L.A., Hahn, W.J., Hoot, S.B., Sweere, J.A., Kuzoff, R.K., Kron, K.A., Chase, M.W., Swensen, S.M., Zimmer, E.A., Chaw, S.-M., Gillespie, L.J., Kress, W.J., and Sytsma, K.J., Angiosperm phylogeny inferred from 18S ribosomal DNA sequences, *Ann. Missouri Bot. Gard.*, 84, 1–49, 1997.

Sy, L.-K., Saunders, R.M.K., and Brown, G.D., Phytochemistry of *Illicium dunnianum* and the systematic position of the Illiciaceae, *Phytochemistry*, 44, 1099–1108, 1997.

Takhtajan, A., Outline of the classification of flowering plants (Magnoliophyta), *Bot. Rev.*, 46, 225–359, 1980.

Thorne, R.F., Classification and geography of the flowering plants, *Bot. Rev.*, 58, 225–348, 1992.

Walker, J.W., Comparative pollen morphology and phylogeny of the Ranalean complex, in *Origin and Early Evolution of Angiosperms*, Beck, C.B., Ed., Columbia University Press, New York, 1976, p. 241–299.

Wodehouse, R.P., *Pollen Grains: Their Structure, Identification and Significance in Science and Medicine,* Hafner, New York, 1959.

Zhang, B.L., Magnoliaceae trib. Illicieae DC, in *Flora Reipublicae Popularis Sinicae*, Vol. 30, Part 1, Law, Y.-W. Ed., Science Press, Beijing, 1996, pp. 198–231.

3 Plant Sources of the Genus *Pimpinella*

María Reyes González-Tejero

CONTENTS

3.1 INTRODUCTION

Pimpinella L., Sp. *pl.*, 263 (1753). Synonymy: *Tragoselinum* Tourn. *ex* Hall., *Enum. stirp. Helv.* **2**, 428 (1742); *Anisum* Hill, *Brit. herb.*, 424 (1756); *Tragium* Spreng., *Pl. Umbell. prodr.*, 26 (1813); *Ledeburia* Link., *Enum. hort. Berol.*, **1**, 286 (1821); *Gymnosciadium* Hochst., *Flora*, **27**, 20 (1844); *Apium* Caruel Parl., *Fl. Ital.*, **8**, 424 (1889), *pro parte*.

The genus *Pimpinella* L. (Apiaceae/Umbelliferae) comprises 150 species distributed in Eurasia and Africa, over 16 of which are present in Europe. The derived condition of the family Apiaceae in which *Pimpinella* is included can be established by certain characters that are frequently found in the group: the generally herbaceous habit of the family; the frequent presence of compound leaves; and the small, inconspicuous flowers, with few floral parts arranged in whorls and grouped in umbel-shaped inflorescences, the result of prolonged coevolution with insects. The presence of inferious ovaries composed of sealed carpels is a further indication of the degree of evolution.

The genus includes herbaceous annual, biannual, or perennial plants (Figure 3.1), generally with a fine hair-covering. The stem is erect, striate, or furrowed lengthways, and it is sometimes woody at the base. Lower leaves are generally simple and rarely pinnatisect, with a toothed or serrated edge, with a (bi)pinnate division of upper leaves. The flowers are bisexual, small, regular; the corolla consist of five white or yellow (sometimes pink or purple) petals that are sometimes slightly emarginate, with the apex bent inward and with the upper surface glabrous or hairy. The calyx is either very small or else absent. The flowers are grouped in umbel-shaped inflorescences (Figure 3.2), which are generally ebracteate, although occasionally one bract and bracteoles are present. The fruit is a morphologically variable schizocarp (Figure 3.3) consisting of two ovoid to subspherical mericarps that are variably compressed laterally and either completely glabrous or have a thick down; primary ribs are inconspicuous to prominent with numerous vittae.

From a pharmacological and economic viewpoint, the most valuable species is *P. anisum* L., an annual plant of up to 50 cm with white flowers and aromatic fruits (Figure 3.4), covered with fine, short, adpressed hairs. The roots of two other species, *P. saxifraga* L. and *P. major* (L.) Hudson, have been traditionally used in medicine. They are perennial herbaceous plants differentiated from *P. anisum* by their glabrous fruits.

0-415-32246-4/04/$0.00+$1.50

FIGURE 3.1 Crop of *P. anisum*

FIGURE 3.2 Flowers of *P. anisum*

FIGURE 3.3 Infrutescences of *P. anisum*

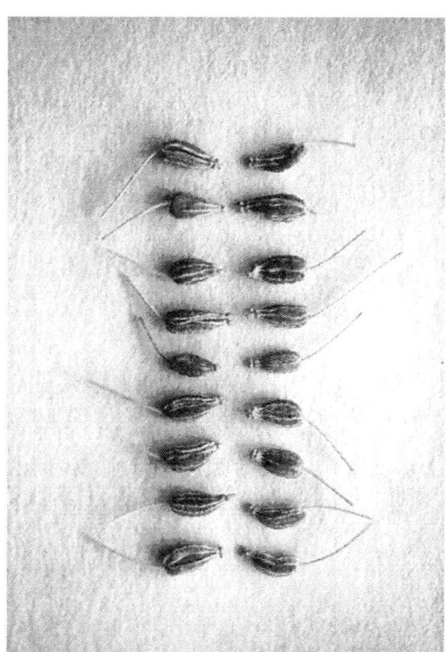

FIGURE 3.4 Fruits of *P. anisum*

3.2 TAXONOMIC HISTORY

The genus was established by Linnaeus in 1753 in *Species Plantarum*, classified within the class *Pentandria dyginia*, which consists of plants with bisexual flowers with five stamens and two carpels. When Linnaeus formally published the generic name *Pimpinella*, he included the pre-Linnaean name *Tragoselinum* Tourn. as a synonym. Linnaeus identified four species: *P. saxifraga*, *P. glauca*, *P. peregrina*, and *P. anisum*.

Various subsequent taxonomists published new generic names for elements of the genus, although all are now reduced to synonymy, namely, *Anisum* Hill, *Tragium* Sprengel, *Ledeburia* Link., and *Gymnosciadium* Hochst. to *Pimpinella*.

In 1827, De Candolle considered the genus *Pimpinella* to be divided into three sections: *Tragoselinum*, characterized by its glabrous fruits and perennial roots; *Tragium*, with hairy fruits, perennial (rarely biennial) roots, and radical leaves uni- to bipinnatisect with ovate segments; and *Anisum*, which included species with down-covered annual fruits.

One of the most important reviews was that of Bentham and Hooker (1867), who included 65 to 70 species under the name *Pimpinella*, classified in six sections according to the habitat of the plant, leaf and fruit morphology, and petal color. Yet it was not until 1910 when Wolff used petal color and morphology and fruit indumentum to undertake a thorough taxonomic revision that to some extent clarified the status of the genus. He divided the genus into three sections: *Reutera*, which included species with yellow flowers and glabrous or hairy fruits; *Tragium*, generally with white flowers and bristly or hairy fruits, granular or tubercled, sometimes nearly glabrous or almost completely smooth; and *Tragoselinum*, also with white flowers and glabrous or totally glabrous fruits. *P. anisum* was included within the section tragium.

3.3 NOMENCLATURE

The Latin name *Pimpinella* was originally used in reference to various species of the genus *Sanguisorba* (Rosaceae), because of the similarity between the leaves of certain Pimpinella species (e.g., *P. saxifraga*) and those of *Sanguisorba* species.

Anise was used and cultivated in ancient Egypt (Hedrick, 1972). Theophrastus, in *Historia Plantarum*, mentions it along with cumin, coriander, dill, or fennel as a condiment herb characterized by its naked seeds, an idea that lasted until the eighteenth century and that led De Jussieu (1748–1836) to the use of this character to separate the Umbelliferae from the Araliaceae, as in the taxa of the latter the seeds are enclosed in a fleshy pericarp.

In the third century, Albertus Magnus referred to anise as *Roman fennel*, a name that was conserved and subsequently used by Ruellius in France in 1536 (Hedrick, 1972). Bahuin (1671) used the specific name *Anisum herbariis*, which was later included by Linnaeus (1753) under *P. anisum*.

3.4 PHYLOGENETIC RELATIONSHIPS

Most authors recognize a phylogenetic relationship between the Apiaceae and the Araliaceae (e.g., Takhtajan, 1980; Cronquist, 1981). Many of the diagnostic features of the Apiaceae are found individually in members of the Araliaceae, and it has been suggested that the Araliaceae were ancestral to the Apiaceae (Cronquist, 1981). Thorne (1973, 1983) broadened the concept of the Araliaceae to include the Apiaceae, although he subsequently returned to a more traditional treatment (Thorne, 1992).

Phytochemical data support the close relationship between the two families (Gershenzon and Mabry, 1983). They both possess similar essential oils, containing volatile monoterpenoids as well as other constituents; triterpene saponins with oleanene- and ursene-type saponins; polyacetylenes;

an unusual fatty acid, petroselinic acid; and caffeic acid esters; they have similar flavonoid patterns and are similar in their absence of proanthocyanidins and ellagic acid derivatives. Recent cladistic analyses of morphological (Judd et al., 1994) and molecular (Plunkett et al., 1996a, 1997) data further support the close relationship between the Apiaceae and Araliaceae.

There has been some debate over the relationship of the Apiaceae and Araliaceae with the family Cornaceae. Takhtajan (1980, 1997) has argued that the two families are difficult to separate based solely on morphological characters. Cronquist (1981), however, noted phytochemical differences between the two families, with the Cornaceae apparently distinct because of the production of iridoids, ellagitannins, and proanthocyanidins and because of the absence of the polyacetylenes, essential oils, and coumarins.

Dahlgren (1980) has emphasized a relationship with the family Asteraceae (Compositae) based on the presence of sesquiterpene lactones, polyacetylenes, coumarins, and triterpenes in both groups. Dahlgren also noted morphological, embryological, and palynological similarities. The relationship with the Asteraceae was not supported by Cronquist (1981), however, who argued that morphological differences between the two families are significant, despite the phytochemical similarities.

Recent molecular investigations (Downie and Katz-Downie 1996; Plunkett et al., 1996b; Downie et al., 1998) have indicated that subfamily Apioideae sensu Drude (1898), which includes *Pimpinella*, is monophyletic, but that lower level tribal classifications within the subfamily are not supported, because of homoplasies in the more traditionally used fruit characters.

References

Bauhin, G., *Pinax Theatri Botanici,* 2nd ed., Basel, 1671.

Bentham, G. and Hooker, J.D., *Genera Plantarum,* Vol. 1, Part 3, L. Reeve, London, 1867.

Cronquist, A., *An Integrated System of Classification of Flowering Plants,* Columbia University Press, New York, 1981.

Dahlgren, R.M.T., A revised system of classification of the angiosperms, *Bot. J. Linnaeus Soc.,* 80, 91–124, 1980.

De Candolle, A.P., *Prodromus Systematis Naturalis Regni Vegetalis, Organographie Vegetale*, Teruttel et Mürtz, Paris, 1873.

Downie, S.R. and Katz-Downie, D.S., A molecular phylogeny of Apiaceae subfamily Apioideae: evidence from nuclear ribosomal DNA internal transcribed spacer sequences, *Am. J. Bot.*, 83, 234–251, 1996.

Downie, S.R., Ramanath, S., Katz-Downie, D.S., and Llanas, E., Molecular systematics of Apiaceae subfamily Apioideae: phylogenetic analyses of nuclear ribosomal DNA internal transcribed spacer and plastid *rpoC1* intron sequences, *Am. J. Bot.*, 85, 563–591, 1998.

Drude, C.G.O., Umbelliferae, in *Die naturlichen Pflanzenfamilien III*, Vol. 8, Engler, A. and Prantl, K., Eds., W. Engelmann, Leipzig, 1898, p. 63–250.

Gershenzon, J. and Mabry, T.J., Secondary metabolites and the higher classification of angiosperms, *Nord. J. Bot.*, 3, 5–34, 1983.

Hedrick, U., Ed., *Sturtevant's Edible Plants of the World*. Dover, New York, 1972.

Hooker, J.D., *Illicium verum*, *Curtis's Bot. Mag.*, 114, pl. 7005, 1888.

Hooker, W.J., *Illicium religiosum*, *Curtis's Bot. Mag.*, 69, pl. 3965, 1842.

Judd, W.S., Sanders, R.W., and Donoghue, M.J., Angiosperm family pairs: preliminary cladistic analyses, *Harvard Pap. Bot.*, 5, 1–51, 1994.

Linnaeus, C., *Species Plantarum*, Salvius, Stockholm, 1753.

Plunkett, G.M., Soltis, D.E., and Soltis, P.S., Higher level relationships of Apiales (Apiaceae and Araliaceae) based on phylogenetic analysis of *rbcL* sequences, *Am. J. Bot.*, 83, 499–515, 1996a.

Plunkett, G.M., Soltis, D.E., and Soltis, P.S., Evolutionary patterns in Apiaceae: inferences based on *matK* sequence data, *Syst. Bot.*, 21, 477–495, 1996b.

Plunkett, G.M., Soltis, D.E., and Soltis, P.S., Clarification of the relationship between Apiaceae and Araliaceae based on *matK* and *rbcL* sequence data, *Am. J. Bot.*, 84, 565–580, 1997.

Takhtajan, A., Outline of the classification of flowering plants (Magnoliophyta), *Bot. Rev.*, 46, 225–359, 1980.

Takhtajan, A., *Diversity and classification of flowering plants,* Columbia University Press, New York, 1997.

Thorne, R.F., Inclusion of the Apiaceae (Umbelliferae) in the Araliaceae, *Notes R. Bot. Gard. Edinburgh*, 32, 161–165, 1973.

Thorne, R.F., Proposed new realignments in the angiosperms, *Nordic J. Bot.*, 3, 85–117, 1983.

Thorne, R.F., Classification and geography of the flowering plants, *Bot. Rev.*, 58, 225–348, 1992.

Wolff, H., Umbelliferae—Apioideae—Amminae—Carinae, Amminae novemjugatae ex genuinae, in *Das Pflanzenreich IV*, Vol. 9, Engler, A., Ed., Wilhelm Engelmann, Berlin, 1910, p. 1–398.

4 Chemical Constituents of the Genus *Illicium*

Yoshiyasu Fukuyama and Jian-mei Huang

CONTENTS

4.1 INTRODUCTION

The genus *Illicium* belongs to the family Illiciaceae and is an evergreen shrub or tree. About 40 species have been disjunctively distributed in eastern North America, Mexico, the West Indies, and eastern Asia. The highest concentration of species is in northern Myanmar and southern China, where nearly 35 species have been described (Saunders, 1995). The genus is comparatively primitive and has close affinities with the Magnoliaceae. Therefore, *Illicium* was classified into the family Magnoliaceae in much of the early taxonomic literature (Torre et al., 1900) but was subsequently excluded from the Magnoliaceae by Smith on the basis of floral morphology and vegetative anatomy (Smith, 1947) and was assigned familial rank as the Illiciaceae.

The fruits of *Illicium* species are distinctive star-shaped follicles that emit a characteristic refreshing flavor. The fruits of *I. verum* Hook., in particular, are the source of the only economically important product derived from the genus: Chinese star anise, which is widely used as a spice for flavoring food and beverages. Hence, essential oils have been the primary subject of chemical research on *Illicium* species, and the presence of volatile phenols have been reported as constituents of various parts of all *Illicium* species so far studied. However, the fruits of *I. anisatum*, Japanese star anise, have been known to be very toxic for several centuries. Many researchers have been involved in the attempt to isolate the toxic substance since the struggle began in the middle of the nineteenth century. In 1952, Lane et al. succeeded in the isolation of the pure toxic principle named anisatin (Lane et al., 1952) for the first time, the complete structure of which was later established by Yamada and Hirata (Yamada et al., 1968). Since anistatin was first recognized as the unprecedented sesquiterpene dilactone containing an unusual β-lactone, several other members of the toxic *Illicium* species have also been the subject of chemical investigation mainly for the purpose of clarifying the relationship between structure and toxicity, resulting in the isolation of a number of unusual sesquiterpenes and neolignans. The chemical studies of *Illicium* have developed rapidly over the last 20 years as a result of more efficient methods of purification and of the use of modern spectroscopic techniques for structure elucidation.

0-415-32246-4/04/$0.00+$1.50

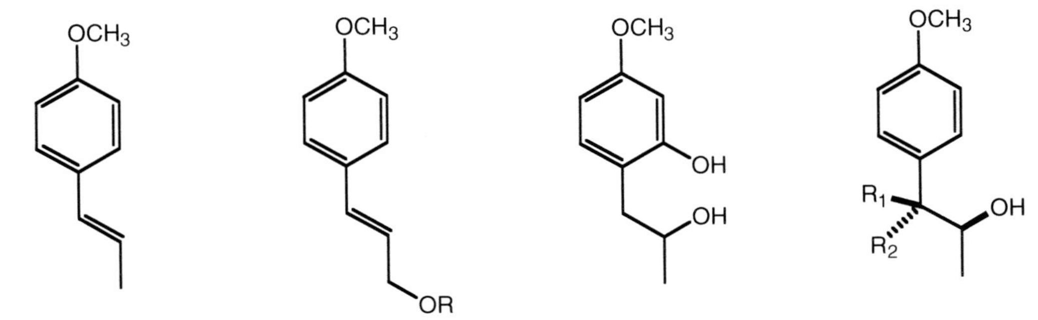

1

2 R=H
3 R=Ac

4

5 R_1=OH, R_2=H
6 R_1=H, R_2=OH

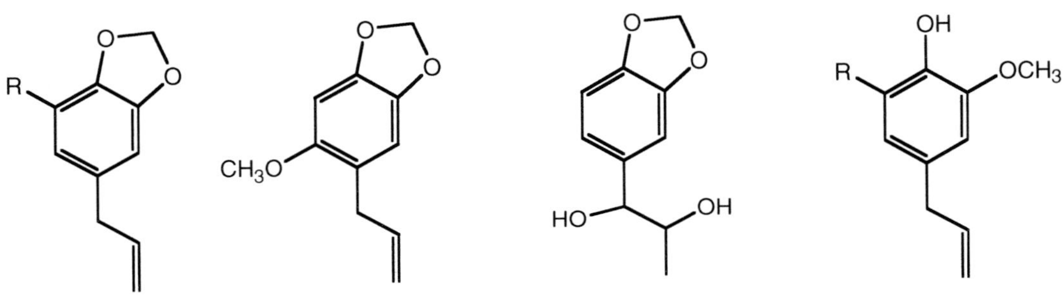

7 R = H
8 R = OCH_3

9

10

11 R = H
12 R = OCH_3

FIGURE 4.1 Phenylpropanoids (compounds **1–12**)

The primary review (*Dictionary of Natural Products* 1996) indicates that *Illicium* is rich in prenylated C_6-C_3 compounds (phenylpropanoid), neolignans, and sesquiterpenes in addition to several common flavonoids, diterpenoids, and triterpenoids. This chapter is intended to be a survey of the literature that reported chemical constituents obtained from plants of the genus *Illicium* during the period, 1965–1999, with the exception of the components of their essential oil.

4.2 C_6-C_3 COMPOUNDS

A very large number of compounds exhibit a characteristic C_6-aromatic-C_3-side chain compounds; for example, aromatic amino acids, hydroxycinnamic acids, phenylpropenes, coumarins, isocoumarins, chromones, and so forth. The biosynthesis of these compounds follows the shikimic acid pathway. Shikimic acid was isolated as early as 1885 by Eijkman from the Japanese plant *I. anisatum*. Its name originally derives from the Japanese *Shikimi,* but shikimic acid was later found to be widespread in nature. Therefore, C_6-C_3 compounds, which are biosynthesized from shikimic acid, are most likely to be characteristic components of *Illicium* species.

Volatile simple phenylpropanoids (Figure 4.1) have been found in leaf oil of most *Illicium* species. *E*-Anethol (compound **1**), safrole (**7**), and eugenol (**11**) were isolated from the leaf oil of *I. anisatum* by J.F. Eijkman a long time ago. *E*-Anethol was then isolated again, as an insecticide (Miyazawa et al., 1993) and as a sweetening substance, from the fruits of *I. verum* (Hussain et al., 1990). In fact, *E*-anethol has been used as a flavoring agent in beverages, candy, baked goods, and chewing gum (Furita and Bellanca, 1975). It is worth noting that the oil of Chinese star anise usually contains over 70% *E*-anethol. *E*-anethol, however, is a present in small amounts or is absent in the oil of other *Illicium* plants (Liu and Yang, 1989; Yang et al., 1992). The analogues of *E*-anethol, **2–6**, were also found in the leaves of *Illicium verum* (Sy and Brown, 1998). However, safrole (**7**) and novel 1,2-methylenedioxy-4-(propane-1,2-diol)benzene (**10**) were found in the fresh leaves of *I. dunnianum* (Sy et al., 1997), and more oxygenated C_6-C_3 compounds, myrsticin (**8**) (Yakushijin et al., 1983), and sarisan (**9**) and methoxyeugenol (**12**) (Shibuya et al., 1978; Sy et al., 1997) were isolated from *I. anisatum* and *I. dunnianum*. Another phenylpropanoid bearing more substituents (**13–20**) (Figure 4.2) was isolated from the bark of *I. difengpi* (Kouno et al., 1992) and the leaves of *I. verum* (Sy and Brown, 1998). Compounds **13–18** appear to be biogenetically derived from threo-anethol glycol (**5**), which was also present as a significant component of the methylene chloride extract of the leaves of *I. verum*. Moreover, phenylpropanoid glycosides (**21–28**) (Figure 4.3) were present in *I. difengpi* and *I. anisatum* (Kouno et al., 1992; Tanaka et al., 1998; Jiang et al., 1999). Glycoside **21** was 6-*p*-coumaroyl glucoside of **19**.

Phenylpropanoids bearing an additional prenyl group (Figure 4.4) are also characteristic phenols in *Illicium* plants. *O*-Prenylated chavicol (**29**), *O*-prenylated eugenol (**30**), and **32** occur in the fresh leaves of *I. dunnianum* (Sy et al., 1997), whereas the leaves of *I. anisatum* contain *O*-prenylmethoxyeugenol (**31**) in addition to **30** (Shibuya et al., 1978). Illicinol (**33**) was the only phenolic component isolated from the dried leaves of *I. arborescens* (Yakushijin et al., 1983). The structure of **33** was confirmed by direct comparison of illicinol with an authentic sample obtained by the reaction of 2-allyl-4,5-methylenedioxyphenol and prenylbromide in the presence of base. In addition, Yakusuijin et al. (1983) reported the presence of compounds **34** and **35** in the wood of *I. anisatum*. Anisoxide (**36**) was first isolated by Jackson and Short (1937) from star anise oil obtained from the seeds of *I. verum*. The structure of **36** was established by Barton et al. (1958) by degradation as well as by synthesis from 1-(2-hydroxyl-4-methylphenyl)-3-methylbutan-2-one. Although anisoxide contains a chiral carbon, it was obtained as a racemate, and thereby might be an artifact formed during the isolation procedure. Okely and Grundon (1971, 1981), who asked this question, proved that anisoxide could originate from an abnormal Claisen rearrangements of feniculin (**36a**), a main component in star anise oil, at 185°C, followed by cyclization of the resultant 1,2-dimethylallylphenol (**36b**) (Scheme 4.1). Later, anisoxide was confirmed to be absent in the star anise oil and the seeds of *I. verum*, using GC analysis. Okely and Grundon concluded that anisoxide had been previously obtained as a result of prolonged distillation of anise oil (Okely and Grundon, 1971, 1981).

Another group of C_6-C_3 compounds bearing an additional prenyl group occurs in *I. tashiroi* and *I. arborescens* (Figure 4.5). Illicinone A (**37**) is a typical example of these compounds, which feature a structure consisting of the units of both lignans [C_6 + C_3] and terpene [C_5] (Yakushijin et al., 1980). (4*S*)-Illicinone A (**37a**), its dihydro-derivative (2*R*, 4*S*)-illicinone B (**38a**), and (4*S*)-illicinone C (**39a**) and (2*R*, 4*S*)-illicinone D (**40a**), having an epoxide ring on a prenyl group, were found in *I. tashiroi* (Yakushijin et al., 1980, 1984; Fukuyama et al., 1992). It is interesting that each isomer, with regard to the C-4 chirality, (4*R*)-illicinone A (**37b**), (2*S*, 4*R*)-illicinone B (**38b**), (2*R*, 4*R*)-illicinone B (**38c**), (4*R*)-illicinone C (**39b**), (2*S*, 4*R*)-illicinone D (**40b**), and (2*R*, 4*R*)-illicinone D (**40c**), was isolated from the leaves of *I. arborescens* (Yakushijin et al., 1984). In addition, a number of illicinone-type compounds having a variety of the modified prenyl units, illicinone G (**41**), (2*R*, 4*S*)-illicinone F (**42a**), and (2*S*, 4*S*)-illicinone F (**42b**), were isolated from *I. tashiroi* (Fukuyama et al., 1994). Illicinone H (**43**) and its methyl derivatives **44** and **45**, which lacked a methylenedioxy group, existed in usual illicinones A–F and were found in *I. tashiroi* (Fukuyama et al., 1994) and *I. anisatum* (Hano et al., 1992; Kaku et al., 1993), as shown in Figure 4.6.

The C-4 chirality of illicinone A was determined by the CD spectra of both α-hydroxycyclohexanones **48** and **49**, derived from **37b** as shown in Scheme 4.2. Two chair conformations **a** and **b** for **48** and **49** seem to be possible. In fact, the CD spectra of **48** and **49** exhibited a positive Cotton effect in a nonpolar solvent; therefore, **48a** and **49a** configurations should dominate. However,

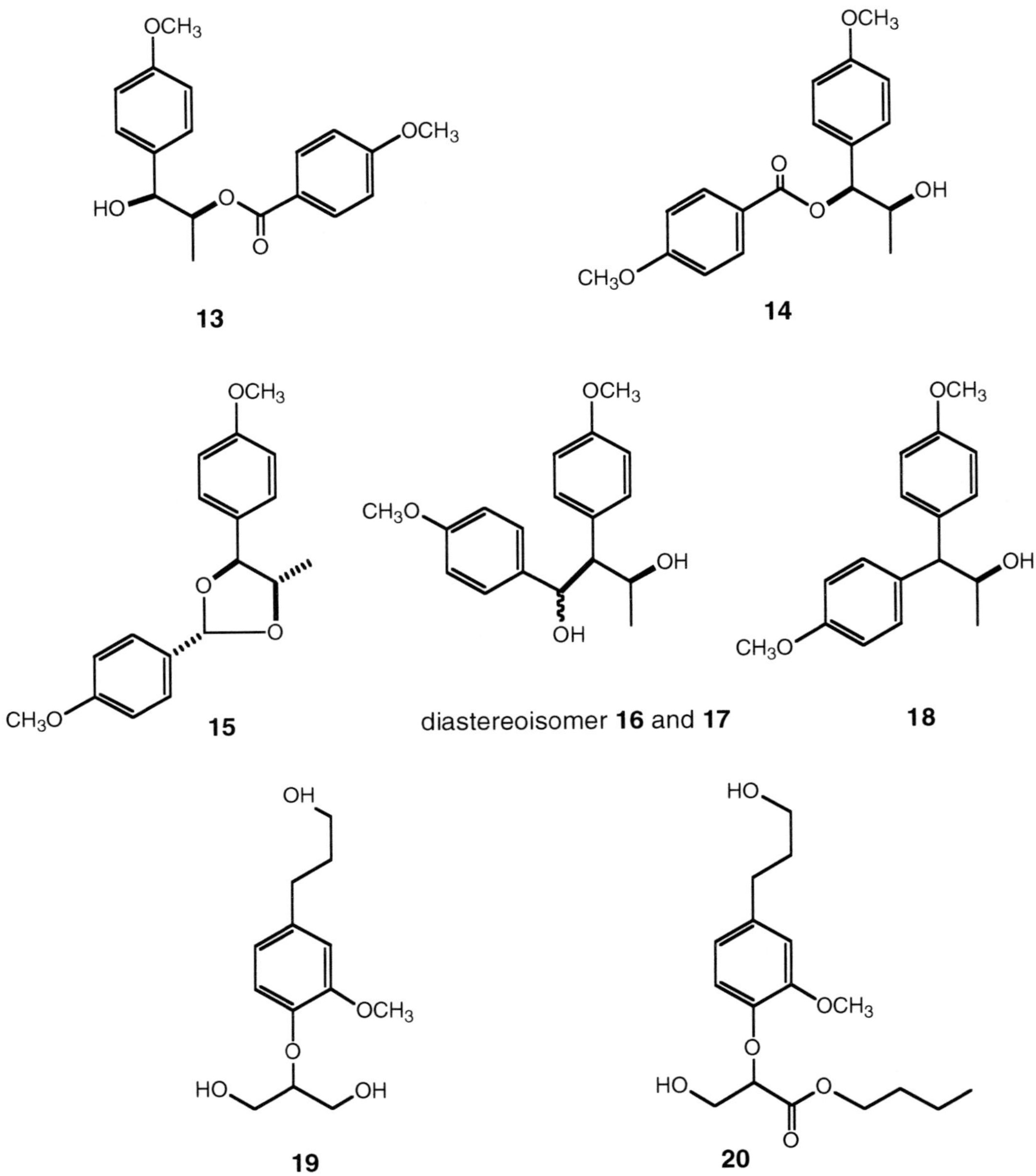

FIGURE 4.2 Phenylpropanoids (compounds **13–20**)

the CD spectrum of **48** still showed only a positive Cotton effect in polar solvents, but inversion of the sign of the Cotton effect in the CD spectrum of **49** was observed on changing the polarity of the solvent. These results indicate that **48** predominantly takes the conformer **48a** in both nonpolar and polar solvents, whereas **49** adopts conformer **49a** in nonpolar solvents and there is an equilibrium between **49a** and **49b** in polar solvents. On the basis of the CD spectra, the absolute configuration of **37b** and **37a** isolated from *I. arborescens* and *I. tashiroi* could be established as *4R* and *4S*, respectively (Yakushijin et al., 1984).

Yakushijin et al. (1984) reported interesting thermal and photochemical transformations of illicinone A (**37**) (Scheme 4.3 and Scheme 4.4). (4*R*)-Illicinone A (**37b**) was heated in a sealed

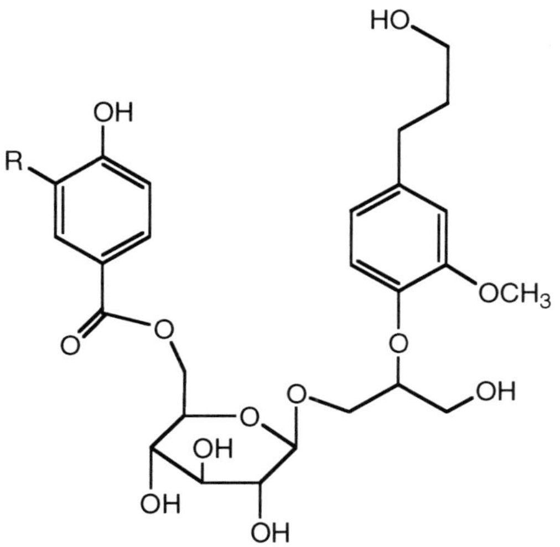

21 R=H
22 R=OCH_3

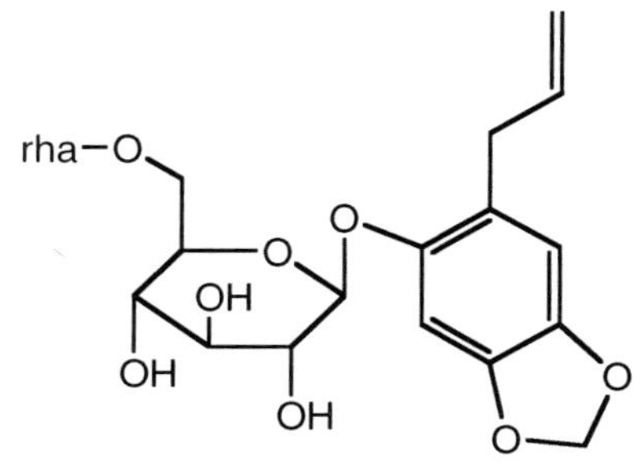

23

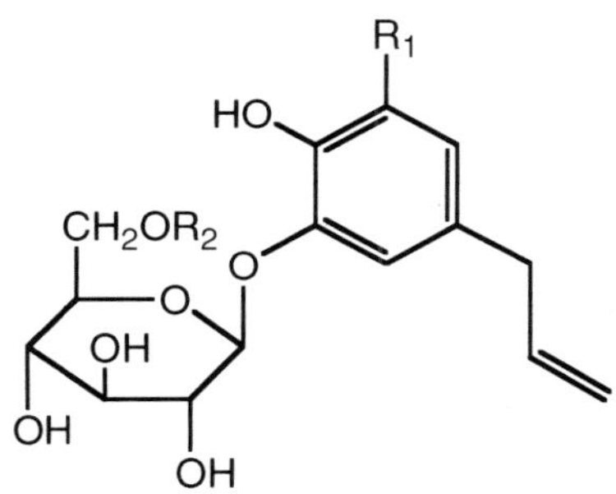

24 R_1=H, R_2=α–arabinofuranosyl
25 R_1=OCH_3, R_2=H
26 R_1=OCH_3, R_2=α–arabinofuranosyl

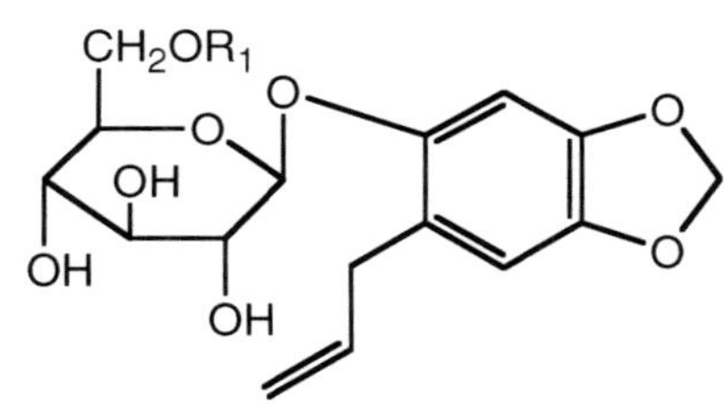

27 R_1=β-D-apiofuranosyl
28 R_1=α–L-arabinofuranosyl

FIGURE 4.3 Phenylpropanoid glycosides (compounds **21–28**)

tube at 190 to 220°C, giving rise to a mixture of **50** (31%), **51** (11%), and illicinol (**33**) (13%), one of the chemical constituents of the leaves of *I. arborescens.* Compounds **50** and **51** were regarded as the products formed by the Claisen rearrangement of **33** derived from **37b**, because **37b** was again heated at the same temperature to yield **50** and **51** in 50 and 12% yields, respectively. Illicinol (**33**) was considered to be formed from **37** irreversibly, but a detailed mechanism was not proposed. In contrast, irradiation of **37b** in ether at 20 to 30°C with a 400-W high-pressure mercury lamp afforded a mixture of novel tricyclic and tetracyclic products **52** (24%), **53** (11%), and **50** (15%). In addition, illicinol (**33**) was irradiated under the same conditions for 30 minutes to give illicinone A (**37**) (19%), **52** (28%), **53** (11%) as racemates, and **50** (16%). Irradiation of **33** for 10 minutes yielded **37** in 44% yield, but after 1 hour, **37** was not detected in the products. Thus, transformation from illicinol (**33**) to illicinone A (**37**) was photochemically correlated, whereas the reverse transformation occurred thermally. These chemical results are most likely to throw light on the biogenesis of nonaromatic prenylated C_6-C_3 compounds from illicinol A.

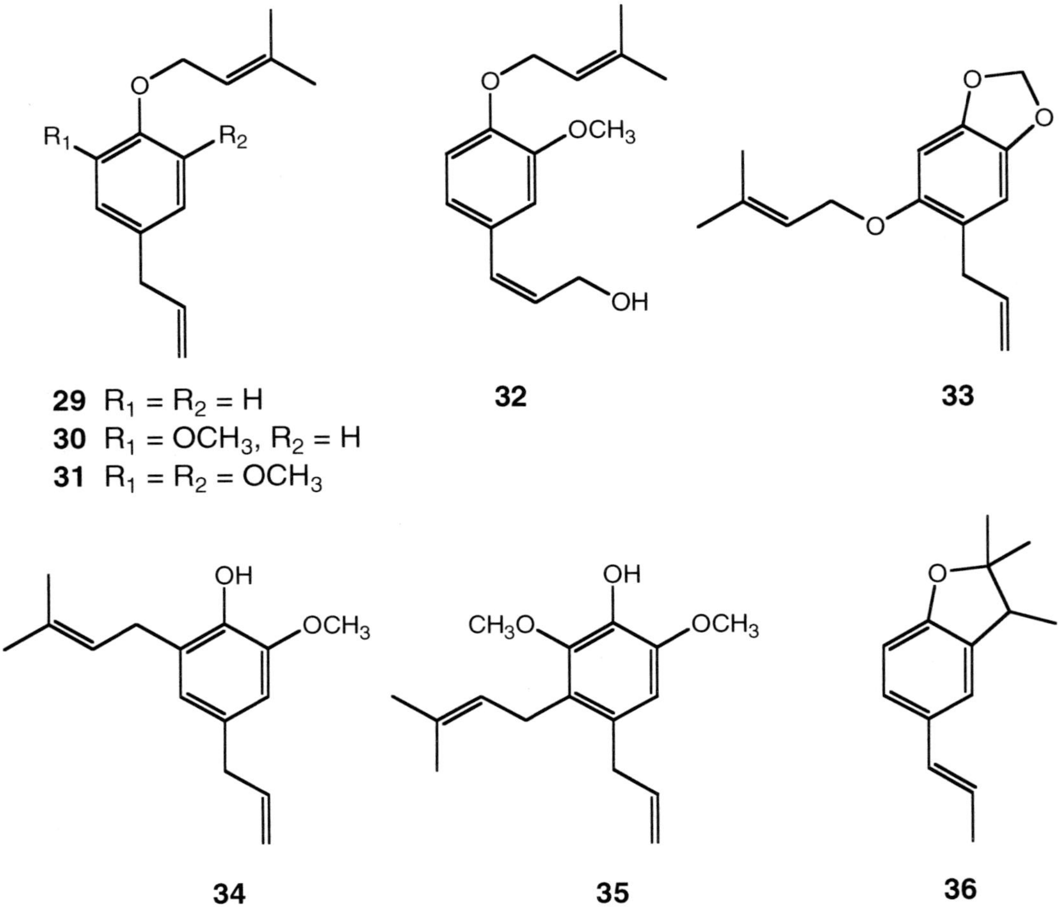

FIGURE 4.4 Prenylated phenylpropanoids (compounds **29–36**)

Bicycloillicinone asarone acetal (**54**) and bicycloillicinone (**55**), which was an enantiomer of the product **52** photochemically derived from 4*R*-illicinone A (**37b**), were isolated from the wood of *I. tashiroi* (Fukuyama et al., 1995, 1997) (Figure 4.7). Bicycloillicinone (**55**) should be photochemically transformed from 4*S*-illicinone A (**37a**) by essentially the same way as that used for the preparation of **52**. The unusual structure of **54** was elucidated by extensive two-dimensional nuclear magnetic resonance analysis and was identified by the acid-degraded product **56** and by 4,5-*O*-dimethyl-γ-asarone (**57**). The relative configurations at C-2 and C-4 were set up by the bicyclo[3,2,1]octane framework, and the relative stereochemistry of the remaining chiral carbons at C-3, C-11 was proved by the NOE (nuclear Overhauser effect) results, as shown in Figure 4.8. These spectral data indicate that **54** consists of a unique stereostructure; the γ-asarone acetal moiety takes an pseudoaxial orientation at the C-3 position, whereas the dimethylcarbinol group at the C-11 position lies inside the molecule. The formation of **57** may be presumably involved in a radical process, but its details have been a puzzle. It is worth noting that both compounds **54** and **55** could increase choline acetyltransferase activity at 30 μM in the culture of P10 rat septal neurons (Hatanaka et al., 1988).

Another example of prenylated C_6-C_3 compounds, illifunones **58–67**, belongs to tetrahydrofuran type, as shown in Figure 4.9 (Yakusijin et al., 1984; Fukuyama et al., 1994; Kouno et al., 1991, 1997). Illifunones **58–67** are probably obtained from illicinones C (**39**) and D (**40**) having an epoxide ring on the prenyl group, either by concerted cyclization with cleavage of the epoxy group

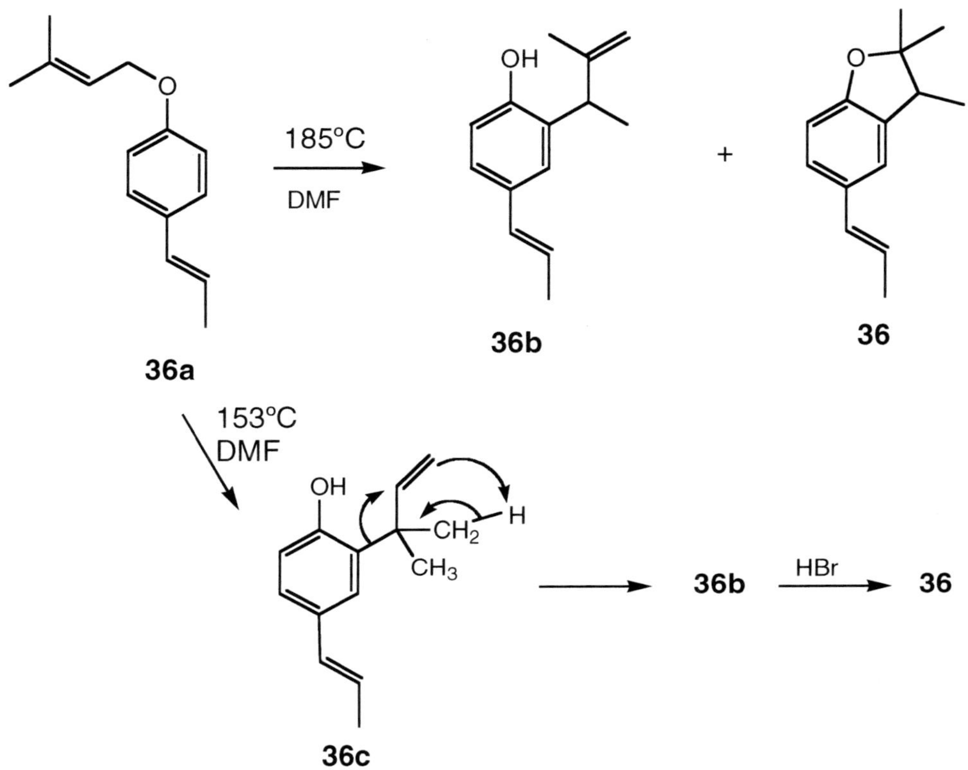

SCHEME 4.1 Chemical correlation of feniculin (**36a**) with anisoxide (**36**) by consecutive thermal rearrangements

(Yakushijin et al., 1984) or by 1,4-addition of the secondary alcohol derived from ring opening of the epoxide followed by hydrolysis of the methylenedioxy ring. The relative spatial relationship between the hydroxyl group at C-4 and the dimethylcarbinol group at C-11 in illifunones can be easily predicted on the basis of the comparison of the chemical shift value for H-11, as summarized in Figure 4.9 (Fukuyama et al., 1994). Namely, H-11 in illifunones A (**59**) and B (**61**), and 12-*O*-methylillifunone A (**60**), and 2,3-dihydro-5,6-di-*O*-methylillifunone E (**67**), having an *anti*-hydroxyl group, appeared at a higher field than δ 4.4 ppm, whereas the presence of a *syn*-hydroxyl group as in illifunones C (**62**) and D (**65**), 12-chlroroillifunone C (**63**), 2,3-dehydroillifunone C (**58**), and 4,12-di-*O*-methylillifunone C (**64**) caused a significant low field shift (δ 4.6~4.8 ppm) for H-11. Among them, illifunone E (**66**) was not consistent with this diagnostic difference value. Thus, the structure of illifunone E (**66**) was confirmed by dehydration of **66** with acid converting to illifunone A (**59**) (Fukuyama et al., 1994).

There are novel prenylated C_6-C_3 compounds such as illicinone E (**68**), as shown in Figure 4.10, which belong to a tetrahydrofurano type with a methylenedioxy group (Fukuyama et al., 1992, 1994; Kouno et al., 1997). These compounds occur so far in only *I. tashiroi* and *I. anisatum*. Taking their biosynthesis into consideration, however, other *Illicium* species should elaborate illicinone E-related compounds.

The absolute configuration of illicinone E (**68**) was established on the basis of CD exciton chirality rule (Harada and Nakanishi 1982) of allylic benzoates **81a** and **81b**, derived from **68** by the following ways, as shown in Scheme 4.5. Illicinone E was reduced with $NaBH_4$-$CeCl_3$ (Luche, 1978) to afford two alcohols, **79a** and **79b**, which were readily separated by silica gel chromatography. The allylic double bond in **79a** and **79b** was selectively hydrogenated over Wilkinson's

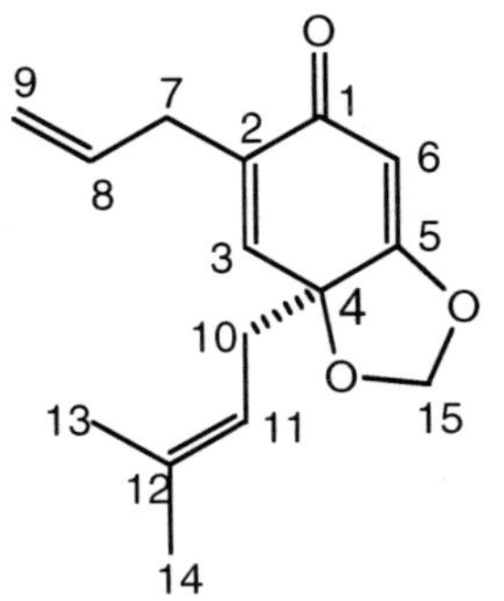

37a

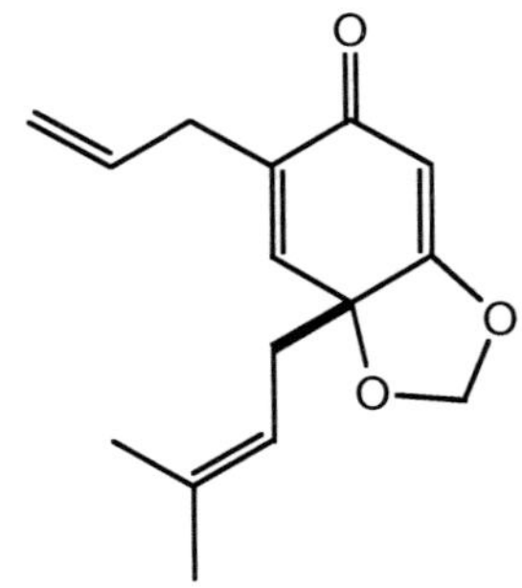

37b

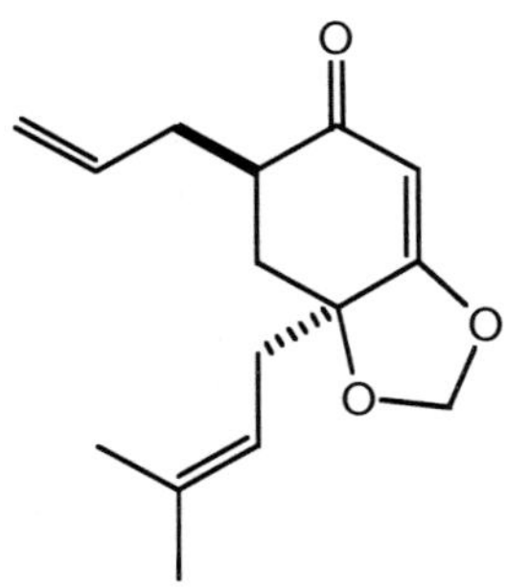

38a

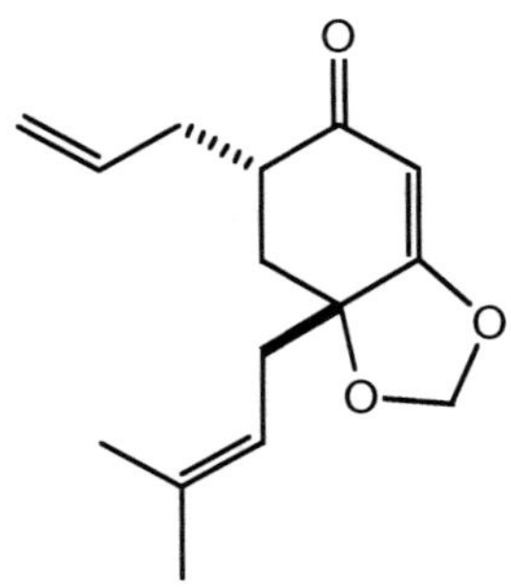

38b

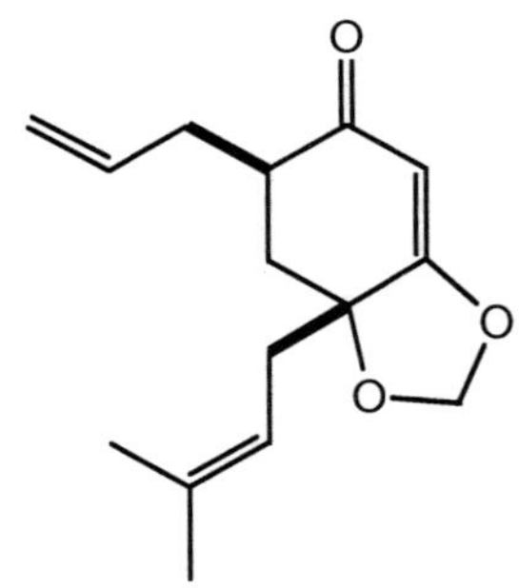

38c

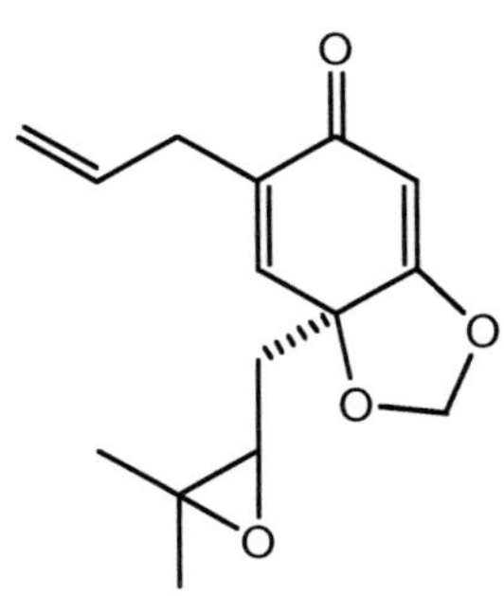

39a

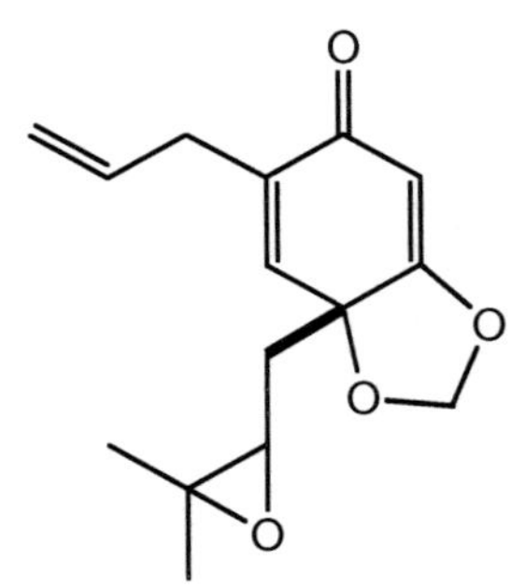

39b

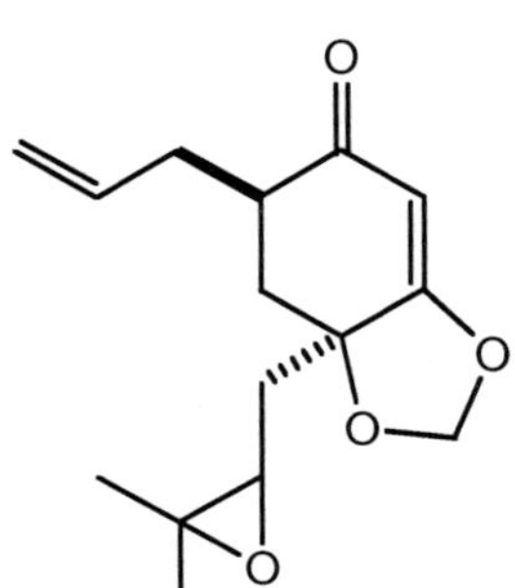

40a

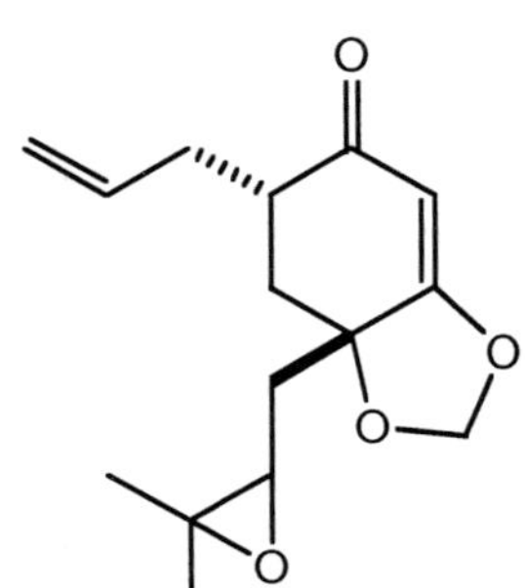

40b

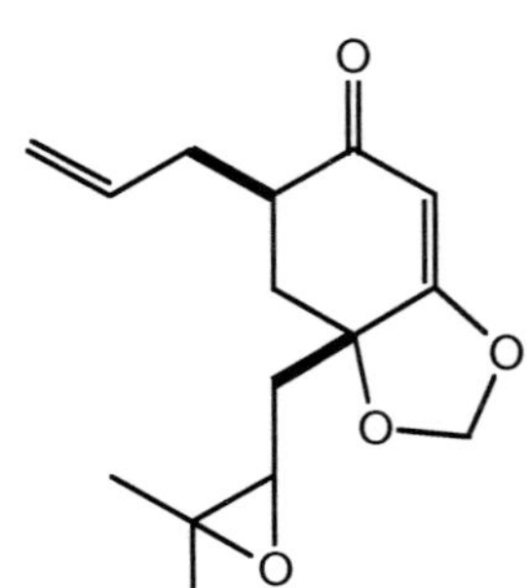

40c

FIGURE 4.5 Prenylated C_6-C_3 compounds **37–40** occurring in *I. tashiroi* and *I. arborescens*

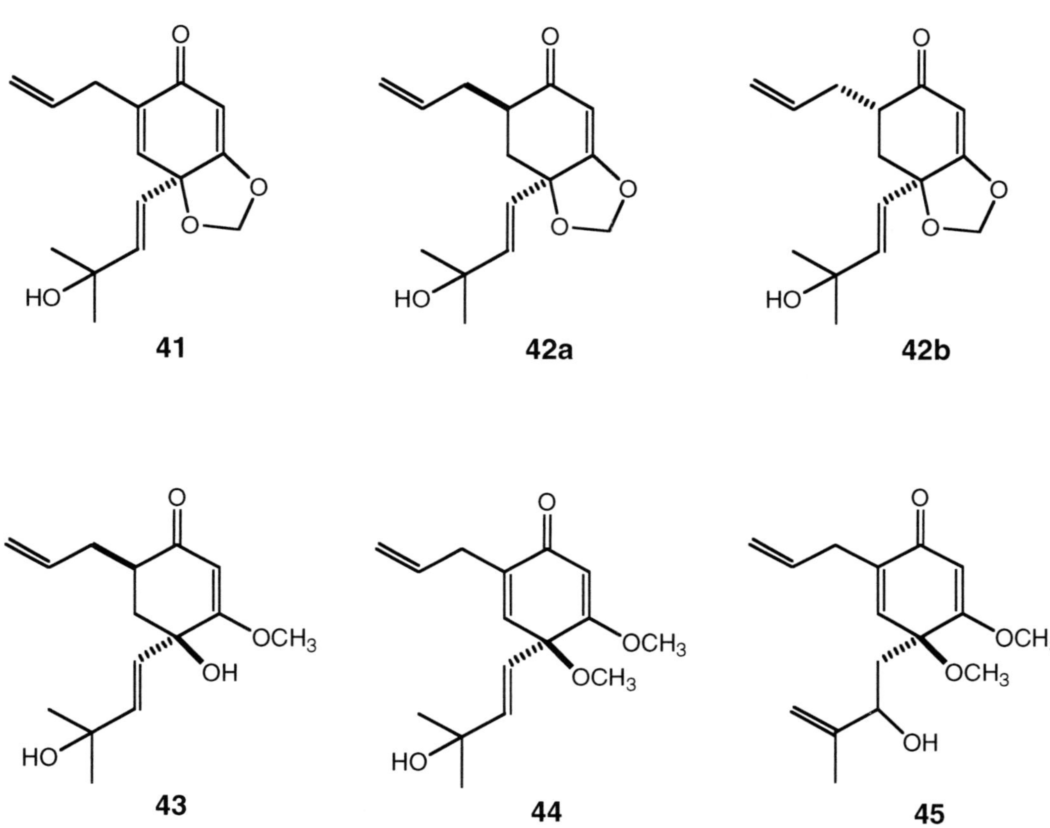

FIGURE 4.6 Prenylated C_6-C_3 compounds **41–45** isolated from *I. tashiroi* and *I. anisatum*

catalyst to yield **80a** and **80b**, which in turn were converted into *p*-bromobenzoyl derivatives **81a** and **81b**, respectively, with *p*-bromobenzoyl chloride-pyridine. The relative configurations of the *p*-bromobenzoyl groups in **81a** and **81b** were determined on the basis of NOEs, as shown in Figure 4.11. These results disclosed that the *p*-bromobenzoyl group is *cis* to the methylenedioxy group in **81a** and *trans* in **81b**. Thus, the relative configurations of **81a** and **81b** are ready to refer to their CD spectra. The CD spectra of **81a** and **81b** showed a negative Cotton effect at 240 nm and a positive Cotton effect at 239 nm, respectively, thereby allowing us to assign the absolute configuration at C-1 in **81a** and **81b** to *R* and *S*, respectively. Accordingly, the absolute configuration for illicinone E was determined to be (4*S*), (5*S*), and (11*R*) (Fukuyama et al., 1992).

(2*R*)-2,3-Dihydroillicinone E (72) and (2*S*)-2,3-dihydroillicinone E (**75**) were hydrogenated to yield the reduced products **72a** and **75a**, which were identical in all respects to each of the tetrahydro-derivatives transformed from illicinone E (**68**), respectively (Scheme 4.6). The chemical correlation indicated that **72** and **75** are stereoisomers with regard to C-2 in 2,3-dihydroillicinone E. Finally, NOE experiments could readily make a stereochemical difference to C-2 between **72** and **75**. However, the structure of illicinol E (**77**) was assigned as 1*R* and 2*R* on the basis of the following data: the relative stereochemistry of the acetate **77a** was elucidated as depicted in Figure 4.12, and the spectral data of **77** were identical with those of one of two products that were obtained by $NaBH_4$ reduction of (2*R*)-2,3-dihydroxyillicinone E (**72**).

Illicium plants biosynthesize not only simple prenylated C_6-C_3 compounds (Kouno et al., 1990; Hano et al., 1992; Kaku et al.1993) but also illicinone E (**68**) and its analogues **69–78**, which contain the methylenedioxy group in the molecule. Illicinone E (**68**) and its variants are chemical

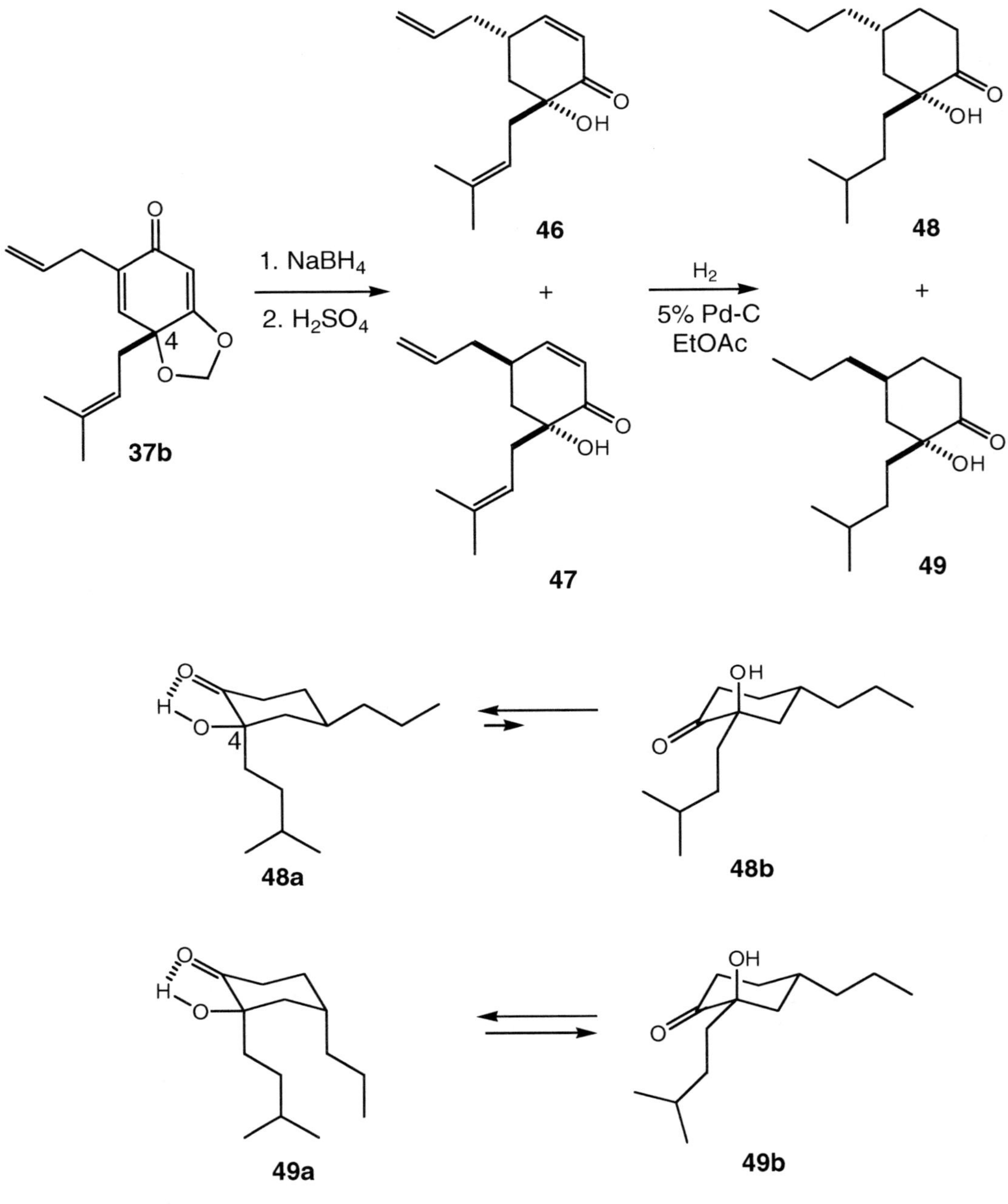

SCHEME 4.2 Absolute configuration at C-4 in illicinone A (**37**)

components characteristic of *Illicium* plants, which have not been so far found in the other plants (Fukuyama et al., 1994).

4.3 LIGNANS

Illicium was classified in the family Magnoliaceae in much of the early taxonomic literature. The chemical constituents of the genus of *Magnolia* are characterized by the presence of a series of unusual biphenyl-type neolignans. Thus, it is natural to search for biphenyl-type neolignans in

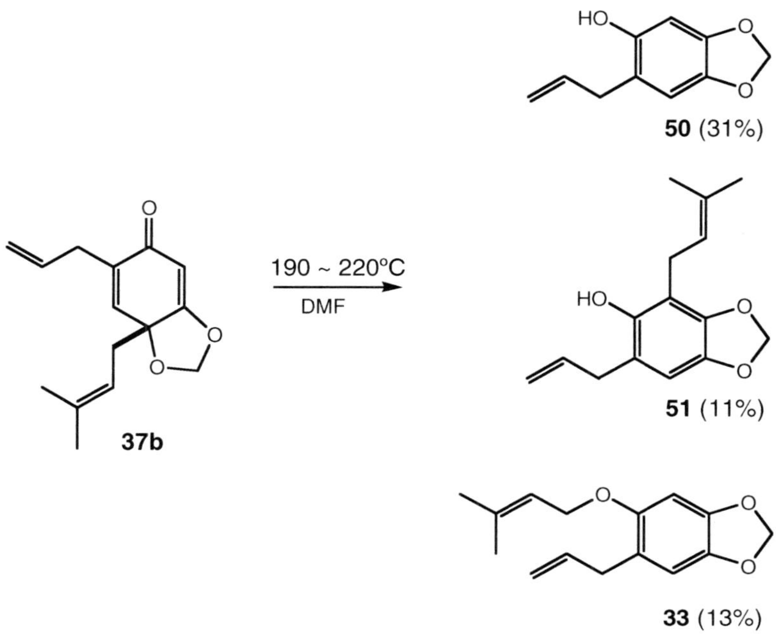

SCHEME 4.3 Thermal reaction of (4*R*)-illicinone A (**37b**)

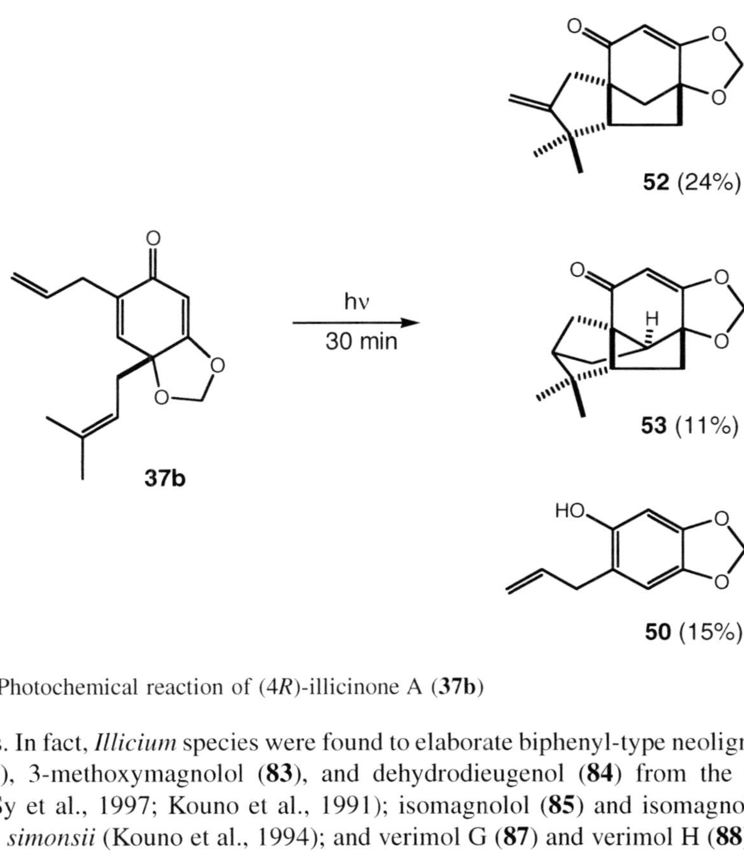

SCHEME 4.4 Photochemical reaction of (4*R*)-illicinone A (**37b**)

Illicium plants. In fact, *Illicium* species were found to elaborate biphenyl-type neolignans as follows: magnolol (**82**), 3-methoxymagnolol (**83**), and dehydrodieugenol (**84**) from the pericarps of *I. dunnianum* (Sy et al., 1997; Kouno et al., 1991); isomagnolol (**85**) and isomagnolone (**86**) from the barks of *I. simonsii* (Kouno et al., 1994); and verimol G (**87**) and verimol H (**88**) from *I. verum*

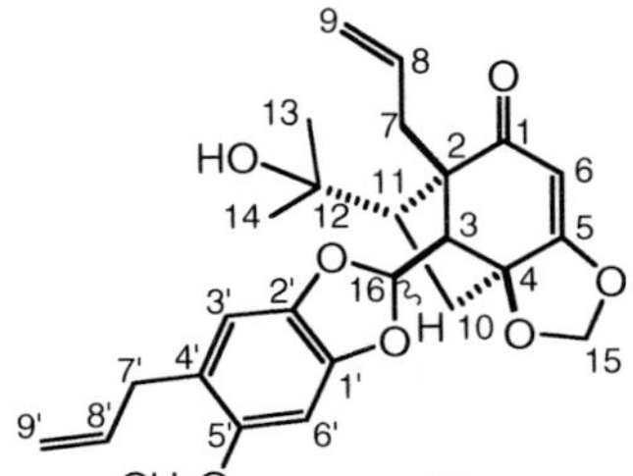

54

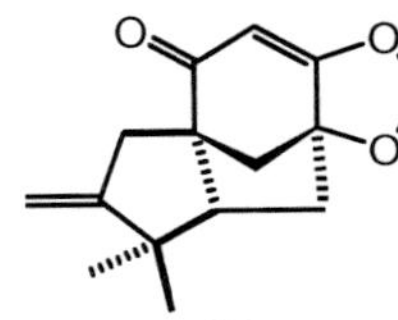

55

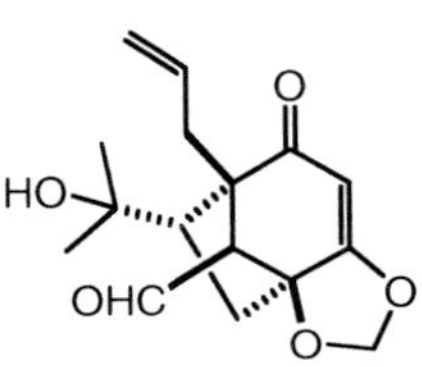

56

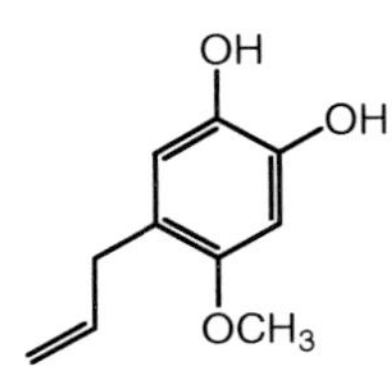

57

FIGURE 4.7 Bicycloillicinone asarone acetal (**54**) and bicycloillicinone (**55**) isolated from *I. tashiroi*

(Sy and Brown 1998) (Figure 4.13). In addition to these neolignans, another neolignan and its glycosides (**89–98**) were obtained from the pericarps of *I. majus* and the barks of *I. difengpi* (Kouno et al., 1991, 1993; Huang et al., 1996) (Figure 4.14). The configurations of C-7 and C-8 of **90–98** have not been determined, and Kouno et al. (1991) suggested that the configurations of **97** and **98** might be the same as that of Theander's compound (Theander et al., 1985) because of their nearly similar optical rotations. Moreover, some small signals in the ^{1}H-NMR of **91** and **92** indicated that they could be obtained as a mixture of diastereoisomers, respectively.

The most fascinating lignans, regarded as characteristic constituents of *Illicium* plants, are a series of triphenyl-type neolignans, called sesquineolignans (Figure 4.15). The first example is macranthol (**99**), from the pericarps of *I. macranthum*, which is presumably formed by both *ortho, ortho-* and *ortho, para*-oxidative coupling of three 4-allylphenol chavicol (Kouno et al., 1989). Because macranthol (**99**) was found in *Illicium macranthum*, some triphenyl-type neolignans, dunnianol (**100**), isodunnianol (**101**), and simonsinol (**102**), were isolated from the bark of *I. dunnianum* (Kouno et al., 1991) and *I. simonsii* (Kouno et al., 1994). Another sesquineolignan (**103**) has been isolated from the aerial part of *I. dunnianum* (Sy and Brown, 1996; Sy et al., 1997).

The compound **103** is of biosynthetic interest because it is derived from three 4-allylphenol-chavicol units linked together by both *ortho, ortho*-coupling and *ortho, para*-coupling. The first *ortho, ortho*-linkage occurs by standard oxidative coupling of the two molecules' chavicol radical (**105**), yielding magnolol (**82**). The second *ortho, para*-coupling (Scheme 4.7) then arises from oxidative coupling of magnolol radical (**104**), with a further molecule of chavicol leading to the formation of a trimer, which is unable to regain aromaticity at the central ring, thereby resulting in the formation of the second ether linkage between the central and right-hand rings by intramolecular Michael-type addition of the hydroxyl anion to the 5,6-double bond of the central ring (Sy and Brown, 1996).

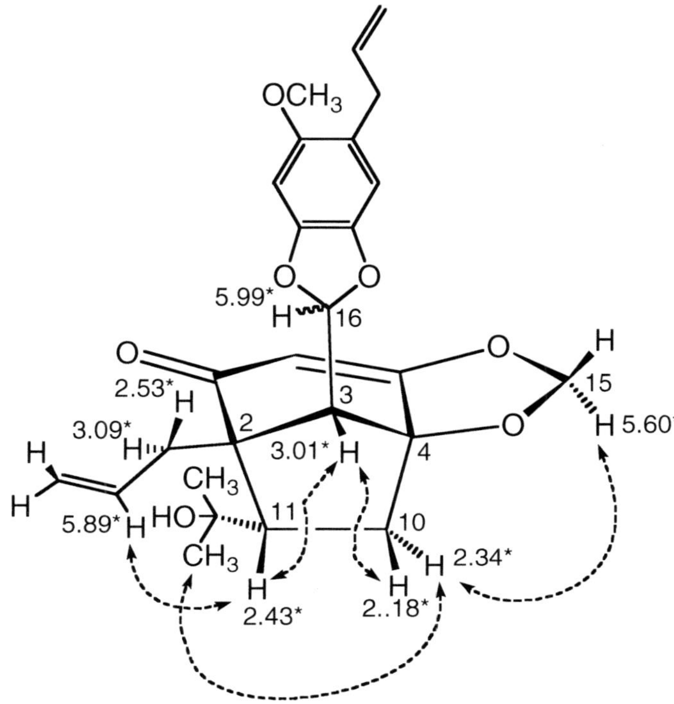

FIGURE 4.8 Relative configuration of compound **54** based on NOEs, indicated by arrows; asterisks show δ_H values

4.4 SESQUITERPENES

Shikimi (*I. anisatum* L.) has been known as a toxic plant since ancient times. It has been documented that people in Japan often were killed by unintentionally eating the fruits of the Japanese star anise (*I. anisatum* L.). Although the convulsive activity of the fruits of *I. anisatum* had been known for several centuries, it was only in 1952 that Lane et al. (1952) succeeded in the first isolation of a pure convulsive principle, which was named *anisatin* (Figure 4.16).

This result was reconfirmed in 1958 by Kawano and Matsuo (1958). In 1968, the structures of anisatin (**107**) and another convulsive principle, neoanisatin (**108**), were determined to be unique cage-shaped sesquiterpenes having a rare spiro β-lactone ring by Yamada and Hirata (Yamada et al., 1968). The structure of **107** was fully assigned on the basis of the both structures of the oxidatively degraded products, noranisatin (**107a**) and noranisatinone (**107b**), which had been already established by the spectral data (Yamada et al., 1965) and x-ray crystallographic analysis of bromonoranisatinone (**107c**) (Sakabe et al., 1965) (Scheme 4.8). Later, the structure of anisatin was confirmed by x-ray crystallographic analysis (Wong et al., 1988). The total syntheses of (-)-anisatin (**107**) and (-)-noranisatin (**107a**) were first accomplished by Niwa et al. (1990, 1991), although several groups challenged synthesis of its complex molecule (Linder et al., 1982; Kende and Chen, 1985).

Anisatin causes picrotoxin-like convulsions in mice and dogs (Kajimoto et al., 1955), and the toxicity of anisatin and neoanisatin is much stronger than that of picrotoxin in the estimation of their LD_{50}: 1 mg/kg for anisatin, in mice (i.p.). The neuropharmacological study of anisatin demonstrates that it is a potent noncompetitive GABA antagonist like picrotoxin (Kudo et al., 1981).

Kouno, one member of Kawano's group who had continued to investigate the toxic substance in *I. anisatum* (Kawano et al., 1958), succeeded in systematic studies on chemical components in *Illicium* plants. A number of anisatin-related sesquiterpenes, which have been reported by his group,

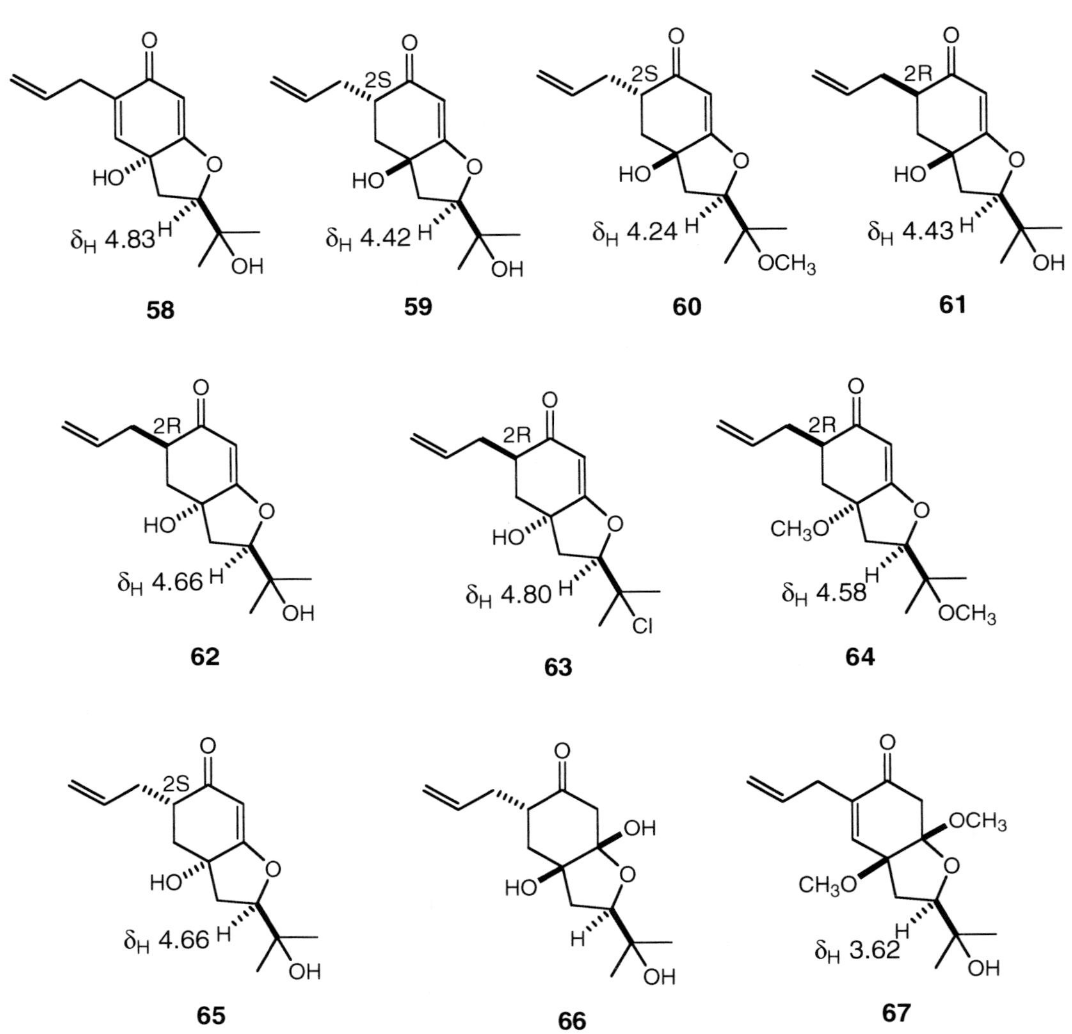

FIGURE 4.9 Tetrahydrofuran-type prenylated C_6-C_3 compounds **58–67** and chemical shift values (δ_H) for H-11 in illifunones

can be classified into some subgroups as follows: anisatin (Figure 4.17), pseudoanisatin (Figure 4.18 and Figure 4.19), minwanensin (Figure 4.20), majucin (Figure 4.21), pseudomajucin (Figure 4.22), and cycloparvifloralone types (Figure 4.23), and some rare carbon skeletons (Figure 4.24).

The presence of very toxic anisatin is notified in many *Illicium* plants, such as *I. anisatum*, *I. majus*, *I. floridanum*, *I. minwanense*, and *I. merrillianum* (Lane et al., 1952; Kouno et al., 1989; Shimidt et al., 1998; Wang et al., 1994; Huang et al., 1999). The examples of anisatin-type sesquiterpenes are shown in Figure 4.17. All of them except for **114** belong to neoanisatin derivatives lacking the hydroxyl group at the C-3 position. 1-Hydroxyneoanisatin (**109**), 6-deoxy-1-hydroxyneoanisatin (**110**), 2-oxo-6-deoxyneoanisatin (**115**), and an epimeric mixture of **116** and **117** were isolated from *I. majus* (Yang et al., 1990; Kouno et al., 1991), whereas 2 α-hydroxyneoanisatin (**111**) was isolated from the pericarps of *I. anisatum* (Kouno et al., 1991). It should be noted that veranisatins A (**112**), B (**113**), and C (**114**) were isolated as new convulsants in little amounts from nontoxic Chinese star anise (*I. verum* Hook. f.), which has been used as a basic spice and in traditional Chinese and Japanese medicines. They caused severe convulsions and death at 3 mg/kg

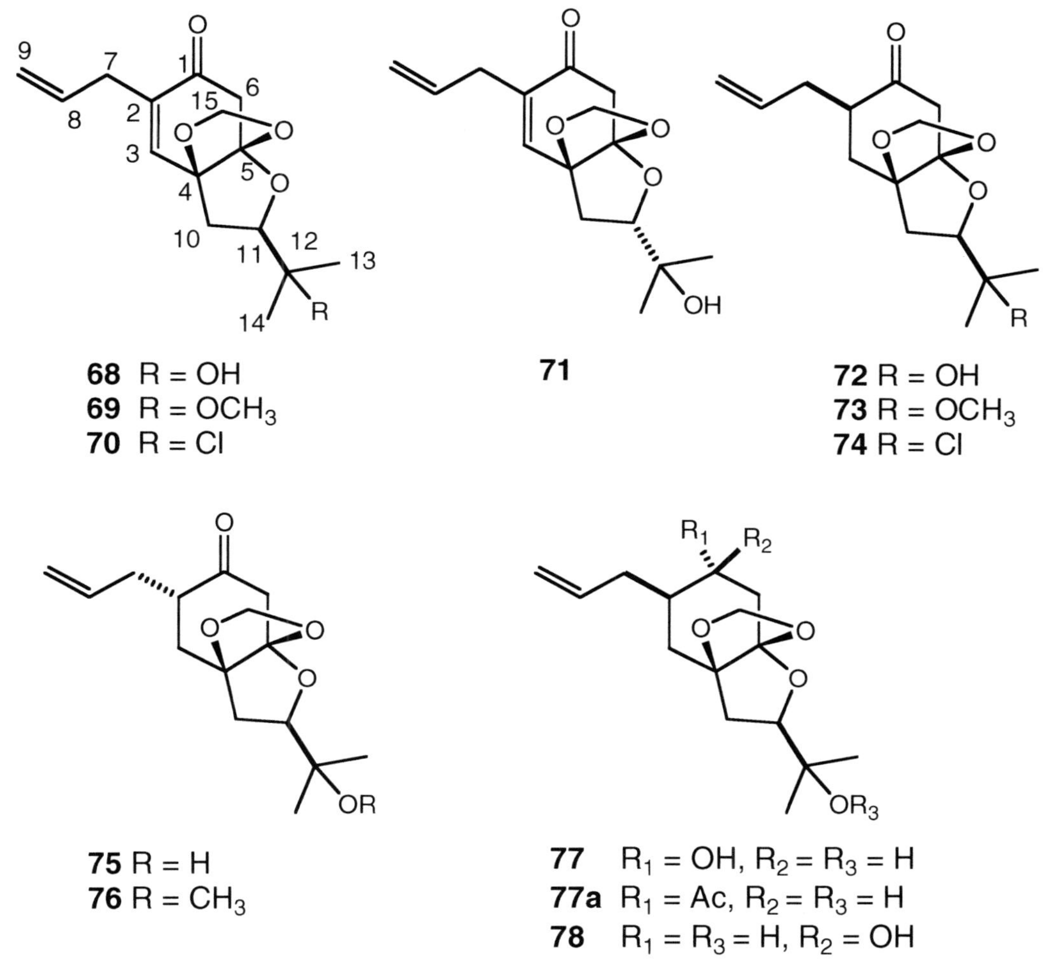

FIGURE 4.10 Illicinone E (**68**) and its related prenylated C_6-C_3 compounds

(p.o.) in mice (three of three) (Okuyama et al., 1993; Nakamura et al., 1996). Compound **115** also exhibited picrotoxin-like convulsion, and its LD_{50} was 1.46 mg/kg in mice (i.p.) (Yang et al., 1990).

Pseudoanisatin was first isolated as a nontoxic compound from *I. anisatum* by Lane et al. (1952). The wrong structure of pseudoanisatin was first proposed on the basis of the spectral data (Okigawa and Kawano, 1971), but later was revised as **118** by x-ray crystal structure determination (Kouno et al., 1983). Treatment of **118** with sodium methoxide in methanol gave rise to a trans-lactonisation product **118a**, being of special interest, with the unusual inversions at C-4 and C-5. The structure of **118a** was unambiguously determined by x-ray crystallographic analysis. The two plausible mechanisms for this base-catalyzed transformation were proposed as shown in Scheme 4.9 (Kouno et al., 1984).

Other pseudoanisatin-type sesquiterpenes, 6-deoxy-pseudoanisatin (**120**) (Kouno et al., 1988) and **121** (Tanaka et al., 1998), occurred in the seeds, fruits, and leaves of *I. anisatum*. The structure of 7-deoxy-7-hydroxypseudoanisatin (**123a**) (Kouno et al., 1991) from the seeds of *I. anisatum* was first elucidated as pseudoanisatin-type sesquiterpene, and then was revised as **123b** and finally corrected as **123** (Figure 4.20), which belongs to the minwanensin type. Other *Illicium* plants, *I. dunnianum, I. tashiroi*, and *I. merrillianum*, elaborated dunnianin (**124**), 6-deoxydunnianin (**126**) (Kouno et al., 1988), 3-benzoylpseudoanisatin (**119**) (Huang et al., 1996), isodunnianin (**125**)

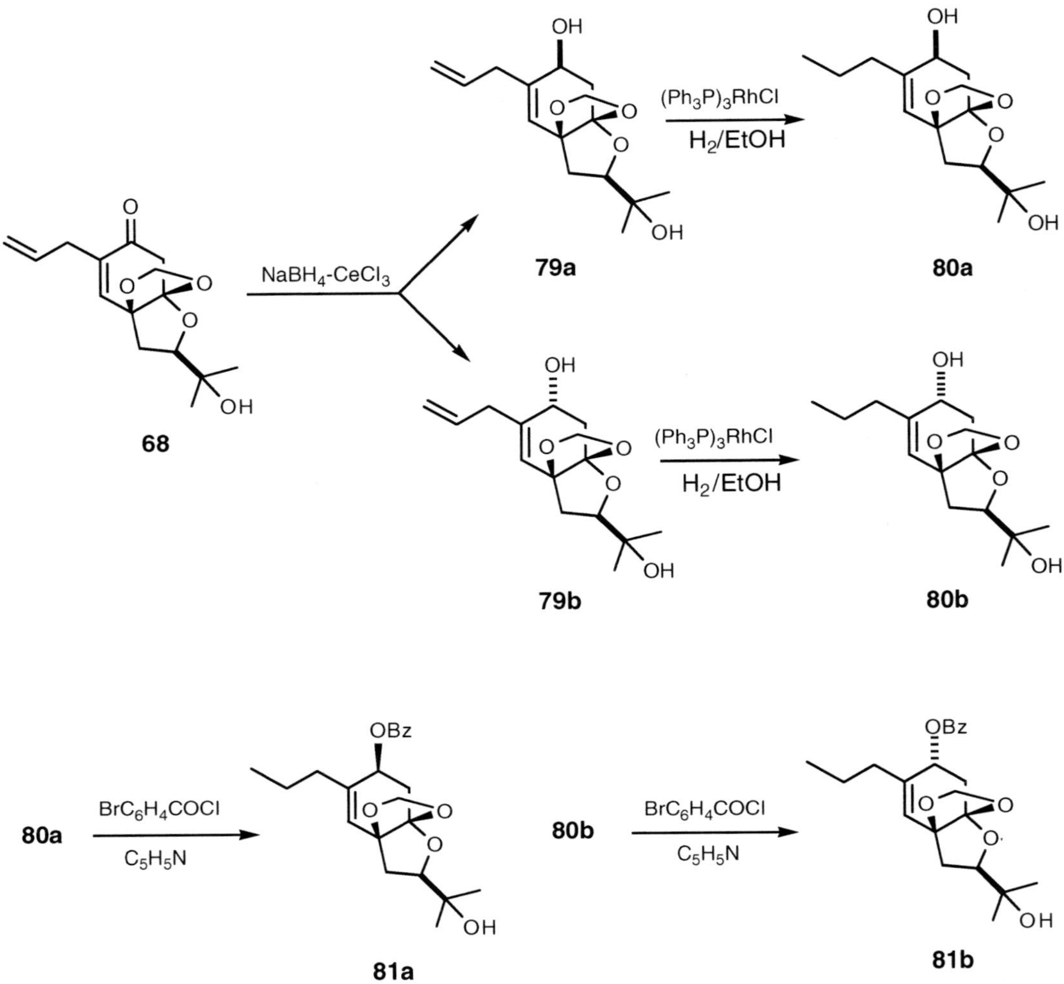

SCHEME 4.5 Chemical conversion from illicinone E (**68**) to *p*-bromobenzoates **81a** and **81b**

(Fukuyama et al., 1993), and 1-hydroxy-3-deoxypseudoanisatin (**122**) (Huang et al., 1999). The structures of these compounds were assigned on the basis of nuclear magnetic resonance spectral data and by comparison with spectral data of pseudoanisatin (**118**) and dunnianin (**124**). Schmidt and Peters (1997) reported that dunnianin (**124**), 7-deoxy-7-hydroxypseudoanisatin (**123a**), and 3-benzoylpseudoanisatin (**119**) were isolated from the American star anise, *I. floridanum*, and their structures were reinvestigated by nuclear magnetic resonance spectroscopic analyses as well as by x-ray crystallographic analysis of dunnianin. As a result, the structures of **119**, **124**, and **123a** were revised as **119a**, **124a**, and **123b**, respectively, which consist of an 11, 3-δ-lactone instead of an 11, 14-ε-lactone. In comparison of their spectral data, the occurrence of the very large geminal coupling constant (near 20 Hz), accounting for the presence of δ-lactone, and the appearance of H-3 as a doublet (about 5 Hz) with essentially no coupling with H-2, were found to be characteristic of sesquiterpenes having an 11, 3-δ-lactone. In light of the spectral data of the other pseudoanisatin-type sesquiterpenes **120**, **125**, and **126**, the structure of isodunnianin (**125**) and 6-deoxydunnianin (**126**) should be revised as **125a** and **126a**. In 1999, the structure of **123b** was revised one more time as **123**. The 11, 3-δ-lactone ring was revised to a δ-lactone closed between C-11 and C-7. The vicinal coupling constant between H-2 and H-3 of about 0 Hz, however, was not agreeable with such a structure. After all, the x-ray crystallographic analysis proved that the structure of **123**

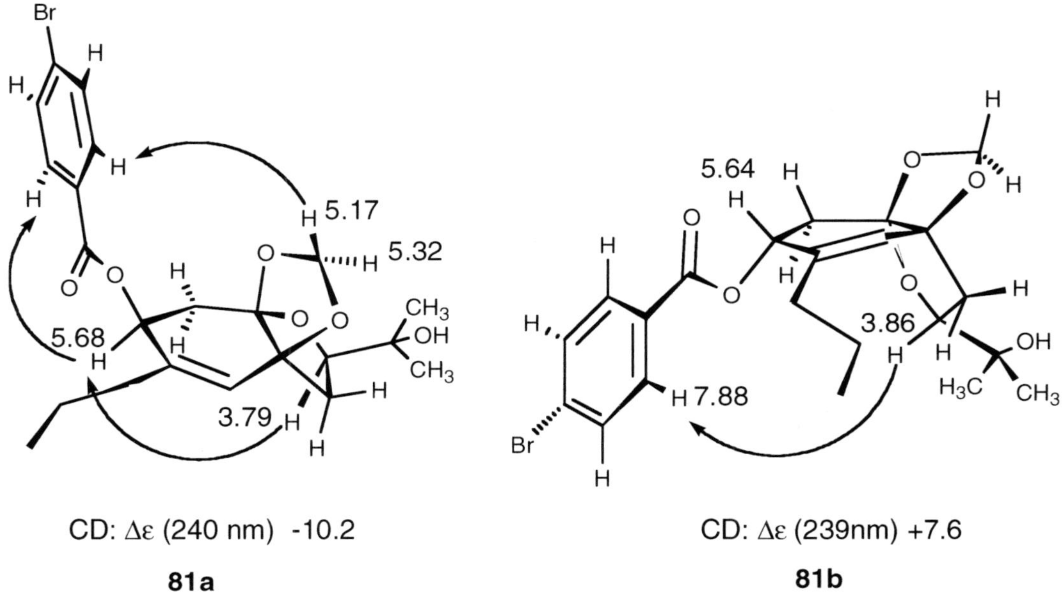

FIGURE 4.11 Conformation of *p*-bromobenzoates **81a** and **81b** based on NOEs, indicated by arrows, and the results of their CD spectra

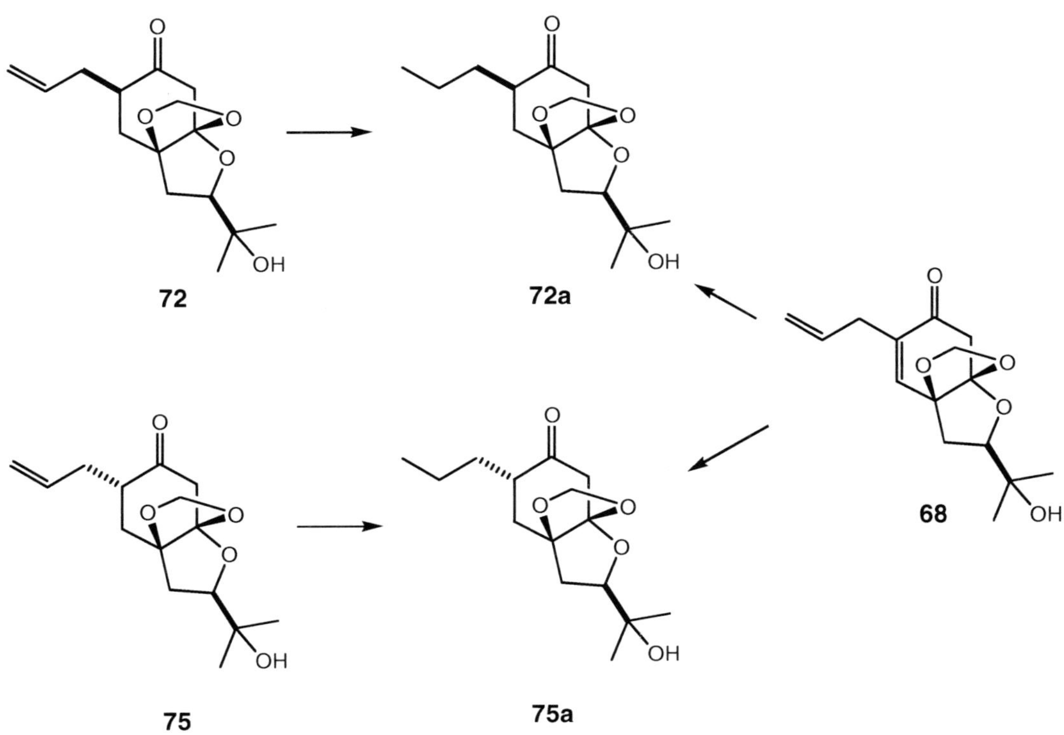

SCHEME 4.6 Chemical correlation of **72** and **75** with **68**

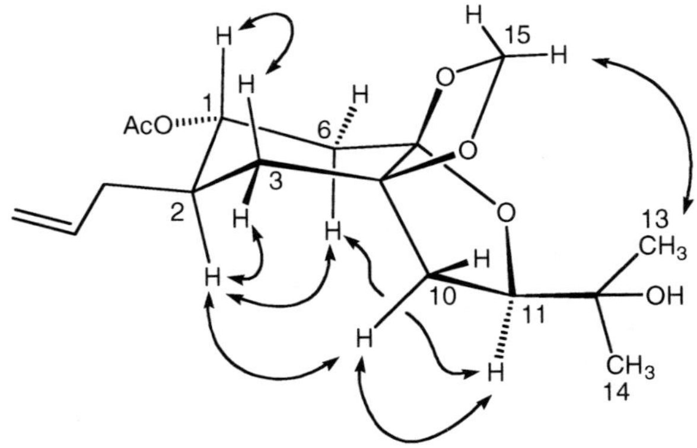

FIGURE 4.12 Relative stereochemistry of **77a** based on NOEs, indicated by arrows

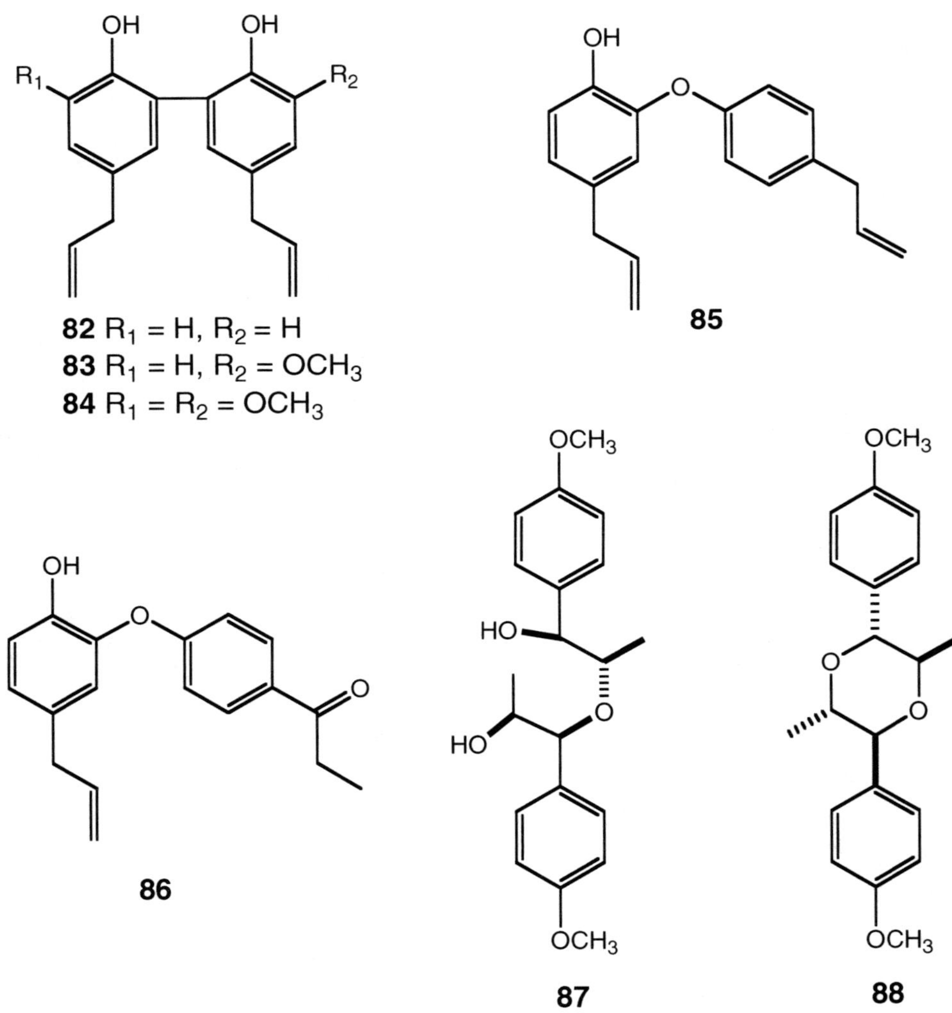

FIGURE 4.13 Neolignans (compounds **82–88**)

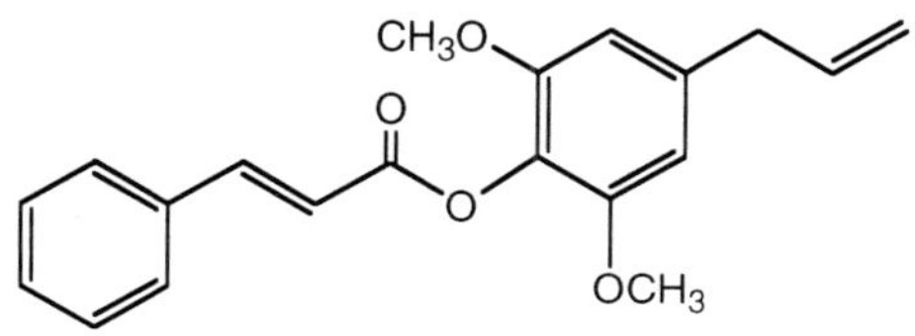

89

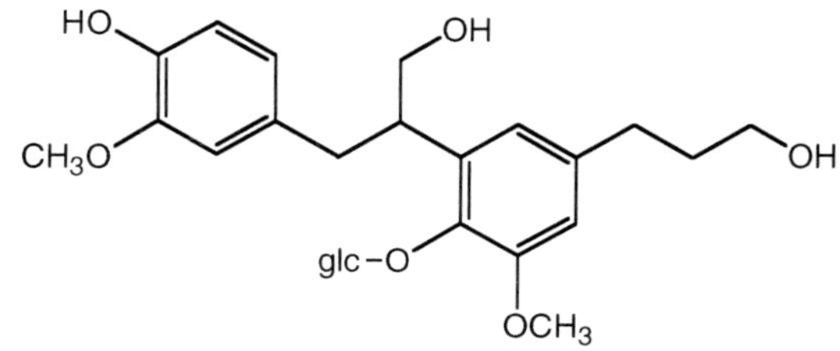

90

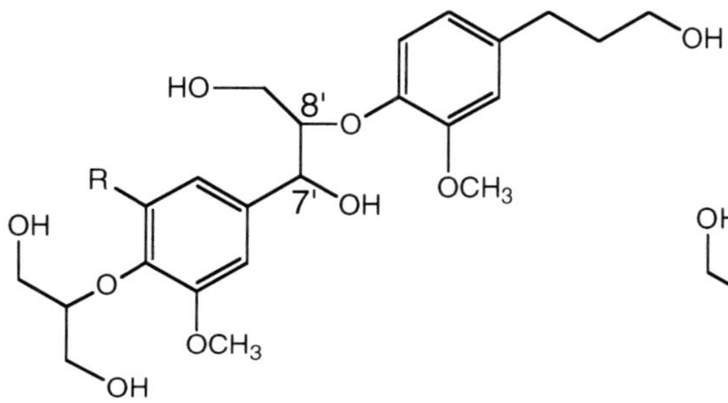

91 R=H
92 R=OCH_3

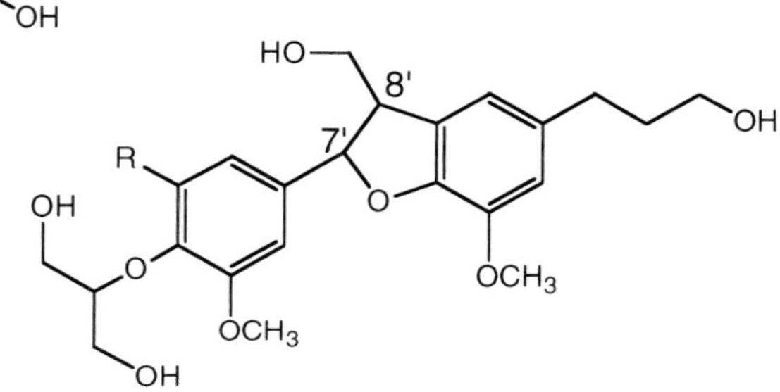

93 R=H
94 R=OCH_3

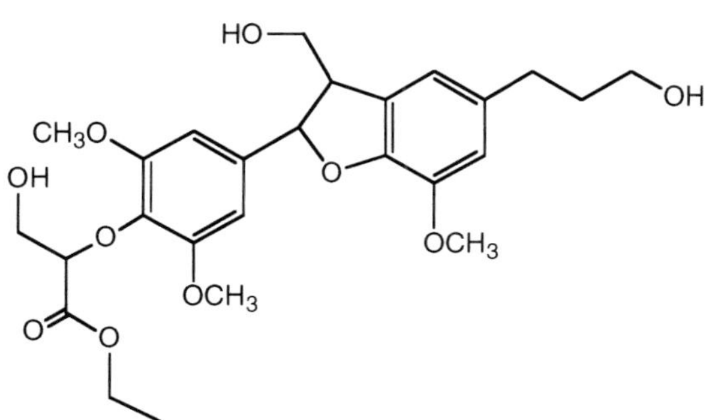

95

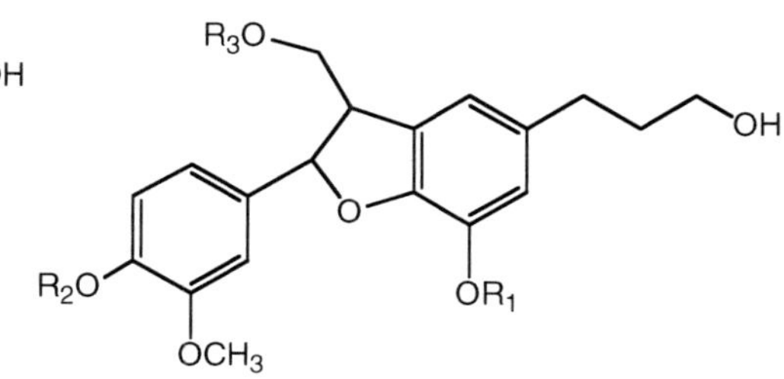

96 R_1=CH_3, R_2=H, R_3=xyl
97 R_1=R_3=H, R_2=rha
98 R_1=CH_3, R_2=rha, R_3=H

FIGURE 4.14 Neolignans (compounds **89–98**)

was correct, and the aforementioned spectral discrepancy could be attributed to a strong intramolecular hydrogen bond between the OH group attached to C-14 and C-3.

Some pseudoanisatin-type sesquiterpenes were found to coexist as a ketone and hemiacetal equilibrium (Figure 4.19), such as pseudoanisatin (**118**) and 4,7-hemiketal of pseudoanisatin (**127**), parviflorolide (**128**), cycloparviflorolide (**129**) (Schmidt et al., 1999), merrillianolide (**130**), and cyclomerrillianolide (**131**) (Huang et al., 1999). In addition, an unusual sesquiterpene neodunnianin (**132**) was isolated from the pericarps of *I. dunnianum* (Huang et al., 1997), featured by an 11,15-δ-lactone. Pseudoanisatin derivatives, which have been so far checked, are nontoxic substances unlike anisatin, but it is worthy of note that isodunnianin (**125**) not only promotes neurite outgrowth

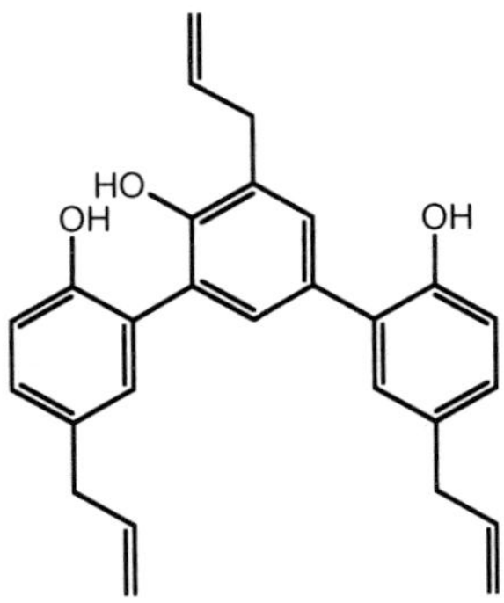

99

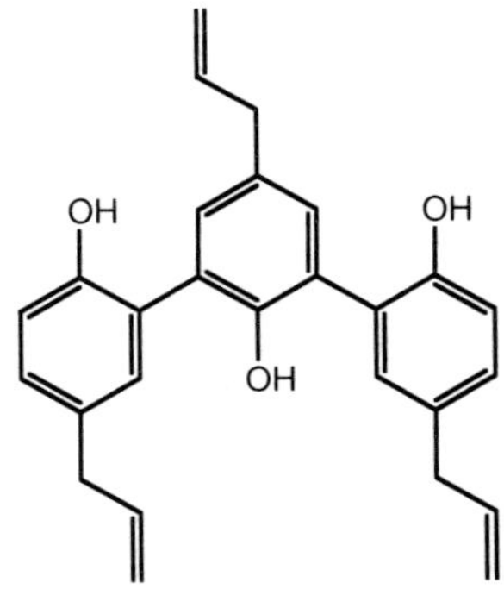

100

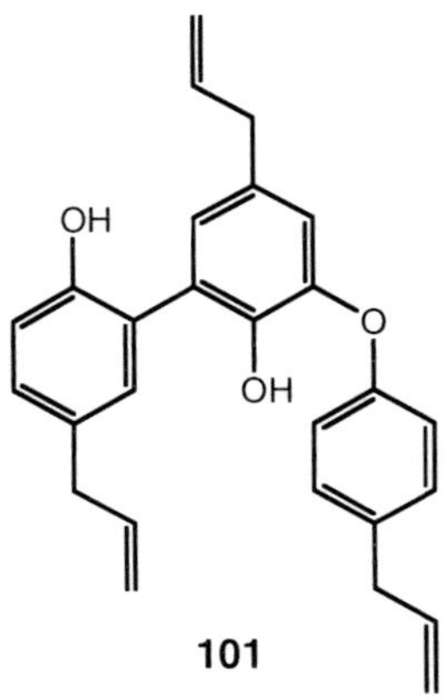

101

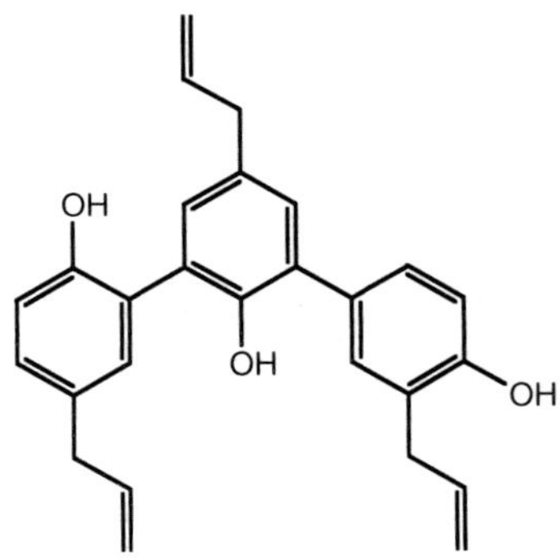

102

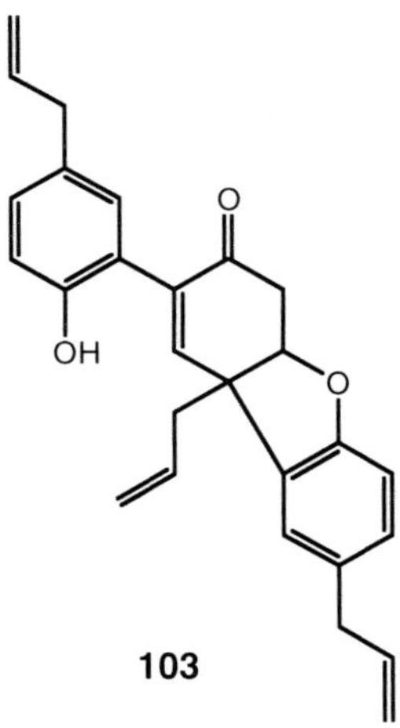

103

FIGURE 4.15 Neolignans (compounds **99–103**)

in the primary culture of fetal rat cerebral hemisphere at 10 μM but also increases choline acetyltransferase activity (Fukuyama et al., 1993).

The opening of the spiro β-lactone of anisatin-type sesquiterpene leads to minwanensin type. After the structure of minwanensin (**133**) from the pericarps of *I. minwanense* was elucidated (Figure 4.20), this type of compound such as 3-acetoxy-14-*n*-butyryloxy-10-deoxyfluoridanolide (**134**), 14-acetoxy-3-oxofloridanolide (**135**), 13-acetoxy-14-(*n*-butyryloxy)fluoridanolide (**136**), and debenzoyldunnianin (**123**) were also found in the fruits of *I. floridanum* (Schmidt et al., 1998). The structure of debenzoyldunnianin was first proposed as **123a** and/or **123b** (Kouno et al., 1991;

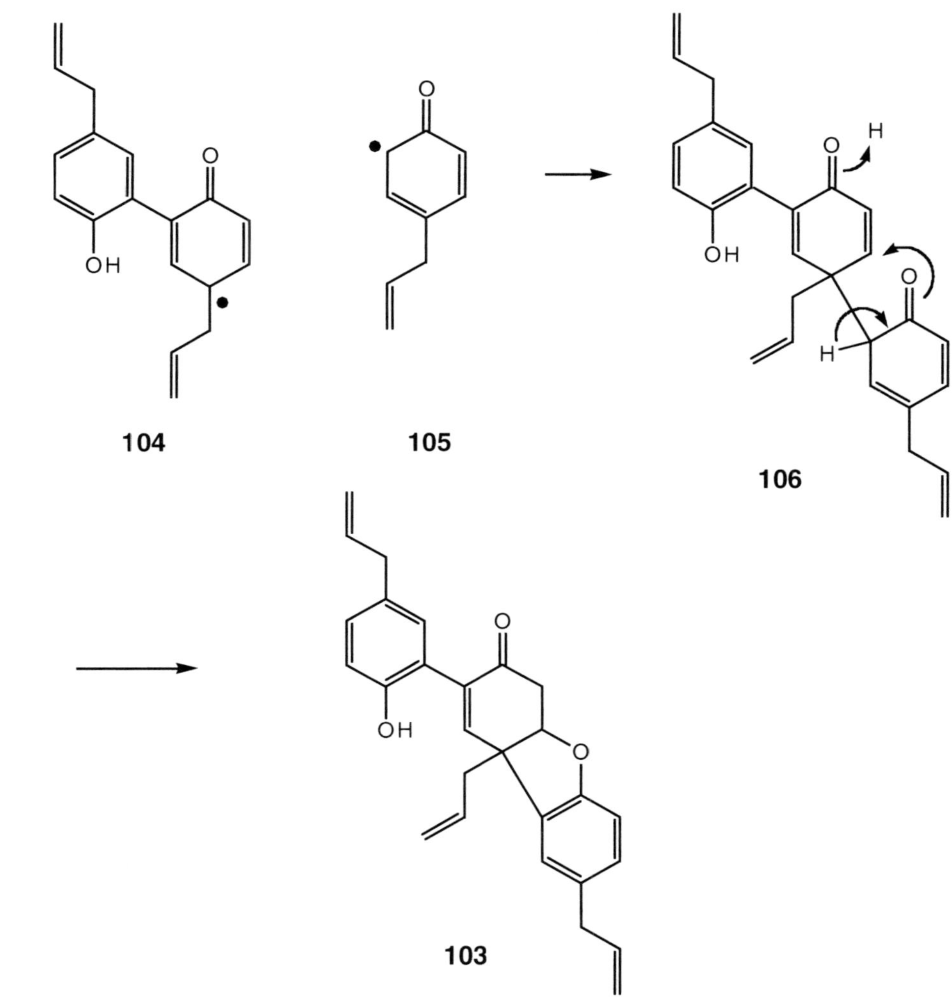

SCHEME 4.7 Possible biosynthesis of **103**

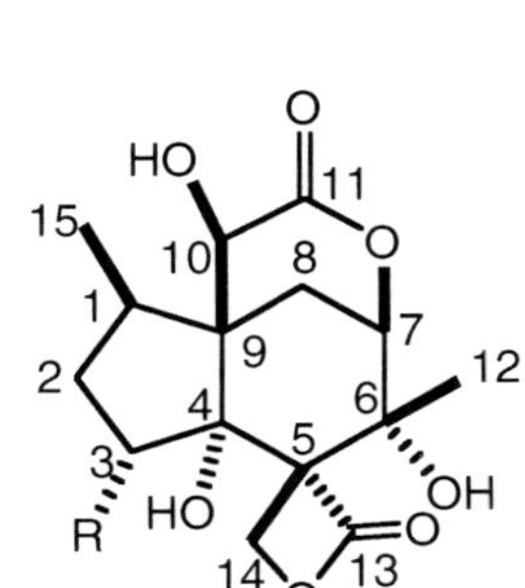

107 R = OH
108 R = H

FIGURE 4.16 Anisatin (**107**) and neoanisatin (**108**)

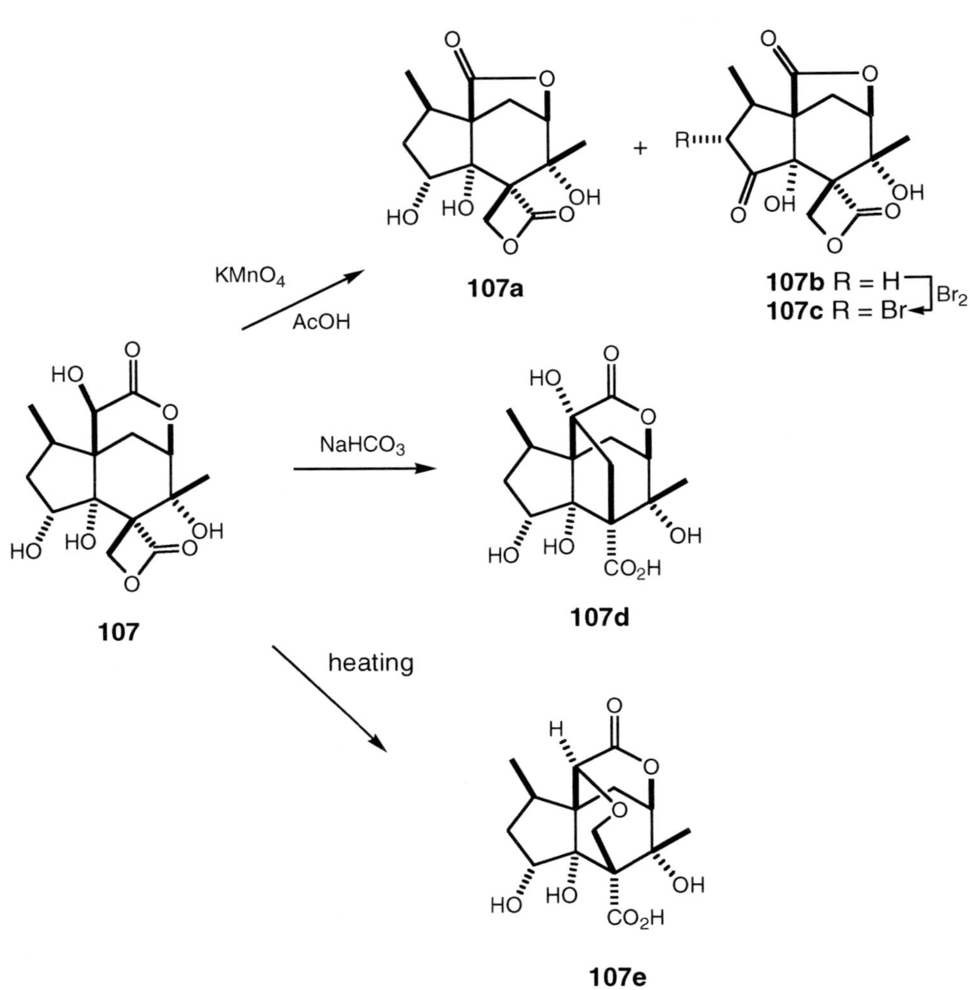

SCHEME 4.8 Chemical characteristics of anisatin (**107**) (Yamada et al., 1968)

Schmidt and Peters, 1997), but the correct structure **123** was finally established by x-ray crystallographic analysis. The structure of minwanensin was considered to be similar to that of anisatin, but it does not show toxicity against mouse at the dose of 50 mg/kg. This indicates that the presence of β-lactone in a molecule may be responsible for toxicity.

A number of new majucin-type sesquiterpenes having a γ-lactone ring, as shown in Figure 4.21, were found in the pericarps of *I. majus,* belong to one of the Chinese *Illicium* plants. Majucin (**137**) was assigned the structure by extensive spectroscopic analysis and by comparing its nuclear magnetic resonance data with those of anisatin as well as of neomajucin (**138**), established by an x-ray diffraction method (Yang et al., 1988; Kouno et al., 1989). More majucin-type sesquiterpene lactones, (2*S**)-hydroxyneomajucin (**139**), 2-oxoneomajucin (**142**), 2,3-dehydromajucin (**143**), (2*R**)-hydroxy-3,4-dehydroneomajucin (**144**), (1*S**)-2-oxo-3,4-dehydroneomajucin (**145**), (1*R**)-2-oxo-3,4-dehydroneomajucin (**146**), and (1*R**, 10 S*) -2-oxo-3,4-dehydroneomajucin (**147**), were isolated (Kouno et al., 1989, 1990). It should be noted in particular that the compound bearing the (10*S**)-hydroxyl group is only **147** among the anisatin-like sesquiterpenes. 6-Deoxyneomajucin (**140**), isolated from the seeds of *I. anisatum,* is the first example of majucin type found in Japanese star anise (Kouno et al., 1988). *I. angustisepalum* also contains majucin-type sesquiterpene like 10-benzoyl ester of neomajucin (**141**) (Sy and Brown, 1998).

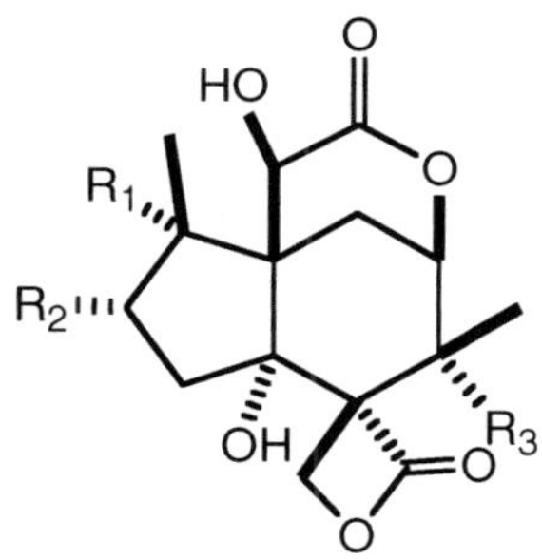

109 $R_1 = R_3 = OH, R_2 = H$
110 $R_1 = OH, R_2 = R_3 = H$
111 $R_1 = H, R_2 = R_3 = OH$

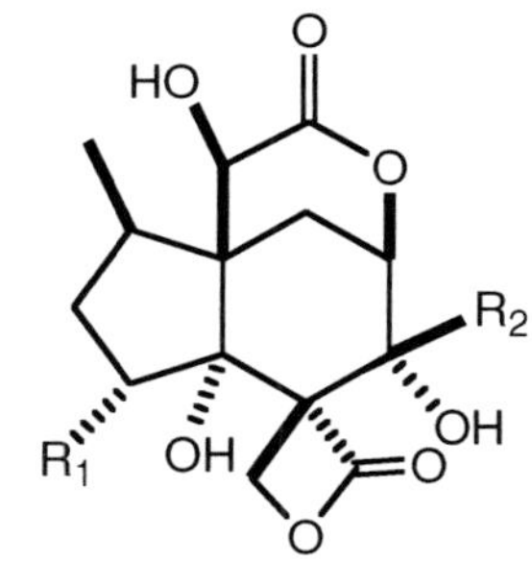

112 $R_1 = H, R_2 = CH_2OCH_3$
113 $R_1 = H, R_2 = CO_2CH_3$
114 $R_1 = OH, R_2 = CO_2CH_3$

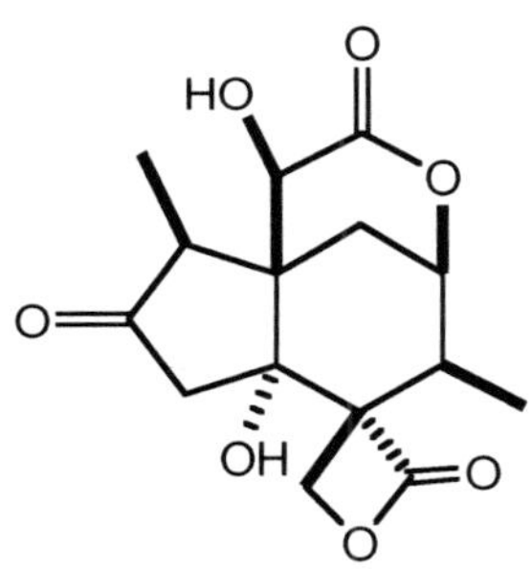

115

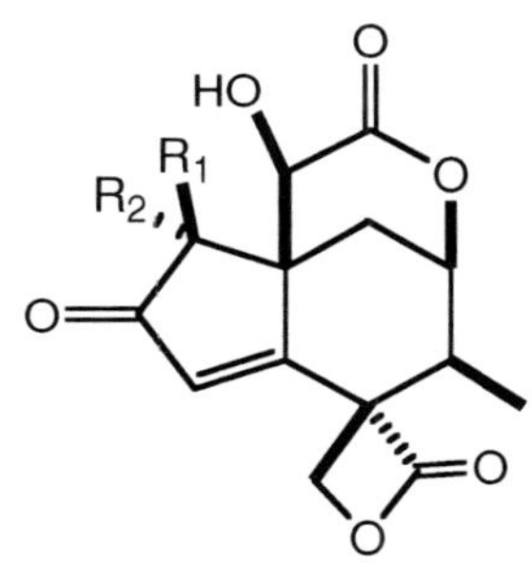

116 $R_1 = H, R_2 = CH_3$
117 $R_1 = CH_3, R_2 = H$

FIGURE 4.17 Anisatin-type sesquiterpenes (compounds **109–117**)

As shown in Scheme 4.9, base treatment of pseudoanisatin (**118**) led to the formation of the γ-lactone **118a** with the inversion at C-4. Pseudomajucin (**148**) and its 7-*O*-β-D-glucoside (**149**), which were isolated from the pericarps of *I. majus* (Kouno et al., 1989), were found to consist of the same carbon skeleton as that of **118a**. The structure of **148** was established by x-ray crystallographic analysis. The glucoside linkage position in **149** was determined as C-7 by glycosylation shift (2.6 ppm) at C-7 compared with that of **148**. No pseudomajucin-type has so far been known as a natural product except **148** and **149** (Figure 4.22).

In 1999, new cycloparvifloralone-type sesquiterpenes were found (Figure 4.23). Cycloparvifloralone (**150**) occurred in the leaves of *I. parviflorum*, and (11)7,14-ortholactone of 14-hydroxy-3-oxofloridanolide (**151**) was isolated from the fruits of *I. floridanum* (Schmidt et al., 1999), whereas *I. merrillianum* produced merrillianone (**152**) (Huang et al., 1999). All of these sesquiterpenes possess a hitherto rare ring system with a cage-like acetal and hemiacetal structure.

New carbon skeletal sesquiterpenes, which are not able to take their place with the known subclasses of *Illicium* sesquiterpenes, have been found as natural products, as shown in Figure 4.24. Anislactones A (**153**) and B (**154**) were isolated from the fruits of *I. anisatum* (Kouno et al., 1990). The structure of anislactone A was established by an x-ray crystallographic analysis (Kouno et al., 1989), whereas anislactone B was an epimer with regard to the 7-hydroxyl group. Both compounds have unique types of carbon-skeleton that have never been recorded as natural compounds.

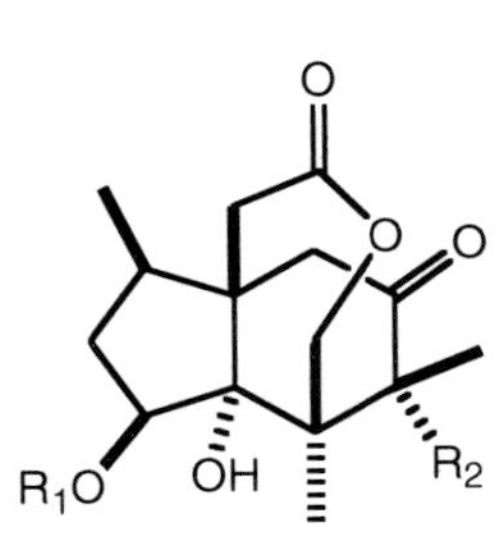

118 R_1 = H, R_2 = OH
119 R_1 = Bz, R_2 = OH
120 R_1 = H, R_2 = H

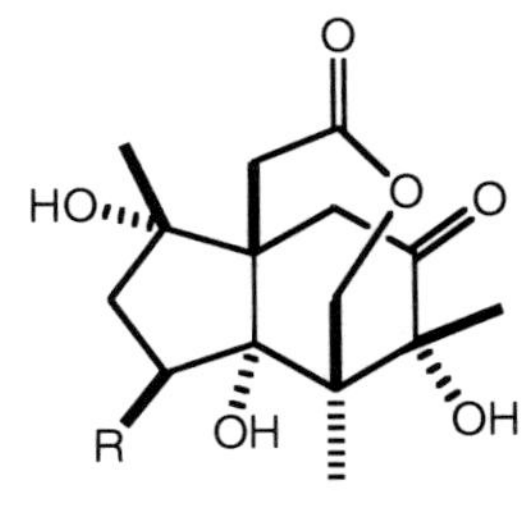

121 R=OH
122 R=H

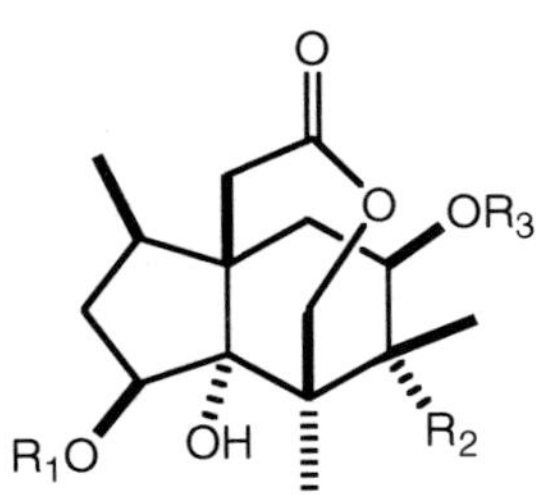

123a R_1 = R_3 = H, R_2 = OH
124 R_1 = Bz, R_2 = OH, R_3 = H
125 R_1 = Ac, R_2 = OH, R_3 =Bz
126 R_1 = Bz, R_2 = R_3 =H

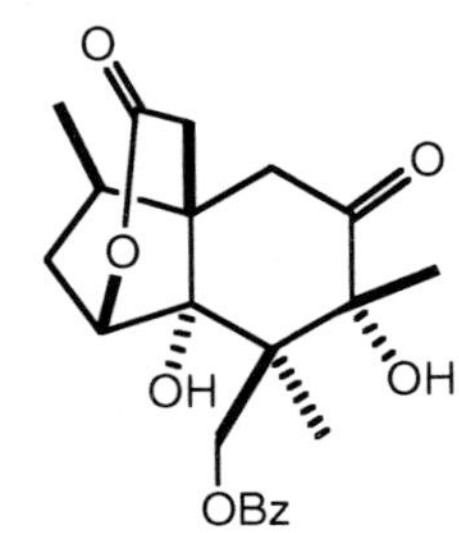

119a

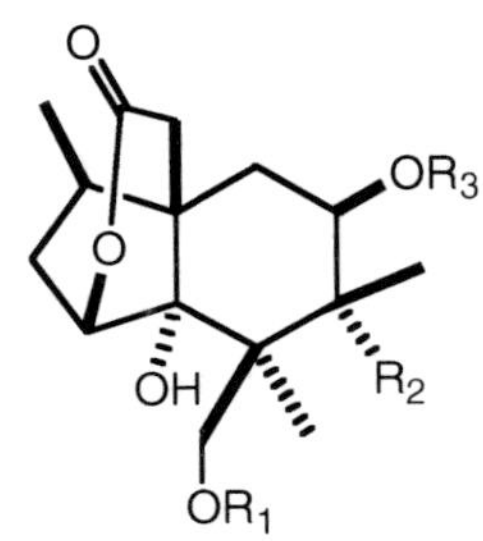

123b R_1 = R_3 = H, R_2 = OH
124a R_1 = Bz, R_2 = OH, R_3 = H
125a R_1 = Ac, R_2 = OH, R_3 =Bz
126a R_1 = Bz, R_2 = R_3 =H

FIGURE 4.18 Pseudoanisatin-type sesquiterpenes (compounds **118–126**)

The compounds are proposed presumably derived from the majucin-type sesquiterpene as follows: the ring construction occurs between C-7 and C-8 in the majucin-type compound, followed by the bond formation of C-6 and C-8, and then hydroxylation at C-8. In this case, C-7 should become a methyl group. Although this seems to be a better explanation for converting from the majucin-type compound to anislactones, inversion of the C-9 configuration still has remained ambiguous.

Other new skeletal sesquiterpenes, named illicinolides A (**155**) and B (**156**) and tashironin (**157**), were isolated from *I. tashiroi*. The structure of illicinolide A was elucidated on the basis of the spectral data, and then its absolute configuration was established by x-ray crystallographic analysis of the *p*-bromobenzoyl derivative **155a** (Fukuyama et al., 1990). The structure of Illicinolide B was assigned as 6α-hydroxyillicinolide A by spectral data compared with those of illicinolide A, and its absolute structure was determined by applying the CD dibenzoate rule to the *p*-bromobenzoyl derivative **156a** (Fukuyama et al., 1992). Although illicinolides A and B appear to be closely related to the previously reported anisatin (**107**) and majucin (**137**), the structural feature that contains the γ-lactone ring closed between C-7 and C-9 is rather similar to noranisatin (**107a**) (Yamada et al., 1965), an oxidatively degraded product of anisatin. This highly oxygenated abnormal structure may be rationalized biogenetically by assuming that the C-11 carbon in compounds **155** and **156** would originate from the C-11 in normal anisatin skeleton through, for example, a tetracyclic intermediate C, as shown in Scheme 4.10.

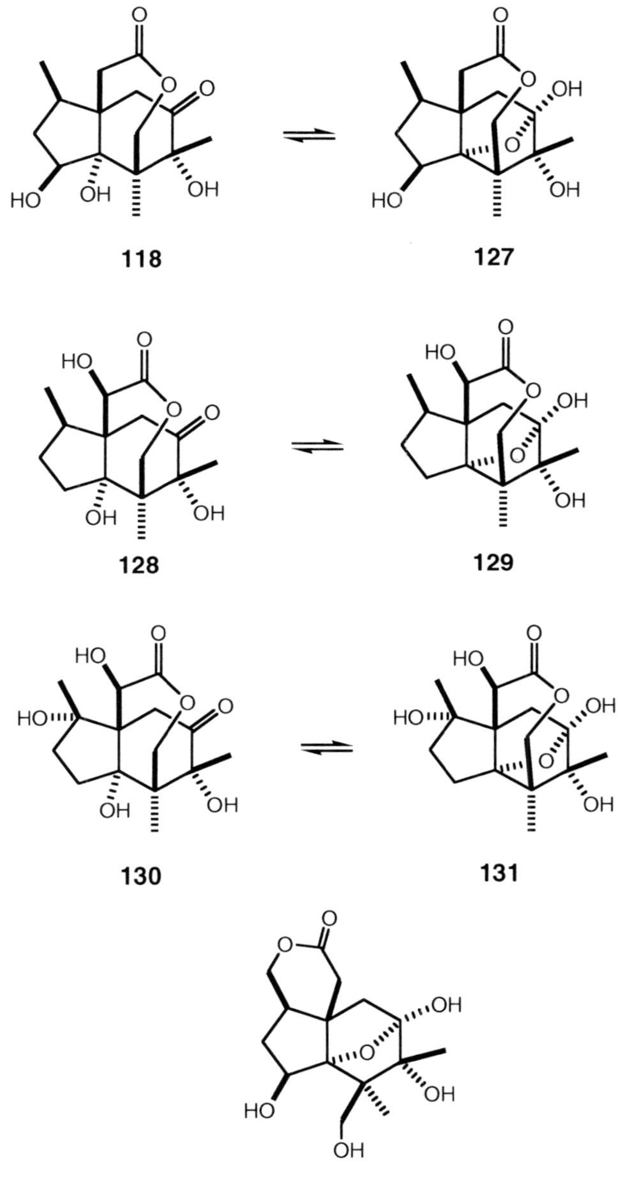

FIGURE 4.19 Ketone and hemiacetal equilibrium of pseudoanisatin-type sesquiterpenes and an unusual sesquiterpene **132**

In contrast, the structure of tashironin (**157**) was elucidated by extensive analysis of spectroscopic data (Fukuyama et al., 1995). Tashironin consists of a 2-oxatricyclo [4.3.1.$0^{4,9}$] heptane skeleton and can be regarded as a rare *allo*-cedrane-type sesquiterpene. Although anisatin (**107**) and its congeners have been postulated to be biosynthesized from a tricarbocyclic precursor A (Tomita and Hirose, 1973), the biosynthetic route leading to an anisatin skeleton has been still obscure because any tricarbocyclic sesquiterpenes made up by the same carbon skeleton of A have not been found in *Illicium* species. Tashironin (**157**) possesses the precise carbon skeleton corresponding

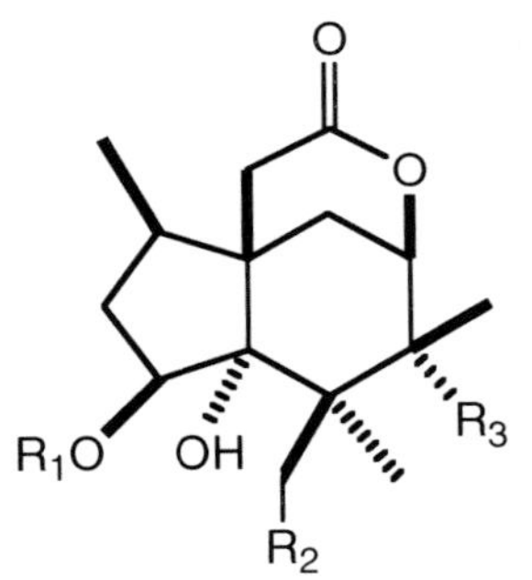

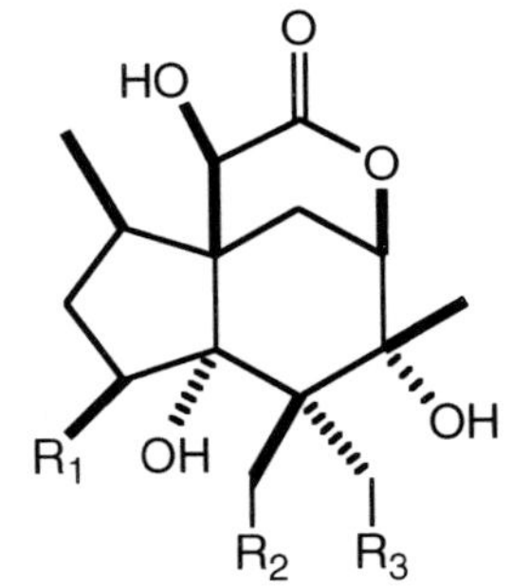

133 R_1 = H, R_2 = OH, R_3 = H
134 R_1 = Ac, R_2 = O-*n*Bu, R_3 = OH

135 R_1 = O, R_2 = OAc, R_3 = H
136 R_1 = H, R_2 = O-*n*Bu, R_3 = OAc

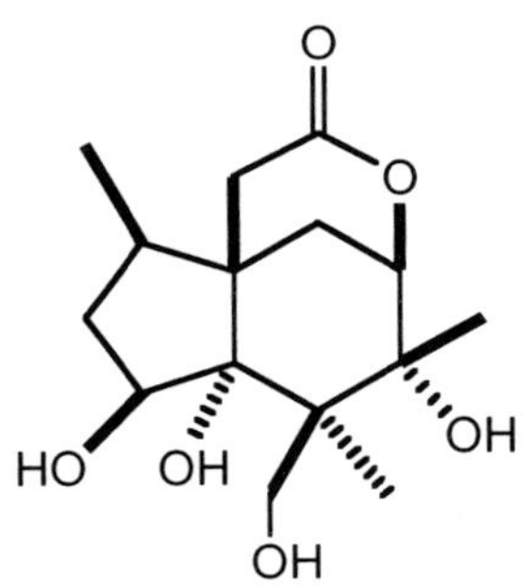

123

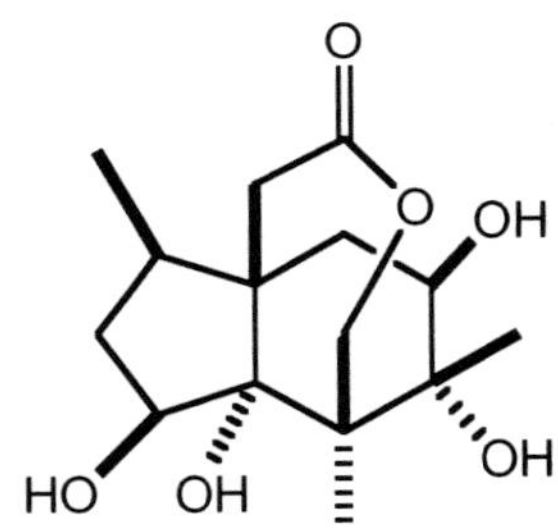

123a

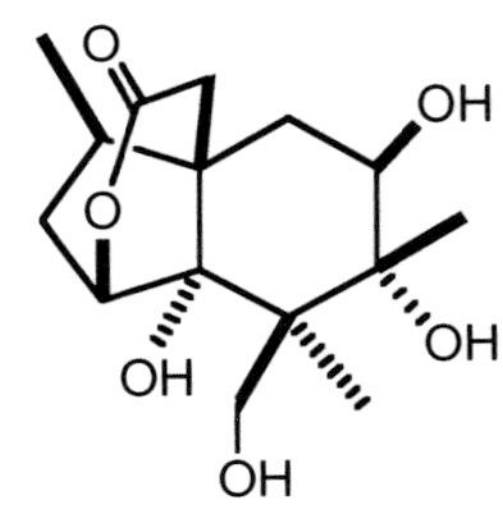

123b

FIGURE 4.20 Minwanensin-type sesquiterpenes **123** and **133–136**

to a biogenetic precursor A of all the *Illicium* sesquiterpenes; that is, anisatin, pseudoanisatin, majucin, and pseudomajucin types. It is generally accepted that anisatin could be biosynthesized from an acrorane through a tricarbocyclic precursor A after breaking the bond between C-6 and C-11, as shown in Scheme 4.10 (Devon and Scott 1972). Thus, the isolation of tashironin, anisatin, and its related sesquiterpenes from the same source is of considerable significance and throws light on the biogenesis of *Illicium* sesquiterpenes.

As discussed previously, anisatin (**107**) and noranisatin (**108**) are convulsive toxic principles in *I. anisatum* and are regarded as picrotoxin-like potent phytotoxins. The neuropharmacological study of anisatin demonstrates that their convulsive toxicity is probably caused by a potent non-competitive GABA antagonist. However, it has remained equivocal, at the present time, which structural part of anisatin is able to cause convulsive activity. A systematic study of the structure and toxicity relationship has not been carried out because of the limited available quantity of compounds, although a number of various anisatin-related compounds have been known as natural products. The toxicity of representative compounds to mice (i.p.) was examined and compared with those of anisatin (**107**) and neoanisatin (**108**) (Kouno et al., 1989; Niwa et al., 1991) (Table 4.1). Another anisatin-type 2-oxo-6-dehydroxyneoanisatin (**115**) and another majucin-type neomajucin (**138**) are also very toxic (Yang et al., 1990; Kouno et al., 1990), whereas majucin (**137**) and its analogues **139**, **142**, **144**, **145**, and **146** could not produce any appreciable behavioral changes at a dose of up to 40 mg/kg (Kouno et al., 1990). As neomajucin (**138**) was recognized as a toxic compound, the presence of a spiro β-lactone moiety in anisatin is not likely to be absolutely

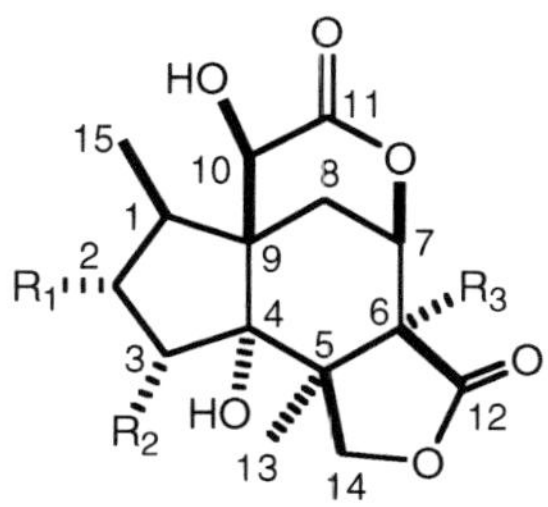

137 R_1 = H, R_2 = R_3 = OH
138 R_1 = R_2 = H, R_3 = OH
139 R_1 = R_3 = OH, R_2 = H
140 R_1 = R_3 = H, R_2 = OH

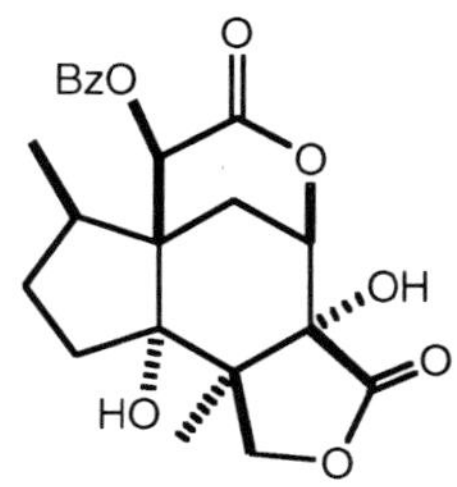

141

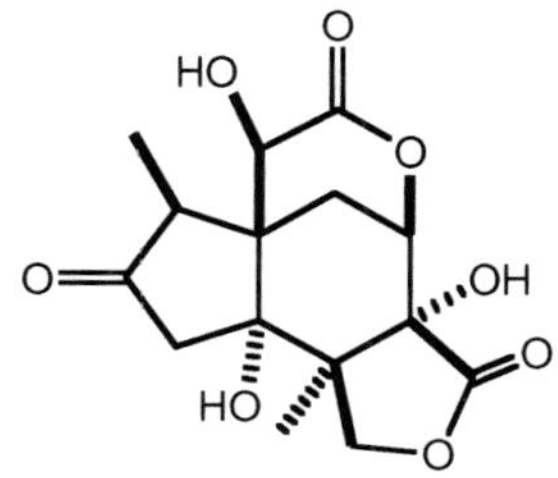

142

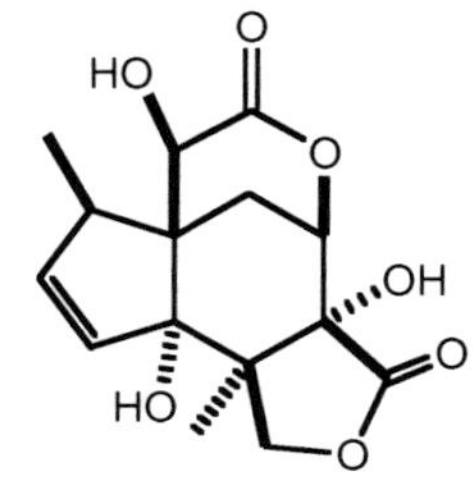

143

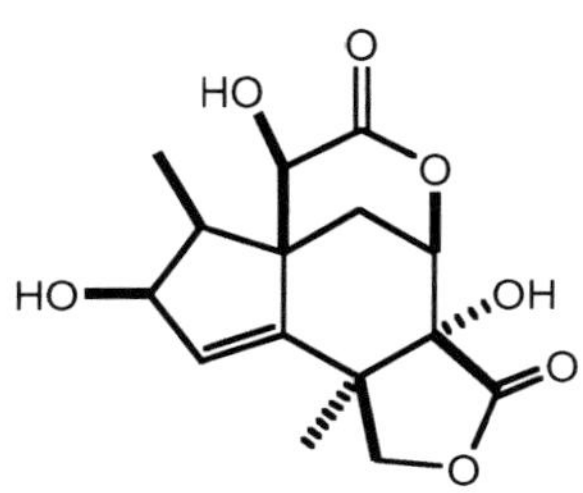

144

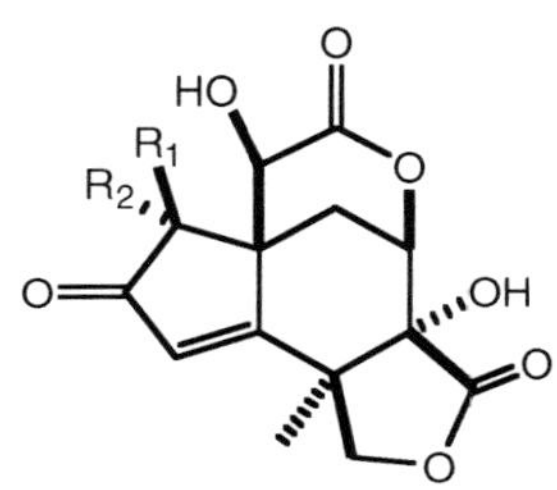

145 R_1 = CH_3, R_2 = H
146 R_1 = H, R_2 = CH_3

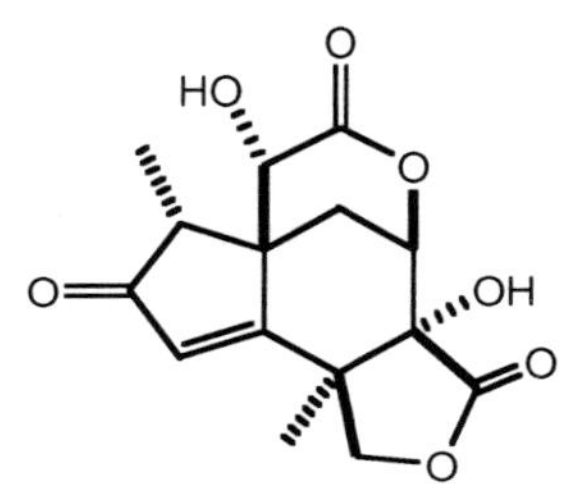

147

FIGURE 4.21 Majucin-type sesquiterpenes (compounds **137–147**)

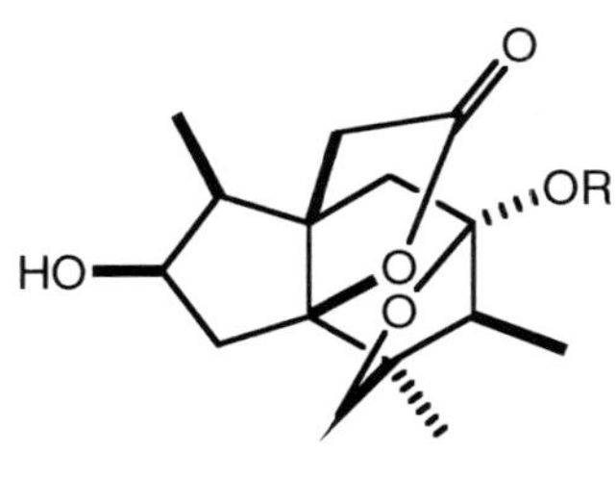

148 R = H
149 R = Glucose

FIGURE 4.22 Pseudomajucin-type sesquiterpenes (compounds **148–149**)

responsible for the convulsive toxicity. We have to wait for further investigation to prove this kind of problem associated with the relationship between the structure and the toxicity in the *Illicium* sesquiterpenes.

4.5 DITERPENES AND TRITERPENES

In contrast to the relatively wide distribution of anisatin-like sesquiterpenes in the genus *Illicium*, few reports concerning about diterpenes and triterpenes have been documented. Two Chinese groups have reported the occurrence of abietane diterpenes in *Illicium* plants. 4-*epi*-Isopimaricacid (**158**),

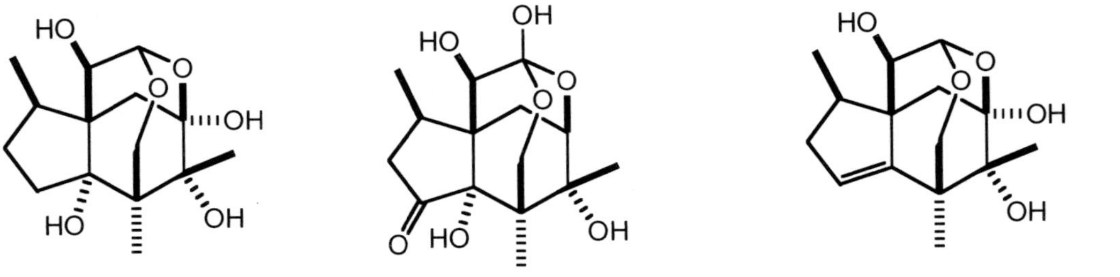

150 **151** **152**

FIGURE 4.23 Cycloparvifloralone-type sesquiterpenes (compounds **150–152**)

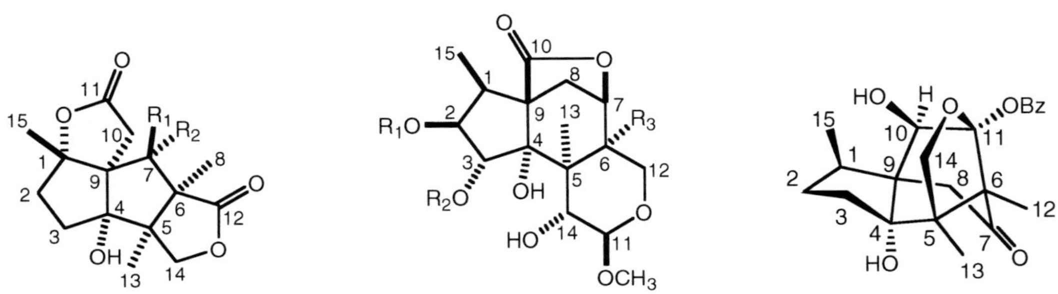

153 R_1 = H, R_2 = OH
154 R_1 = OH, R_2 = H

155 R_1 = R_2 = R_3 =H
155a R_1 = R_2 = *p*-BrBz, R_3 = H
156 R_1 = R_2 = H, R_3 = OH
156a R_1 = R_2 = *p*-BrBz, R_3 = OH

157

FIGURE 4.24 New carbon skeletal sesquiterpenes (compounds **153–157**)

15-hydroxydehydroabietic acid (**159**), and 8,11,13,15- abietatetran-19-oic acid (**160**) were isolated from the bark of *I. jiadifengpi*, which is indigenous to China (Huang et al., 1996). Eleven novel diterpenes (**161–164**, **166–170**, and **172–174**) and two known abietane diterpenes (**165**, **171**) were found from the aerial part of *I. angustisepalum* (Sy and Brown, 1998) (Figure 4.25), but angustanoic acid E, reported as a new compound in this chapter, has to be the known diterpene **160**. Sy and Brown (1998) emphasized that these abietane diterpenes are unusual in the points such that they are oxygenated at the axial C-19 methyl rather than at the equatorial C-18 methyl group. The same type of diterpenes found in *I. jiadifengpi* and *I. angustisepalum* seem to serve as a chemical proof supporting the closely taxonomic relationship between these two species, as both species have been taxonomically arguable.

Although triterpenes are widely distributed in nature, there are very few reports on the triterpenes found in *Illicium* plants. New *seco*-cycloartane triterpenes **175** was isolated together with the previously known schizandronic acid (**176**), schizandrolic acid (**177**), and magniferolic acid (**178**) from *I. dunnianum* (Zhang, 1989; Sy et al., 1997). In addition to **178**, magniferonic acid (**179**), betulic acid (**180**), and a new cycloartane triterpene, 3-O-acetylmagniferolic acid (**181**) was obtained from the bark of *I. difengpi* (Huang et al., 1997) (Figure 4.26). Sy et al. (1997) suggested that these cycloartane-type triterpenes might become one of the phytochemical markers characteristic of the *Illicium* species.

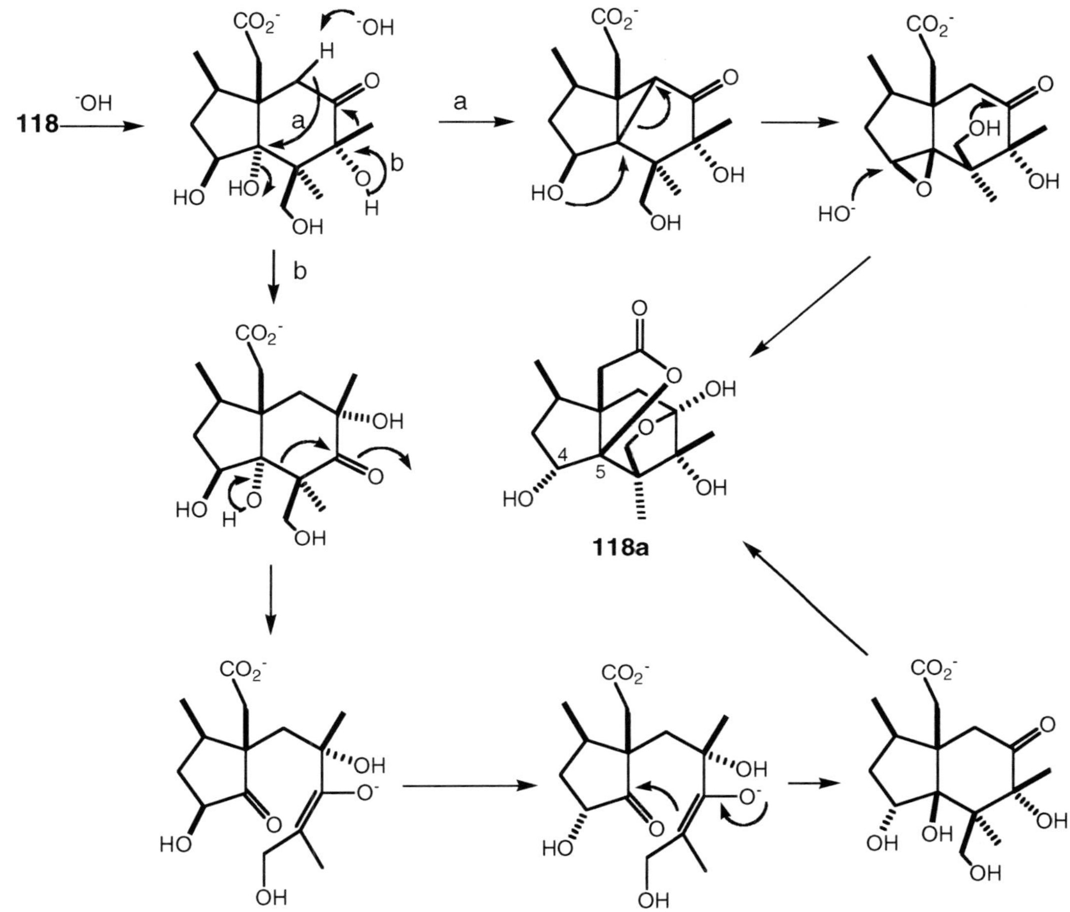

SCHEME 4.9 Two plausible mechanisms for the base-catalyzed transformation of pseudoanisatin (**118**) to **118a**

4.6 FLAVONOIDS AND TANNINS

The common flavonols widely occurring in the plant kingdom were also detected in the fruits of *I. verum* (Knacksted and Herrmann, 1981). They are kaempferol (**182**), and its glycosides **183–185**, and quercetin (**186**), and its glycosides (**187–191**), as shown in Figure 4.27.

In addition to the flavonoids, the occurrence of tannins had been suggested in *I. anisatum* because of the astringency of its bark. In fact, (+)-catechin (**192**) was detected in *I. anisatum* (Morimoto et al., 1988) and *I. henryi* (Xie et al., 1990). Morimoto et al. (1988) investigated the polar parts of the bark of *I. anisatum*, resulting in the isolation of two new prenylated catechins (**193**) and (**194**) and a new trimeric procyanidin (**195**), along with procyanidins B-1 (**196**) and B-7 (**197**), epicatechin-(4β → 8)-epicatechin-(4β → 8)-catechin (**198**), epicatechin-(4β → 8)-epicathechin-(4β → 6)-catechin (**199**), and epicatechin-(4β → 6)-catechin-(4β → 8)-catechin (**200**) (Kashiwada et al., 1986) (Figure 4.28 and Figure 4.29). In contrast to the relatively wide distribution of prenylated flavones, prenylated catechins (**193**) and (**194**) are not often found and thus seem to be components characteristic of the *Illicium* plants.

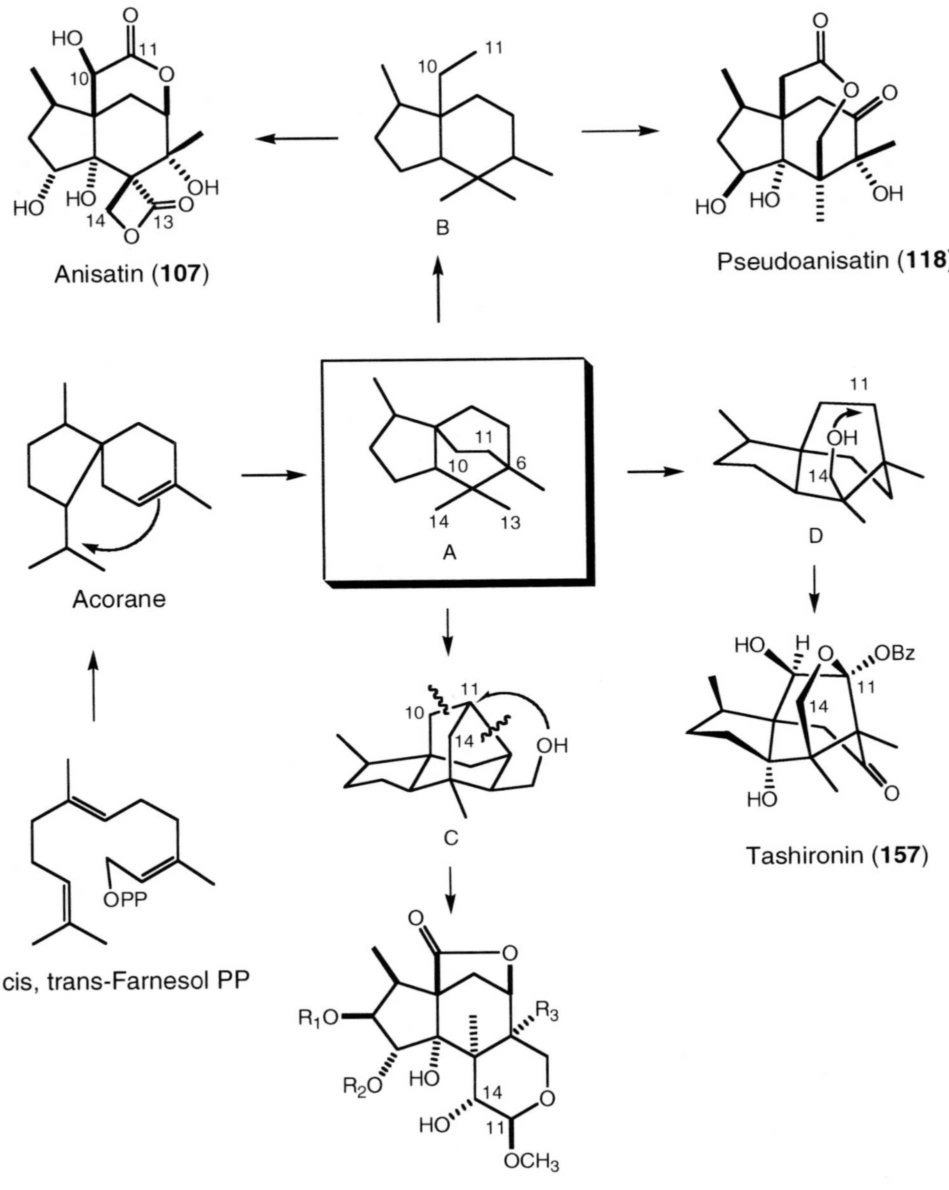

SCHEME 4.10 Plausible biosynthesis for illicinolides A (**155**) and B (**156**), and tashironin (**157**)

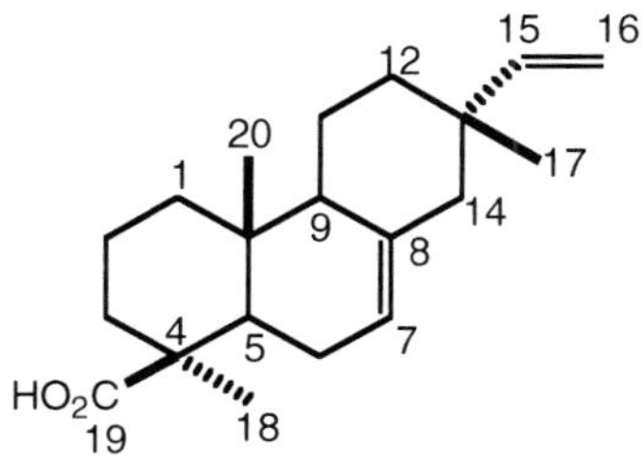

158

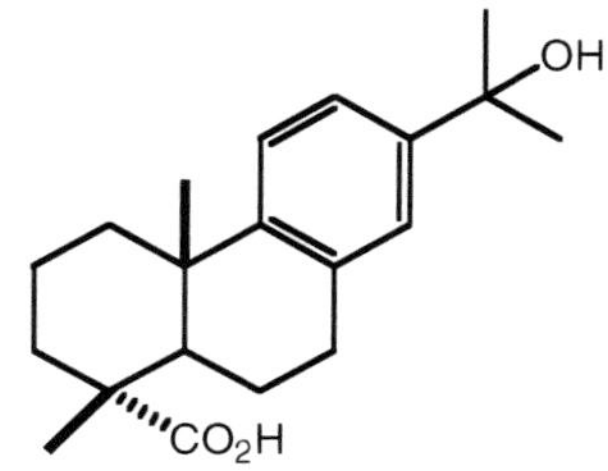

159

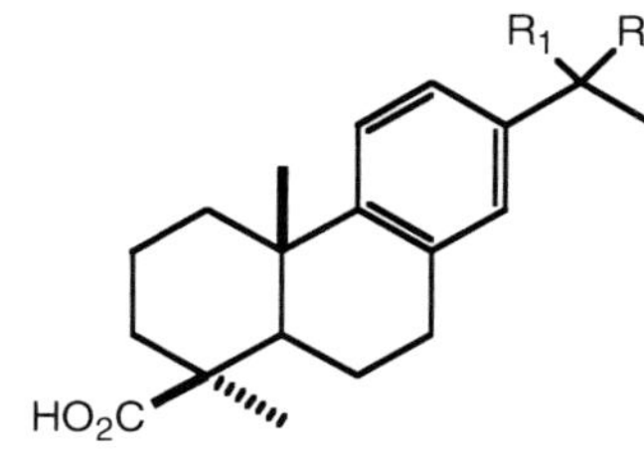

160 R_1, R_2 = CH_2
161 R_1 = Me, R_2 = OH
162 R_1, R_2 = O

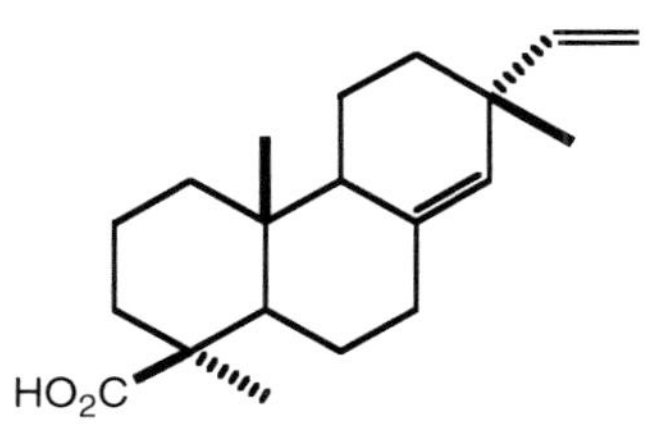

163

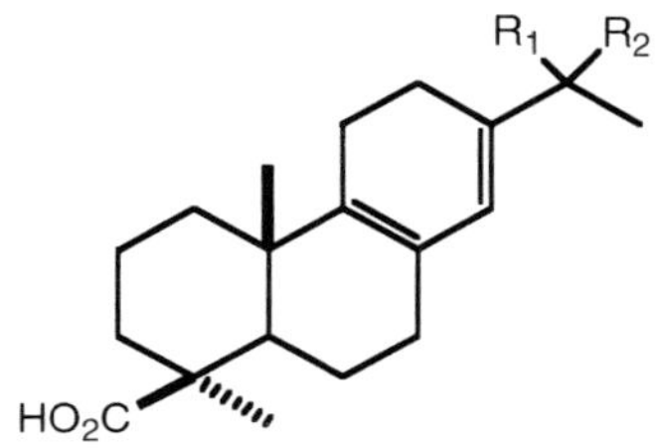

164 R_1, R_2 = CH_2
165 R_1 = Me, R_2 = H

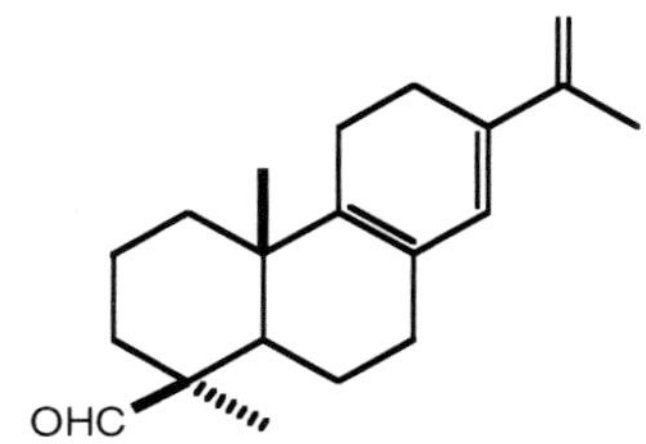

166

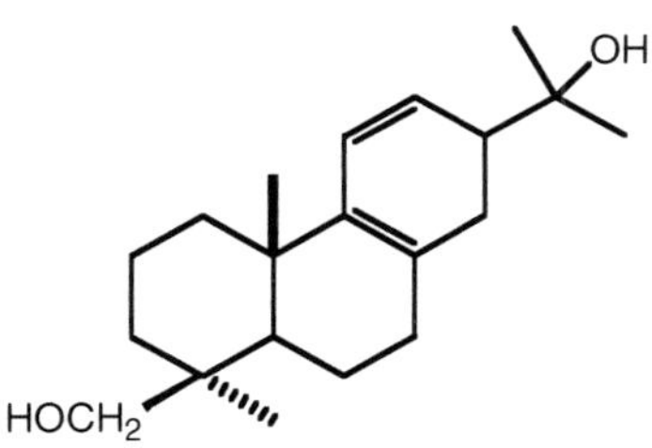

167

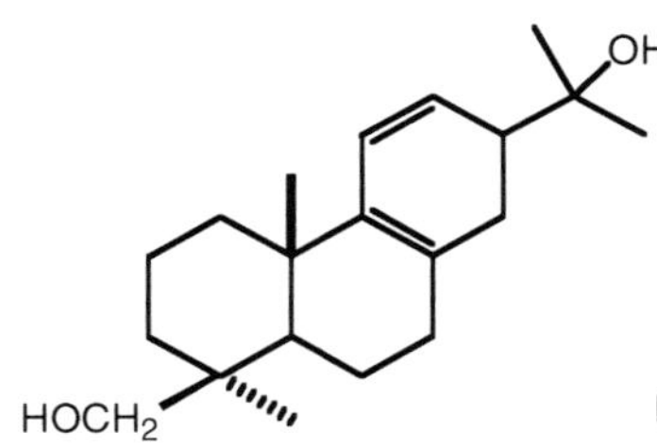

168

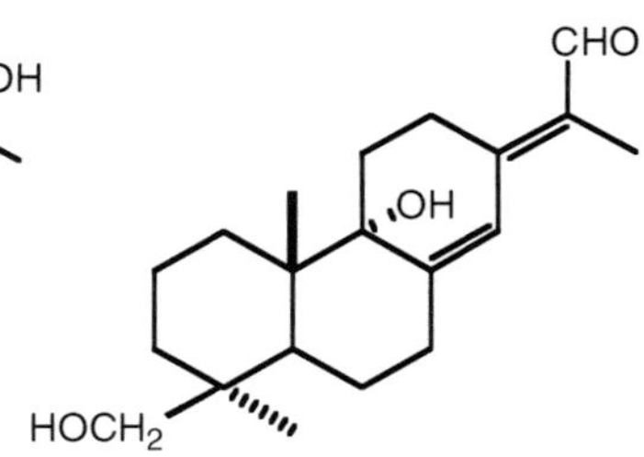

169

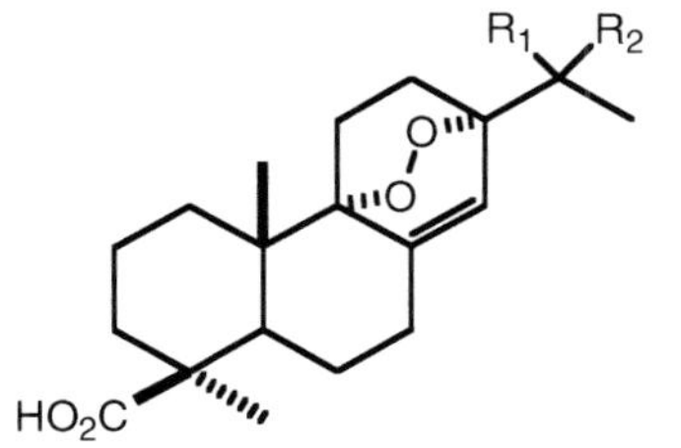

170 R_1, R_2 = CH_2
171 R_1 = Me, R_2 = H
172 R_1 = Me, R_2 = OH

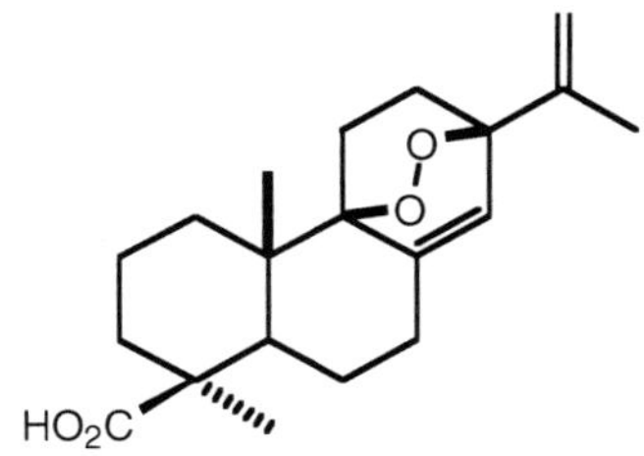

173

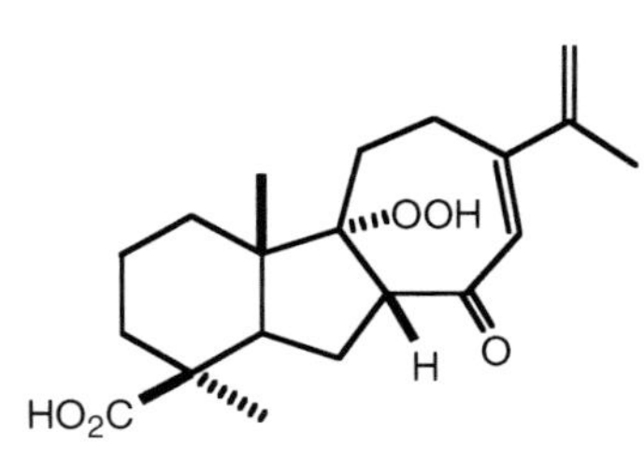

174

FIGURE 4.25 Diterpenes (compounds **158–174**)

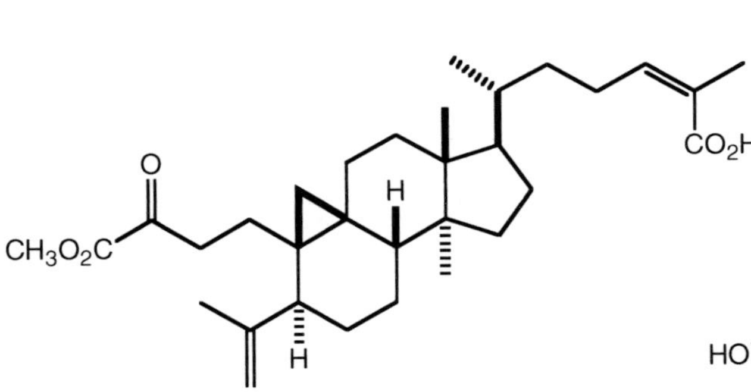

175

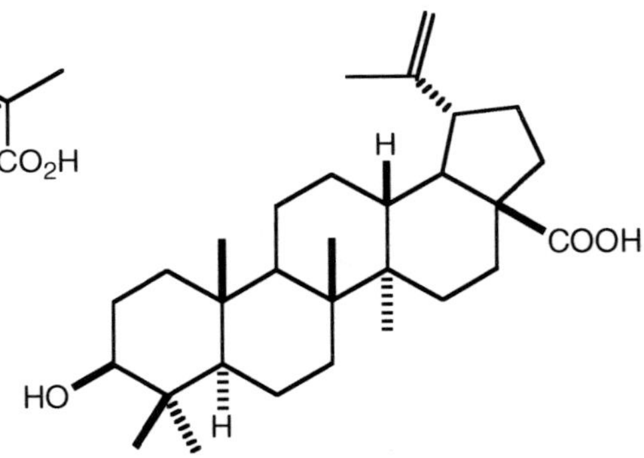

180

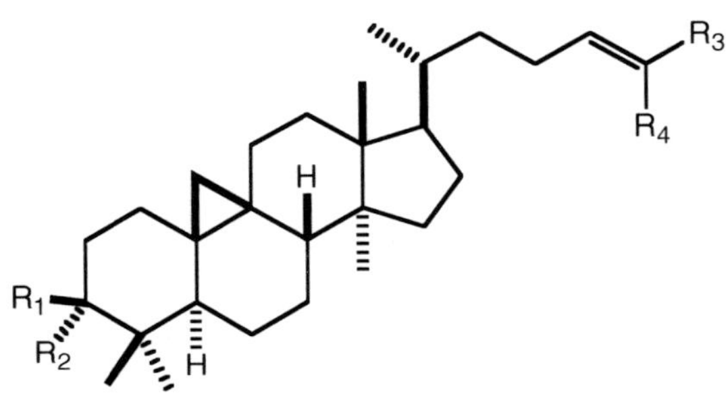

176 R_1, R_2 = O, R_3 = CH_3, R_4 = CO_2H
177 R_1 = OH, R_2 = H, R_3 = CH_3, R_4 = CO_2H
178 R_1 = OH, R_2 = H, R_3 = CO_2H, R_4 = CH_3
179 R_1, R_2 = O, R_3 = CO_2H, R_4 = CH_3
181 R_1=OAc, R_2 =H, R_3 = CO_2H, R_4 = CH_3

FIGURE 4.26 Triterpenes (compounds **175–181**)

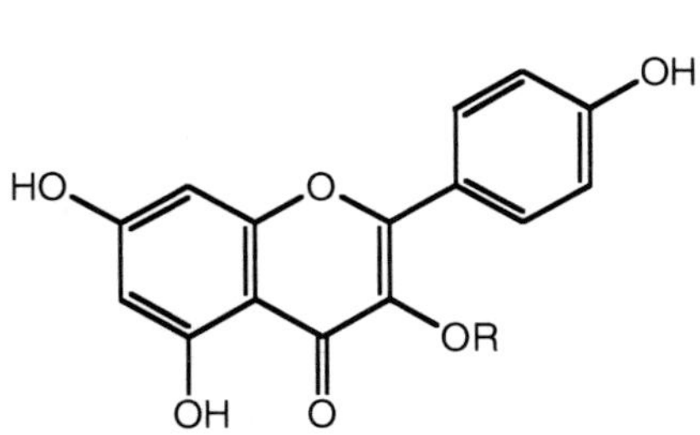

182 R = H
183 R = 6-*O*-α-L-rhamnosyl-D-glucosyl
184 R = D-glucosyl
185 R = D-glactosyl

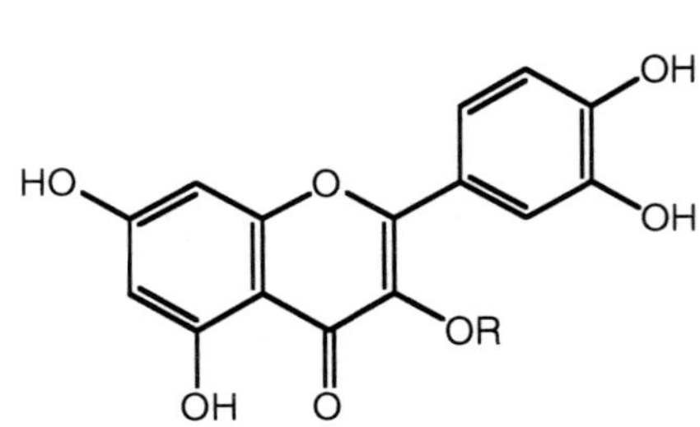

186 R = H
187 R = 6-*O*-α-L-rhamnosyl-D-glucosyl
188 R = D-glucosyl
189 R = D-glactosyl
190 R = D-xylosyl
191 R = L-rhamnosyl

FIGURE 4.27 Flavonoids **182** and **186** and their glycosides

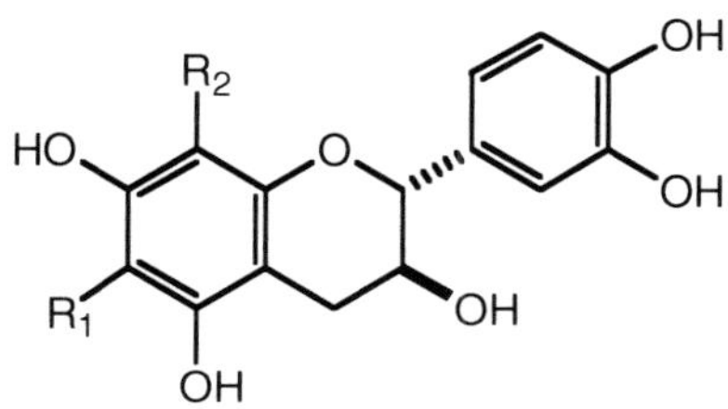

192 $R_1 = R_2 = H$
193 R_1 = prenyl, $R_2 = H$
194 $R_1 = H$, R_2 = prenyl

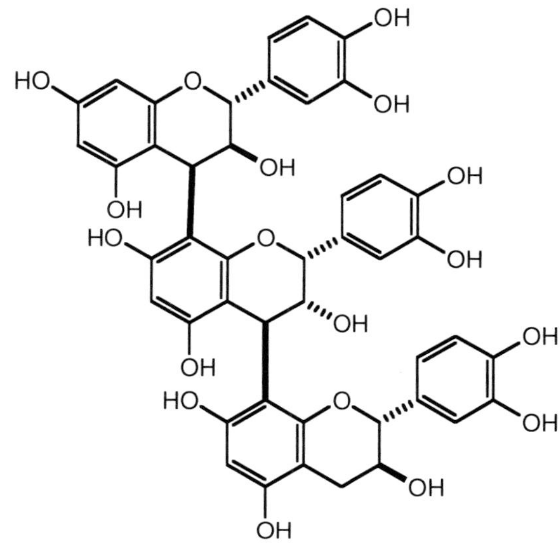

195

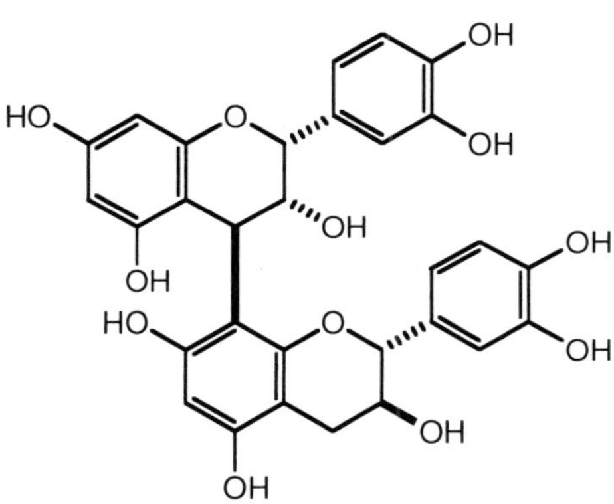

196

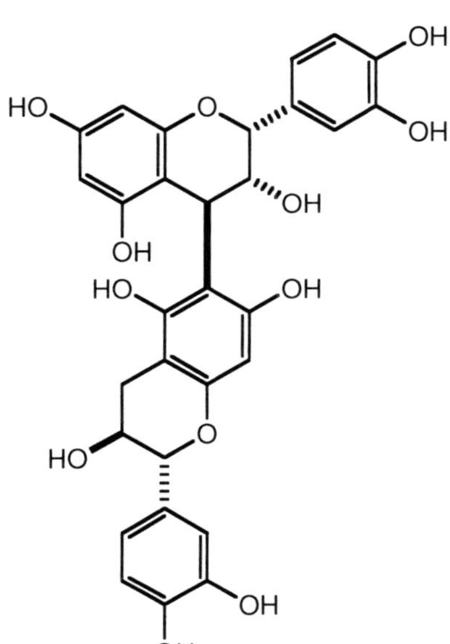

197

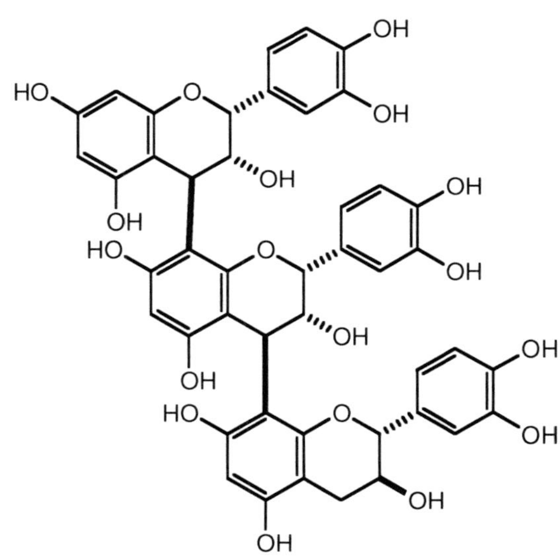

198

FIGURE 4.28 Tannins **192–198** isolated from *I. anisatum*

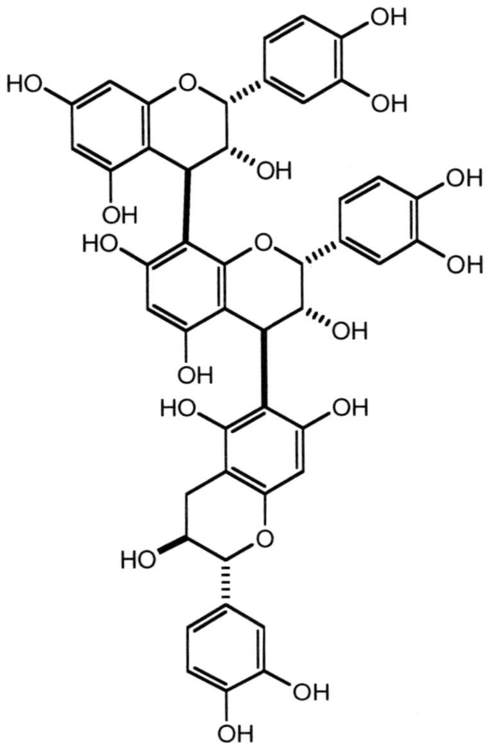

199

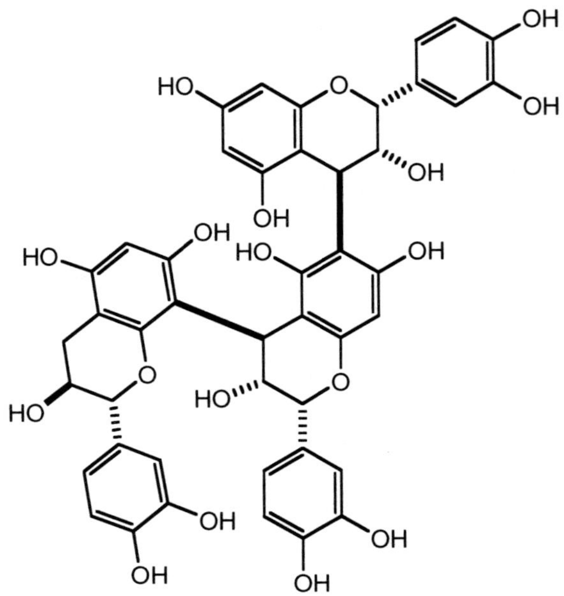

200

FIGURE 4.29 Tannins **131** and **132** isolated from *I. anisatum*

TABLE 4.1
Lethality Induced by *Illicium* Sesquiterpenes

Sesquiterpenes	LD_{50} (mg/kg)
Anisatin (**107**)	1.03[a]
Neoanisatin (**108**)	1.62[a]
Majucin (**137**)	>40[a]
Neomajucin (**138**)	12.2[a]
Pseudoanisatin (**118**)	>100
Minwanensin (**133**)	>50
2-oxo-6-dehydroxyneoanisatin (**115**)	1.46
(2*S**)-hydroxyneomajucin (**139**)	>40
2-oxoneomajucin (**142**)	>40
(2*R**)-hydroxy-3,4-dehydroneomajucin (**144**)	>40
(1*S**)-2-oxo-3,4-dehydroneomajucin (**145**)	>40
(1*R**)-2-oxo-3,4-dehydroneomajucin (**146**)	>40
107a	180
107b	Nontoxic
107c	Nontoxic

[a] Litchfield-Wilcoxon method.

References

Barton, D.H., Bhati, A., Mayo, P.D., and Morrison, G.A., The constitution and synthesis of anisoxide, *J. Chem. Soc.,* 4393–4398, 1958.

Devon, T.K. and Scott, I.A., *Handbook of Naturally Occurring Compounds, Vol. II, Terpenes,* Academic Press, New York, 1972, p. 56.

Dictionary of Natural Products on CD-ROM, version 4.2, Chapman and Hall, London, 1996.

Fukuyama, Y., Okamoto, K., Kubo, Y., Shida, N., and Kodama, M., New chlorine-containing prenylated C_6-C_3 compounds increasing choline acetyltransferase (ChAT) activity in culture of postnatal rat spetal neurons from *Illicium tashiroi, Chem. Pharm. Bull.,* 42, 2199–2201, 1994a.

Fukuyama, Y., Shida, N., Hata, Y., Anzaki, S., and Kodama, M., Prenylated C_6-C_3 compounds related to illicinone E from *Illicium tashiroi, Phytochemistry,* 37, 1653–1657, 1994b.

Fukuyama, Y., Shida, N., Hata, Y., and Kodama, M., Prenylated C_6-C_3 compounds from *Illicium tashiroi, Phytochemistry,* 36, 1497–1503, 1994c.

Fukuyama, Y., Shida, N., and Kodama, M., Structure of illicinolide A, a novel sesquiterpene lactone from the bark of *Illicium tashiroi, Tetrahedron Lett.,* 31, 5621–5622, 1990.

Fukuyama, Y., Shida, N., and Kodama, M., Isodunnianin: a new sesquiterpene enhancing neurite outgrowth in primary culture of fetal rat cerebral hemisphere from *Illicium tashiroi, Planta Med.,* 59, 181–182, 1993.

Fukuyama, Y., Shida, N., and Kodama, M., Tashironin, a plausible biosynthetic precursor of anisatin-type sesquiterpens, *Tetrahedron Lett.,* 36, 583–586, 1995a.

Fukuyama, Y., Hata, Y., and Kodama, M., Bicycloilliocinone asarone acetal: a novel prenylated C_6-C_3 compound increasing choline acetytransferase (ChAT) activity from *Illicium tashiroi, Planta Med.,* 63, 2199–2201, 1997.

Fukuyama, Y., Shida, N., Kodama, M., Chaki, H., and Yugami, T., Tricycloillicinone, a novel prenylated C_6-C_3 compound increasing choline acetyltransferase (ChAT) activity, isolated from *Illicium tashiroi, Chem. Pharm. Bull.,* 43, 2270–2272, 1995b.

Fukuyama, Y., Shida, N., Kodama, M., Kido, M., Nagasawa, M., and Sugawara, M., Illicinolides A and B, novel sesquiterpene lactones from the wood of *Illicium tashiroi, Tetrahedron,* 48, 5847–5854, 1992a.

Fukuyama, Y., Shida, N., Sakurai, T., and Kodama, M., Prenylated C_6-C_3 compounds from *Illicium tashiroi, Phytochemistry,* 31, 3975–3979, 1992b.

Furita, T.E. and Bellanca, N., Eds., *Fenaroli's Handbook of Flavor Ingredients,* 2nd ed., Vols. 1 and 2, CRC Press, Cleveland, OH, 1975.

Hano, Y., Kaku, H., and Nomura, T., *The 112th Annual Meeting of Pharmaceutical Society of Japan,* Fukuoka, abstract no. 2, 1992, p. 200.

Hatanaka, H., Tsukui, H., and Nihonmatsu, I., Developmental change in the nerve growth factor action from induction of choline acetyltransferase to promotion of cell survival in cultured basal forebrain cholinergic neurons from postnatal rats, *Dev. Brain Res.,* 39, 85–95, 1988.

Harada, N. and Nakanishi, K. *Circular Dichroic Spectroscopy—Exciton Coupling in Organic Stereochemistry,* University Science Books, New York, 1982.

Huang, J.-M., Wang J.-L., and Yang C.-S., Sesquiterpene lactones from the pericarp of *Illicium dunnianum, Phytochemistry,* 46, 777–780, 1997.

Huang, J.-M. and Yang C.-S. Pseudoanisatin-like sesquiterpene lactones from the pericarps of *I. dunnianum. Phytochemistry,* 42, 1375–1376, 1996.

Huang, J.-M., Yang, C.-S., Wang, H., Wu, Q.-M., Wang, J.-H., and Fukuyama, Y., Structures of novel sesquiterpenes from the pericarps of *Illicium merrillianum, Chem. Pharm. Bull.,* 47, 1749–1752, 1999.

Huang, P., Xuezhong, Z., Mighi, H., and Nakanishi, T., Chemical studies on the bark of *Illicium jiadifengipi, Zhongguo Zhongyao Zazhi,* 21, 551–553, 1996a.

Huang, P., Yang, M., Lai, M.-X., Zheng, X.-Z., Nishi, M., and Nakanishi, T., Chemical studies on the barks of *Illicium difengpi, Zhongguo Yaoxae Zazhi,* 31, 278–281, 1996b.

Huang, P., Nishi, M., Zheng, X.-Z., Lai, M.-X., and Nakanishi, T., Triterpenoid acids from *Illicium difengpi, Zhongguo Yaoxae Zazhi,* 32, 704–707, 1997.

Hussain, R.A., Poveda, L.J., Pezzuto, J.M., Soejarto, D.D., and Kinghorn, A.D., Sweeting agents of plant origin: phenylpropanoid constituents of seven sweet-tasting plants, *Econ. Bot.,* 44, 174–182, 1990.

Jackson, R.W. and Short, W.F., Anisol. Part I, *J. Chem. Soc.*, 513, 1937.

Jiang, Z.-H., Tanaka, T., and Kouno, I., New phenylpropanoid glucosides from the fruits of *Illicium anisatum*, *Chem. Pharm. Bull.*, 47, 421–422, 1999.

Kajimoto, Y., Fujimori, H., and Hirota, K., Study on the toxic components of "Sikimi." I. Chemical study, *Tokushima J. Exp. Med.*, 2, 104–107, 1955a.

Kajimoto, Y., Fujimori, H., and Hirota, K., Study on the toxic components of "Sikimi." II. Chemical study, *Tokushima J. Exp. Med.*, 2, 108–110, 1955b.

Kaku, H., Hano, Y., and Nomura, T., *The 113th Annual Meeting of Pharmaceutical Society of Japan*, Osaka, abstract no. 2, p. 216, 1993.

Kashiwada, Y., Nonaka, G., and Nishioka, I., Tannins and related compounds. XLVIII. Rhuharb. 7. Isolation and characterization of new dimeric and trimaric procyanidins, *Chem. Pharm. Bull.*, 34, 4083–4091, 1986.

Kawano, N. and Matsuo, A., Studies on the toxic components of *Illicium anisatum* L. I, *Yakugakuzasshi*, 78, 1220–1223, 1958.

Kende, A.S. and Chen, J., Stereoselective total synthesis of 8-deoxyanisatin, *J. Am. Chem. Soc.*, 107, 7184–7186, 1985.

Knackstedt, J. and Herrmann, K., Flavonol glycosides of bay leaves (*Laurus nobilis* L.) and star anise fruits (*Illicium verum* Hook. fil.), *Z. Lebenm. Unters. Forsch.*, 173, 288–290, 1981.

Kouno, I., Akiyama, T., and Kawano, N., Minor constituents from the seeds of Japanese star-anise, *Chem. Pharm. Bull.*, 36, 2990–2992, 1988.

Kouno, I., Baba, N., Hashimoto, M., Kawano, N., Takahashi, M., Kaneto, H., Yang, C.-S., and Sato, S., Isolation of three new sesquiterpene lactones from the pericarps of *Illicium majus*, *Chem. Pharm. Bull.*, 37, 2448–2451, 1989a.

Kouno, I., Baba, N., Hashimoto, M., Kawano, N., Yang, C.-S., and Sato, Y., A new sesquiterpene lactone and its glucoside from the pericarps of *Illicium majus*, *Chem. Pharm. Bull.*, 39, 1773–1778, 1989b.

Kouno, I., Baba, N., Hashimoto, M., Kawano, N., Takahashi, M., Kaneto, H., and Yang, C.-S., Sesquiterpene lactones from the pericarps of *Illicium majus*; 2-oxy derivatives of neomajucin and 3,4-dehydroxy-neomajucin, *Chem. Pharm. Bull.*, 38, 422–425, 1990.

Kouno, I., Hashimoto, M., Enjoji, S., Takahashi, M., Kaneto, H., and Yang, C.-S., Isolation of neoanisatin derivatives from the pericarps of *Illicium majus* with other constituents, *Chem. Pharm. Bull.*, 39, 1773–1778, 1991a.

Kouno, I., Hashimoto, A., Kawano, N., and Yang, C.-S., New sesquineolignan from the pericarps of *Illicium macranthum*, *Chem. Pharm. Bull.*, 37, 1291–1292, 1989c.

Kouno, I., Irie, H., and Kawano, N., The structure of pseudoanisatin and its novel base-catalysed rearrangement product, *J. Chem. Soc. Perkin Trans. I*, 2511–2515, 1984.

Kouno, I., Irie, H., Kawano, N., and Katsube, Y., The revised structure of pseudoanistatin by x-ray crystal structure determination, *Tetrahedron Lett.*, 24, 771–772, 1983.

Kouno, I., Iwamoto, C., Kameda, Y., Tanaka, T., and Yang C.-S., A new triphenyl-type neolignan and a biphenylneolignan from the bark of *Illicium simonsii*, *Chem. Pharm. Bull.*, 42, 112–114, 1994.

Kouno, I. and Kawano, N., New pseudoanisatin-like sesquiterpene lactones from the bark of *Illicium dunnianum*, *J. Chem. Soc. Perkin Trans. I*, 1537–1539, 1988.

Kouno, I., Mori, K., Akiyama, T., and Hashimoto, M., Two sesquiterpene lactones from *Illicium anisatum*, *Phytochemistry*, 30, 351–353, 1991b.

Kouno, I., Mori, K., Kawano, N., and Sato, S., Structure of anislactone A; a new skeletal type of sesquiterpene from the pericarps of *Illicium anisatum*, *Tetrahedron Lett.*, 30, 7451–7452, 1989d.

Kouno, I., Mori, K., Okamoto, S., and Sato, S., Structure of anislactones A and B; novel type of sesquiterpene lactones from the pericarps of *Illicium anisatum*, *Chem. Pharm. Bull.*, 38, 3060–3063, 1990.

Kouno, I., Morisaka, T., Hara, Y., and Yang, C.-S., Two new sesquineolignans from the bark of *Illicium dunnianum*, *Chem. Pharm. Bull.*, 39, 2606–2608, 1991c.

Kouno, I., Shimamoto, S., Jiang, Z.-H., and Tanaka, T., Prenylated C_6-C_3 compounds from root bark of *Illicium anisatum*, *Phytochemistry*, 46, 1389–1392, 1997.

Kouno, I., Yanagida, Y., Shimamoto, S., Shintomo, M., Ito, Y., and Yang C.-S., Neolignans and a phenylpropanoid glucoside from *Illicium difengpi*, *Phytochemistry*, 32, 1573–1577, 1993.

Kouno, I., Yanagida, Y., Shimamoto, S., Shintomo, M., and Yang, C.-S., Phenylpropanoids from the barks of *Illicium difengpi*, *Chem. Pharm. Bull.*, 40, 2461–2464, 1992.

Kudo, Y., Oka, J., and Yamada, K., Anisatin, a potent GABA antagonist, isolated from *Illicium anisatum*, *Neurosci. Lett.*, 25, 83–88, 1981.

Lane, J.F., Koch, W.T., Leeds, N.S., and Gorin, G., On the toxin of *Illicium anisatum*. I. The isolation and characterization of a convulsant principle: Anisatin, *J. Am. Chem. Soc.*, 74, 3211–3214, 1952.

Liu, H. and Yang, C.-S., Component analysis of essential oils from the fruits of seven species of *Illicium* (Illiciaceae) in China, *Acta Phytotaxon. Sinica*, 27, 317–320, 1989.

Lindner, D.L., Doherty, J.B., Shoham, G., and Woodward, R.B., Intermolecular ene reaction of glyoxylate esters: an anisatin model study, *Tetrahedron Lett.*, 23, 5111–5114, 1982.

Luche, J.-L., Lanthanides in organic chemistry. 1. Selective 1,2 reductions of conjugated ketones, *J. Am. Chem. Soc.*, 100, 2226–2227, 1978.

Miyazawa, M., Ota, H., Ishikawa, Y., and Kameoka, H., An insecticidal compound from *Illicium vernum*, *Chem. Express.*, 8, 437–490, 1993.

Morimoto, S., Tanabe, H., Nonaka, G., and Nishioka, I., Prenylated flavan-3-ols and procyanidins from *Illicium anisatum*, *Phytochemistry*, 27, 907–910, 1988.

Nakamura, T., Okuyama, E., and Yamazaki, M., Neurotropic compounds from star anise, *Chem. Pharm. Bull.*, 44, 1908–1914, 1996.

Niwa, H., Ito, S., Wakamatus, K., Mori, T., and Yamada, K., Synthesis of (-)-noranisatin, an oxidation product of neurotoxic sesquiterpene anisatin having a novel spiro β-lactone, *Tetrahedron Lett.*, 32, 1329–1330, 1991.

Niwa, H., Nishiwaki, M., Tsukada, I., Ishigaki, T., Ito, S., Wakamatsu, K., Mori, T., Ikagawa, M., and Yamada, K., Stereocontrolled total synthesis of (-)-anisatin: a neurotoxic sesquiterpenoid possessing a novel spiro β-lactone, *J. Am. Chem. Soc.*, 112, 9001–9003, 1990.

Niwa, H. and Yamada, K., Synthesis of plant neurotoxin, anisatin, *Pharmacia*, 27, 924–927, 1991.

Okely, H.M. and Grundon, M.F., Anisoxide: an artifact from star anise oil formed by Claisen rearrangement, *Chem. Commun.*, 1157–1158, 1971.

Okely, H.M. and Grundon, M.F., Thermal rearrangement of 1-(3-methylbut-2-enyloxy)-4-prop-1-enyl-benzene (Feniculin): a new synthesis of anisoxide and re-investigation of the constituents of star anise oil, *J. Chem. Soc. Perkin I*, 897–899, 1981.

Okigawa, M. and Kawano, N., The structure of pseudoanisatin, *Tetrahedron Lett.*, 75–78, 1971.

Okuyama, E., Nakamura, T., and Yamazaki, M., Convulsants from star anise (*Illicium vernum* Hook. f.), *Chem. Pharm. Bull.*, 41, 1670–1671, 1993.

Saunders, R.M.K., Systematics of the genus *Illicium* L. (Illiciaceae) in Meleasia, *Bot. J. Linn. Soc.*, 117, 333–352, 1995.

Sakabe, N., Hirata, Y., Furusaki, A., Tomiie, Y., and Nitta, I., X-ray structure determination of bromonoranisatinone, a derivative of anisatin, *Tetrahedron Lett.*, 4795–4796, 1965.

Schmidt, T.J., Novel *seco*-prezizaane sesquiterpene from North American *Illicium* species, *J. Nat. Prod.*, 62, 684–687, 1999.

Schmidt, T.J. and Peters, W., Revision of the structure of dunnianin and related sesquiterpene lactone from *Illicium* species, *J. Nat. Prod.*, 60, 783–787, 1997.

Schmidt, T.J., Schmidt, H.M., Muller, E., and Peters, W., New sesquiterpene lactones from *Illicium floridanum*, *J. Nat. Prod.*, 61, 230–236, 1998.

Shibuya, M., Abe, K., Nakahashi, Y., and Kubota, S., Phenolic components from leaf oil of *Illicium anisatum* L, *Chem. Pharm. Bull.*, 26, 2671–2673, 1978.

Smith, A.C., The families Illiciaceae Schisandraceae, *Sargentia*, 7, 1–224, 1947.

Sy, L.-K. and Brown, G. D., A sesquilignan from *Illicium dunnianum*, *Phytochemistry*, 45, 1417–1419, 1996.

Sy, L.-K. and Brown, G.D., Abietane diterpenes from *Illicium angustisepalum*, *J. Nat. Prod.*, 61, 907–912, 1998a.

Sy, L.-K. and Brown, G.D., Novel pheylpropanoids and lignans from *Illicium verum*, *J. Nat. Prod.*, 61, 987–992, 1998b.

Sy, L.-K. and Brown, G.D., A prezizaane sesquiterpene from *Illicium angustisepalum*, *Phytochemistry*, 49, 1715–1717, 1998c.

Sy L.-K., Saunders, R.M.K., and Brown, G.D., Phytochemistry of *Illicium dunnianum* and the systematic position of the Illiciaceae, *Phytochemistry*, 44, 1099–1108, 1997.

Tanaka, T., Ueda, K., Kawojiri, M., Yamasa, H., and Kouno, I., The chemical constituents of the leaf of *Illicium anisatum*, The 118th Annual Meeting of Pharmaceutical Society of Japan, Kyoto, Abstract No. 2, 1998, p. 153.

Theander, O., Lundgren, L.N., and Shen, Z., The constituents of conifer needles. Dilignol glycosides from *Pinus massoniena* Lamb., *Acta Chem. Scan. Ser. B*, 39, 241, 1985.

Tomita, B. and Hirose, Y., *Allo*-Cedrol: a new tricarbocyclic sesquiterpene alcohol, *Phytochemistry,* 12, 1409–1414, 1973.

de Torre, D.C.G. and Harms, H., *Genera Siphonogamarum ad Systema Englerianum Conscripta,* Wilhelm Engelmann, Leipzig, 1900.

Wang, J.-L., Yang, C.-S., Yan, R.-N., Yao, B., and Yang, X.-B., Study on the chemical constituents of the fruits of *Illicium minwanense*, *Zhongguo Yaoxue Zazhi*, 29, 693–696, 1994.

Wong, M.G., Gulbis, J.M., Mackay, M.F., Crail, D.J., and Andrews, P.R., Anisatin: a crystallographic, N.M.R. and theoretical conformation study, *Aust. J. Chem.,* 41, 1071–1086, 1988.

Xie, D.-L., Wang, S., Cheng, Z.-W., Song, Y.-S., and Kong, D.-Y., Chemical constituents of the root-cortex of *Illicium henryi*, *Zhong Caoya.*, 21, 447–449, 1990.

Yakushijin, K., Furukawa, H., and McPhall, A.T., Studies on the constituents of the plants of *Illicium* species. IV. Thermal and photochemical reactions of illicinone A, a novel phytoquinoid from *Illicium* plants, *Chem. Pharm. Bull.,* 32, 23–30, 1984a.

Yakushijin, K., Sekikawa, J., Suzuki, R., Morishita, Furukawa, H., and Murata, H., Novel phytoquinoids from *Illicium tashiroi* MAXIM, *Chem. Pharm. Bull.,* 28, 1951–1954, 1980.

Yakushijin, K., Tohshima, Kitagawa, E., Suzuki, R., Sekikawa, J., Morishita, T., and Furukawa, H., Studies on the constituents of the plants of *Illicium* species. III. structure elucidations of novel phytoquinoids, illicinones and illifunones from *Illicium tashiroi* MAXIM. and *I. arborescens* HATAYA, *Chem. Pharm. Bull.,* 32, 11–22, 1984b.

Yakushijin, K., Tohshima, T., Suzuki, R., Murata, H., Lu, S.-T., and Furukawa, H., Studies on the constituents of the plants of *Illicium* species. II. Structures of phenolic components, *Chem. Pharm. Bull.,* 31, 2879–2983, 1983.

Yamada, T., Takada, S., Nakamura, S., and Hirata, Y., The structure of noranisatin, an oxidation product of anisatin. *Tetrahedron Lett.,* 4785–4794, 1965.

Yamada, K., Takada, S., Nakamura, S., and Hirata, Y., The structure of anisatin and neoanisatin, toxic sesquiterpenes from *Illicium anisatum* L., *Tetrahedron,* 24, 199–229, 1968.

Yang, C.-S., Hashimoto, M., Baba, N., Takahashi, M., Kaneto, H., Kouno, I., and Kawano, N., A new toxic neoanisatin derivative from the pericarps of *Illicium majus*, *Chem. Pharm. Bull.,* 38, 291–292, 1990.

Yang, C.-S., Kouno, I., Kawano, N., and Sato, S., New anisatin-like sesquiterpene lactones from pericarps of *Illicium majus*, *Tetrahedron Lett.,* 29, 1165–1168, 1988.

Yang C.-S., Liu, H., and Wu, X.-G., GC-MS analysis of essential oils of the fruits of *Illicium* plants in China, *Zhongguo Yaoxue Zazhi*, 27, 206–209, 1992.

Zhang, J., Chemical constituents of the stems and leaves of *Illicium dunnianum*, *Zhonngguo Zhongyao Zazhi,* 14, 36–37, 1989.

5 Chemical Constituents of the Genus *Pimpinella*

Jürgen Reichling and Enza Maria Galati

CONTENTS

5.1 INTRODUCTION

The genus *Pimpinella* belongs to the family Apiaceae (Umbelliferae) and includes about 90 species. The dominant features of the family are the umbellate inflorescence, the smaller flower size, and the specialized fruit, consisting of two one-seeded mericarps suspended from a common stalk, the carpophore. According to the following characteristic features, the genus *Pimpinella* is divided into two sections:

1. Section *Anisum* Ludwig DC.: Ovary and fruit hairy; plants annual and biennial; both medicinal and agricultural plants belong to this section; for example, *P. anisum* L.
2. Section *Tragoselinum* Mill. DC.: Ovary and fruit bare; plants perennial; both medicinal and agricultural plants belong to this section; for example, *P. major* L. Huds., *P. peregrina* L., *P. saxifraga* L.

The genus *Pimpinella* comprises annual or perennial herbs, which are mainly distributed in Asia, Europe, and Africa. A few species can also be found in South America and one in the western part of North America (Staesche et al., 1994; Pradhan and Banerji, 1998). From a medical and agricultural point of view, only few species are economically significant. Some of these species will be described in more detail later in this chapter.

0-415-32246-4/04/$0.00+$1.50

P. anisum L. (anise) is an annual herbaceous plant that originated in the eastern Mediterranean region and is native to Asia Minor, Greece, and Egypt. Anise is primarily grown for its fruits, commercially known as "seeds." The fruits are greenish-grey to brown in color, ovoid, and somewhat laterally compressed, ranging from 2 to 5 mm in length. The plant, which is 50 to 60 cm high, is cultivated in Argentina, Chile, China, France, India, Japan, Mexico, Spain, and the United States. As a medicinal herb and aromatic plant, anise is one of the oldest cultigens. The anise fruit and its essential oil are used mainly as a spicy seasoning, as a flavor additive in toothpastes and gargles, in the confectionary industry, and for the production of herb liqueurs or anise brandy. In medicine, the carminative, spasmolytic, and expectorant effects of the fruits and essential oil are of interest (Ernst, 1989; Staesche et al., 1994; Karaali and Basoglu, 1995).

P. major L. Hudson, a perennial plant, is locally common on grassy roadsides and at the edges of woods. The plant grows in most of Europe, except for in the extreme north and south. *P. major* is variable in size, pubescence, leaf dissection, and color of petals (Bohn et al., 1989). There are numerous varieties of *P. major* L. Hudson; however, only *var. rubra* (Hoppe et Schleicher) Fiori et Paoletti und *var. vulgaris* (Moritz) Thell are seen frequently.

P. saxifraga L. is a perennial plant widely distributed over Europe and Asia. The roots of the plant, together with those of the *P. major* (herbal drug Pimpinellae radix), which is closely related to *P. saxifraga*, show expectorant, bronchosecretolytic, and antiphlogistic properties. The species *P. saxifraga* is divided into three subspecies: *ssp. alpestris* (Spr.) Vollm., *ssp. eusaxifraga* Thell., and *ssp. nigra* (Mill.) Gaud. The dissection of the leaves and leaf tips vary greatly among all of the subspecies (Staesche et al., 1994).

P. peregrina L. is a biennial plant with hairy fruits. The plant is native to southern Europe, extending to Asia Minor and Egypt. In Germany, *P. peregrina* has sometimes been mistakenly cultivated as a medicinal plant to yield the herbal drug Pimpinellae radix (Kubeczka et al., 1986). Two varieties of *P. peregrina* are distinguished, according to the density of the hairs; namely, *var. peregrina* DC. and *var. hispida* (Loisel.) Thell. (Staesche et al., 1994).

P. diversifolia DC is found at altitudes of 1500 to 3000 m in the Kumaun Region of the Himalayas. The ethanol extract of its fruits has been reported to be strongly fungitoxic, and the extract of the whole plant has been reported to possess spermicidal activity in rat semen (Bottini et al., 1986).

Although it is beyond the scope of this chapter to review all chemical constituents ever found in the genus *Pimpinella*, the progress that has been made in chemical analysis will be documented by the most important results published in the last 25 years.

5.2 C_6-C_3 COMPOUNDS

In 1976, Stahl and Herting isolated several phenylpropanoids from the roots of *P. saxifraga* and *P. major* (Pimpinellae radix). The authors described these compounds as esters of epoxyisoeugenol with isobutyric, 2-methylbutyric, and tiglic acid. However, a reinvestigation of Pimpinellae radix revealed that all these compounds belong to a new type of phenylpropanoid with a 1,2,5-trisubstituted benzene skeleton. The 1-(E)-propenyl-2-hydroxy-5-methoxy benzene skeleton of these compounds was named pseudoisoeugenol (Figure 5.1) (Martin et al., 1985). In addition to the pseudoisoeugenol derivatives, diesters of 1-(E)-propenyl-2-hydroxy-5-hydroxy benzene (2-(E-3-methyloxiranyl) hydroquinone) as well as derivatives of 1-(E)-propenyl-4-hydroxy benzene (Figure 5.2) in some *Pimpinella* species could also be described (Bottini et al., 1986; Dev et al., 1989; Kubeczka et al., 1986). Rare phenylpropanoids of the pseudoisoeugenol type, such as *trans*-pseudoisoeugenol-2-methylbutyrate, were first detected in the essential fruit oil of *P. anisum* (Kubeczka et al., 1976). Pseudoisoeugenol derivatives have also been identified *in vitro*, namely, in shoot, root, and callus cultures of *P. anisum* (Reichling et al., 1985; Martin and Reichling, 1992; Kemmerer and Reichling, 1996; Santos et al., 1998; Andarwulan and Shetty 2000). As of yet, pseudoisoeugenol derivatives have been found exclusively in the genus *Pimpinella* and not in any

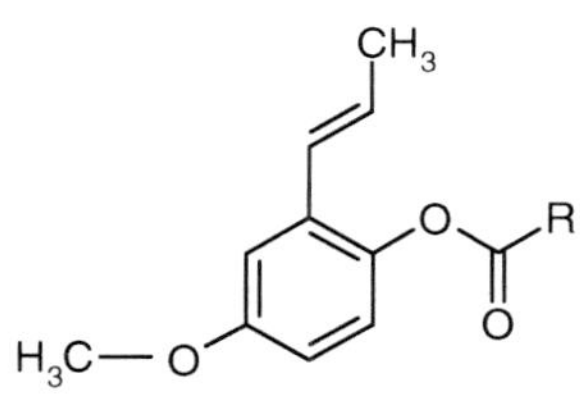

Pseudoisoeugenolderivates

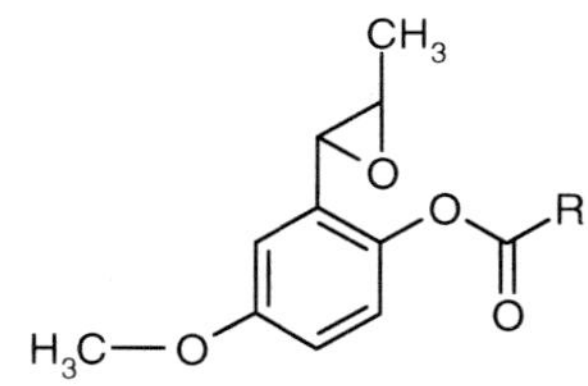

Epoxypseudoisoeugenolderivates

	R=
2-Methylbutyrate	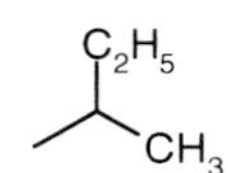
Angelicate	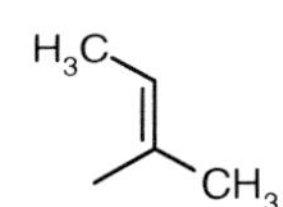
Tiglate	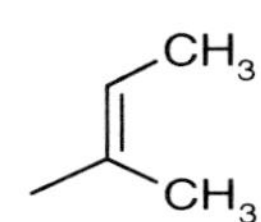
2-Methylproprionate	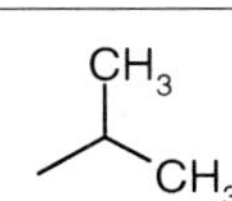

FIGURE 5.1 Derivatives of pseudoisoeugenol and epoxypseudoisoeugenol

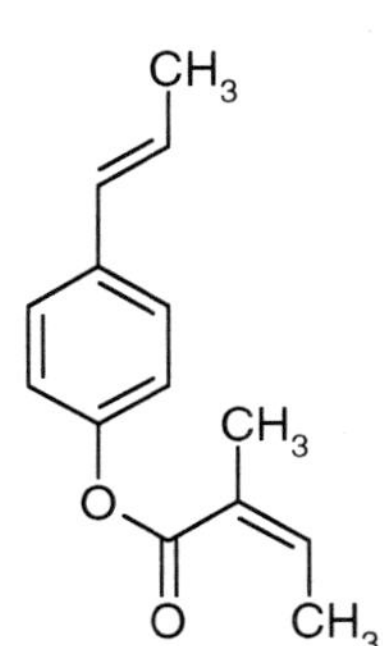

FIGURE 5.2 Derivative of 1-(E)-propenyl-4-hydroxy benzene; for example, isochavicolangelate

other Apiaceae or other plant families (Table 5.1) (Kubeczka et al., 1976; Martin et al., 1985; Bottini et al., 1986; Dev et al., 1989).

The biosynthesis of pseudoisoeugenol derivatives cannot simply be explained by the normal phenylpropanoid pathway, as this pathway involves a hydroxylation step in the para-position next

TABLE 5.1
Occurrence of Various 1-(E)-propenyl-2-hydroxy-5-hydroxy benzene Derivatives in Different Species of the Genus *Pimpinella*

	Pimpinella							
Compound	***anisum***	***diversifolia***	***major***	***nigra***	***peregrina***	***saxifraga***	***tragium***	***villosa***
Epoxypseudoisoeugenol-2-methylbutyrate	•	•	•	•	•	•	•	
Epoxypeusoisoeugenol isobutyrate		•			•	•		
Epoxypseudoisoeugenol tiglate			•			•	•	
Epoxypseudoisoeugenol angelicate		•						+
Pseudoisoeugenol-2-methylbutyrate	•		•	•	•	•		
Pseudoisoeugenol isobutyrate					•			
Pseudoisoeugenol tiglate			•			•		
1-Angelyloxy-2-(E-3-methyloxiranyl)-4-isobutyryloxy benzene		•						
1-Isobutyryloxy-2-(E-3-methyloxiranyl)-4-angelyloxy benzene		•						
1,4-Diangelyloxy-2-(E-3-methyloxiranyl) benzene		•						+
1-Angelyloxy-2-(E-3-methyloxiranyl)-4-epoxyangelyloxy benzene								+

Sources: Merkel and Reichling, 1990; Reichling et al., 1989.
Note: • = definitely identified.

to the C_3 aryl side chain. This position, however, is not substituted in these compounds, which exhibit a typical 2-acyloxy-5-methoxy substitution pattern. To learn more about the biosynthesis of these compounds, a leaf-differentiating callus culture of *P. anisum* was established. The *in vitro* culture selectively promoted the production of epoxypseudoisoeugenol-2-methylbutyrate. This compound served as the final molecule of the biosynthetic pathway in all of the labeling experiments conducted. The biosynthetic sequence found for pseudoisoeugenol derivatives was L-phenylalanine, *trans*-cinnamic acid, p-coumaric acid, p-coumaric aldehyde, p-coumaric alcohol, anol, and *trans*-anethol. The biosynthetic step leading from *trans*-anethol to the pseudoisoeugenol skeleton involves migration of the side chain during introduction of the second OH group in the molecule (NIH shift). The final biosynthetic steps required to form the different compounds must be acylation and epoxidation of the propenyl double bond of pseudoisoeugenol (Martin and Reichling, 1992; Reichling et al., 1995a,b; Kemmerer and Reichling, 1996).

Epoxypseudoisoeugenol-2-methylbutyrate and related compounds have been suggested for use in foods as antioxidants or as modulators of seed germination to obtain seed-based functional phenolics for nutraceutical applications (Anderwulan and Shetty, 2000).

In *P. anisum*, *trans*-anethol is clearly the main component of the fruit's essential oil (Table 5.2). Minor amounts of three other phenylpropanoids could also be detected: *cis*-anethol, pseudoisoeugenol-2-methylbutyrate, and epoxypseudoisoeugenol-2-methylbutyrate. In contrast to the fruit's essential oil, the content of *trans*-anethol in the herb's essential oil only amounted to about 30%. Pseudoisoeugenol-2-methylbutyrate and epoxypseudoisoeugenol-2-methylbutyrate were also present in the herb's essential oil, whereby a relatively high concentration of pseudoisoeugenol-2-methylbutyrate was present. In contrast, epoxypseudoisoeugenol-2-methylbutyrate is one of the major components of the essential oil isolated from the roots of *P. anisum*, and only small amounts of *trans*-anethol and pseudoisoeugenol-2-methylbutyrate were found in essential root oil (Table 5.2) (Martin et al., 1985; Kubeczka et al., 1986).

In the essential oil from roots of *P. diversifolia*, epoxypseudoisoeugenol angelate (54% in oil) and epoxypseudoisoeugenol isobutyrate (21% in oil) have been identified as the major constituents. In addition to these, both major compounds have three minor components, and diesters of 1-propenyl-2-hydroxy-5-hydroxy benzene (2-(E-3-methyloxiranyl)hydroquinone), were present as well: 1-angeloyloxy-2-(E-3-methyloxiranyl)-4-isobutyroyloxy benzene (0.8% in oil), 1-isobutyroyloxy-2-(E-3-methyloxiranyl)-4-angeloyloxy benzene (3.6% in oil), and 1,4-diangeloyloxy-2-(E-2-methyloxiranyl) benzene (1.2% in oil) (Figure 5.3) (Bottini et al., 1986; Dev et al., 1989).

In the roots of *P. major*, the main phenylpropanoid derivative is epoxypseudoisoeugenol tiglate, whereas only traces of epoxypseudoisoeugenol-2-methylbutyrate, pseudoisoeugenol-2-methylbutyrate, and pseudoisoeugenol tiglate are present (Table 5.3). Other than the pseudoisoeugenol derivatives, small amounts of two other phenylpropanoids can also be found: epoxyanol tiglate and anol tiglate, derivatives of 1-(E)-propenyl-4-hydroxy benzene (Martin et al., 1985; Kubeczka and Bohn, 1985; Kubeczka et al., 1989; Merkel and Reichling, 1989). To examine the capacity of various plant tissues of *P. major* to accumulate phenylpropanoids, the occurrence of these compounds was studied in plant material of different origin. As the result showed, the phenylpropanoids accumulated in the plant mainly in an organ-specific manner. For example, a high accumulation of epoxypseudoisoeugenol tiglate was found in the root (11 mg/g fr. wt.), and smaller quantities, (0.4–0.6 mg/g fr. wt.), could be measured in stems, leaves, and flowers. In the inflorescent region, rates of up to 3 mg/g fr. wt. epoxypseudoisoeugenol tiglate could be identified in the umbel pedicels and in the ovary directly after blooming. The same compound could not be identified in the fruit essential oil. Otherwise, both anol derivatives had accumulated in the fruit and root traces. Whereas anol tiglate could only be found in the fruits, epoxyanol tiglate was also identified in the shoot and in the inflorescence in far smaller quanitities (0.1–2.1 mg/g fr. wt.) (Merkel and Reichling, 1989).

The main constituent of the root essential oil of *P. saxifraga* has been shown to be epoxypseudoisoeugenol-2-methylbutyrate. In addition to this, epoxypseudoisoeugenol tiglate, epoxypseudoisoeugenol-2-methylpropionate, and pseudoisoeugenol-2-methylbutyrate have been detected

TABLE 5.2
Percentage Composition of Essential Oil of *P. anisum* in Fruits and Roots

Components	Fruit Oil (%)	Root Oil (%)
n-Heptanal		0.20
Benzaldehyde		0.70
Sabinene		0.20
n-Octanal		0.10
Myrcen		0.10
α-Terpinene		traces
p-Cymene		traces
Limonene		traces
γ-Terpinene		0.10
Undecane		traces
Geijerene		1.80
Borneol		traces
Terpinene-4-ol		traces
α-Terpineol		0.50
Estragol	2.20	
Anisaldehyde	1.90	
cis-Anethole	traces	
trans-Anethole	92.50	2.00
Pregeijerene		6.00
Dodecanol		0.30
β-Caryophyllene		0.20
trans-β-Farnesene		traces
Zingiberene		0.20
β-Bisabolene		6.40
trans-Pseudoisoeugenol-2-methylbutyrate	0.10	4.70
cis-Epoxypseudoisoeugenol-2-methylbutyrate		traces
trans-Epoxypseudoisoeugenol-2-methylbutyrate	traces	70.20

Source: Santos, P.M., Figueiredo, A.C., Oliveira, M.M., Barroso, J.G., Pedro, L.G., Deans, S.G., Younus, A.K.M., and Scheffer, J.C., Essential oils from hairy root cultures and from fruits and roots of *Pimpinella anisum*, *Phytochemistry*, 48, 455–460, 1998.

in small amounts (Table 5.4). Kubeczka et al. (1986) also mentioned the occurrence of an anol ester, likely anol-2-methylbutyrate, in the essential oil of the root of *P. saxifraga*.

P. thellungiana is known as a Chinese folk drug. From the ether extract of the root of this plant, four new phenolic compounds were isolated and the chemical structures determined: 2-(1′,2′-dihydroxy)propyl-4-methoxyphenol, 2-(1′-ethoxy-2′-hydroxypropyl)-4-methoxyphenol, known as thellungianin A; 2-(1′-ethoxy-2′-hydroxy)propyl-4-methoxyphenyl-2-methylbutyrate, known as thellungianin B; and 2-(1′-methoxy-2′-hydroxy)propyl-4-methoxy phenol, known as thellungianin E (Figure 5.4) (Qiao et al., 1997; Li et al., 1998; Shi et al., 1998).

Epoxypseudoisoeugenol angelate, epoxypseudoisoeugenol epoxyangelate, and diesters of 2-(E-3-methyloxiranyl) hydroquinones: 1-epoxyangelyloxy-2-(E-3-methyloxiranyl-4)-angelyloxy benzene, 1,4-diangelyloxy-2-(E-3-methyloxiranyl) benzene, and 1,4-diepoxyangelyloxy-2-(E-methyloxiranyl) benzene (Figure 5.3) — have been reported to accumulate in the roots of *P. villosa* (Macias et al., 1994).

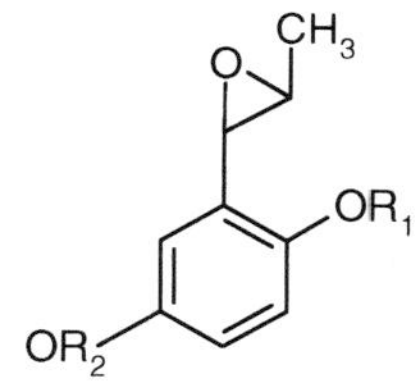

R_2	R_1
Ang	Ep Ang
Ang	Ang
Ep Ang	Ep Ang

Ang =

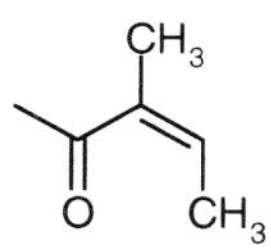

Ep Ang =

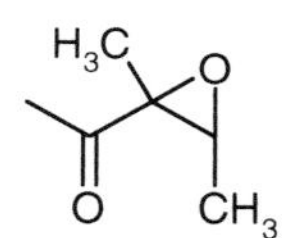

FIGURE 5.3 Diesters of 2-(E3-methyloxiranyl) hydroquinones

TABLE 5.3
Percentage Composition of Essential Oil of *P. major* in Roots

Components	Root Oil (%)	Components	Root Oil (%)
p-Cymene	traces	γ-Elemene	traces
Geijerene	2.33	α-Humulene	0.46
Pregeijerene	9.75	Germacrene-A	1.27
δ-Elemene	12.05	Germacrene-C	7.83
β-Caryophyllene	0.45	*trans*-Pseudoisoeugenol-2-methylbutyrate	traces
trans-β-Farnesene	0.32	*trans*-Epoxypseudoisoeugenol-2-methylbutyrate	traces
Germacrene D	1.87	*trans*-Pseudoisoeugenol tiglate	3.41
β-Bisabolene	2.58	*trans*-Epoxypseudoisoeugenol tiglate	19.54

Source: Bohn et al., 1989.

5.3 COUMARINS

From the aerial part of *Pimpinella diversifolia*, two coumarins, iso-angenomalin and oxypeucedanin, have been isolated (Bottini et al., 1986).

5.4 DITERPENES

The diterpene hydrocarbon neophytadiene (Figure 5.5) was isolated from the hydrocarbon fraction of anise shoots. This compound accounted for about 15% of the hydrocarbon fraction (Burkhardt et al., 1986). Neophytadiene is known to occur in higher plants (Aoki et al., 1982) as well as in zooplankton (Blumer and Thomas, 1965).

TABLE 5.4
Percentage Composition of Essential Oil of *P. saxifraga* in Roots

Components	Root Oil (%)	Components	Root Oil (%)
α-Pinene	0.07	α-Bergamotene	0.10
β-Pinene	0.21	Carvacryl methylether	0.46
Camphene	0.06	*trans*-β-Farnesene	3.15
Sabinene	0.10	Germacrene D	0.49
Myrcene	0.16	β-Bisabolene	0.91
Limonene	0.45	β-Sesquiphellandrene	2.61
β-Phellandrene	0.07	Germacrene B	5.44
cis-Ocimene	0.03	Germacrone	2.25
γ-Terpenene	0.30	Pseudoisoeugenol-2-methylbutyrate	3.56
p-Cymene	0.26	Epoxypseudoisoeugenol-2-methylpropionate	0.64
Geijerene	3.29	Epoxypseudoisoeugenol-2-methylbutyrate	46.24
Pregeijerene	9.18	Epoxypseudoisoeugenol tiglate	Traces

Source: Kubecza, K.H., Bohn, I., and Formacek, V., New constituents from essential oils of *Pimpinella* species, in *Progress in Essential Oil Research*, Brunke, E.J., Ed., Walter de Gruyter, Berlin, 1986, pp. 279–298.

FIGURE 5.4 Derivatives of 2-(1′,2′-dihydroxy)propyl-4-methoxyphenol

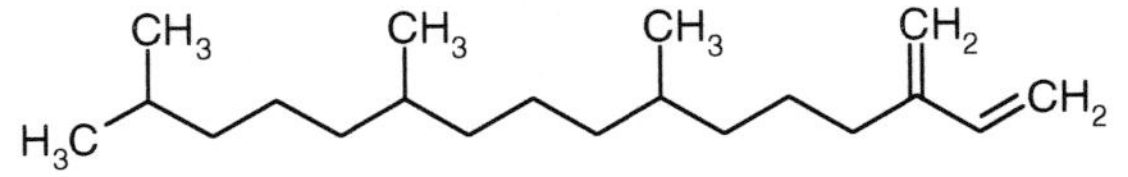

FIGURE 5.5 Neophytadiene, a diterpene hydrocarbone isolated from shoots of *P. anisum*

5.5 ESSENTIAL OIL

Essential oils of the genus *Pimpinella* are complex mixtures of numerous components that contain sesquiterpenes, phenolic compounds (C_6-C_3 compounds), alkenes, and so forth. The essential oil is located in the schizogenetic oil ducts of fruits, shoots and roots. The composition of the essential oils usually varies in the different organs of an individual species. The presence of high amounts (3 to 48%) of pregeijerene and its Cope-rearrangement product geijerene (Figure 5.6) in the essential oils of the roots of all *Pimpinella* species studied seems to be species-specific. Traces of polyenes are also widely found in the essential oil of the roots of the *Pimpinella* species. In the genus *Pimpinella*, it is thought that they largely replace the furanocoumarins otherwise usually found in Apiaceae.

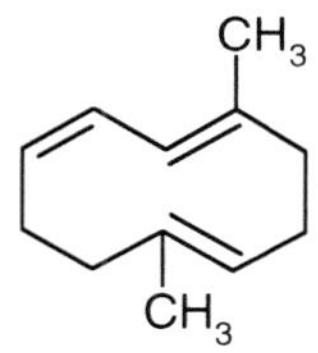

Pregeijerene

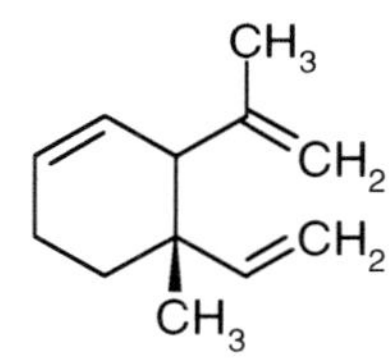

Geijerene

FIGURE 5.6 Pregeijerene and geijerene

TABLE 5.5
Percentage Composition of the Sesquiterpene Hydrocarbone Fraction Derived from Essential Oil of *P. anisum* in Fruits and Shoots

Compounds	Fruit Oil (%)	Shoot Oil (%)	Compounds	Fruit Oil (%)	Shoot Oil (%)
α-Himachalene	5.32		δ-Cadinene		0.11
β-Himachalene	6.95		Humulene		0.10
γ-Himachalene	73.99	23.00	Caryophyllene		0.18
β-Bisabolene	5.06	46.05	β-Farnesene		0.63
δ-Elemene	1.09	0.35	Bergamotene		0.14
ar-Curcumene	1.32		Geijerene		0.42
Copaene	0.32		Pregeijerene		0.71
Sesquiphellandrene		1.62	Germacrene-D		8.71

Source: Reichling, J., Martin, R., Burkhardt, G., and Becker, H., Comparative study on the production and accumulation of essential oil in the whole plant and in tissue cultures of *Pimpinella anisum*, in *Progress in Essential Oil Research*, Brunke, E.J., Ed., Walter de Gruyter, Berlin, 1986, pp. 421–428.

The fruits of *P. anisum* contain about 2 to 6% and the roots about 0.05% of an essential oil (Becker, 1971, Lee *et al.*, 1997). Smaller amounts are found in leaves and stems. Oil of anise is a colorless to pale-yellow liquid with a strong, sweet-spicy, licorice-like odor and a characteristic, sweet, aromatic flavor. The main component (comprising 70 to 90% of the oil) and typical odor and flavor carrier of the fruit oil of *P. anisum* (anise) oil is *trans*-anethol (1-propenyl-4-methoxy benzene). *Trans*-anethol is of medicinal importance because of its carminative and expectorant effect. The isomeric methylchavicol accounts for 4% of the oil composition; it possesses an anise-like smell but lacks the sweetish taste (Ernst, 1989; Santos et al., 1998). Other substances found in anise oil are anisaldehyde, dianethole, anisketone, anisic acid, and p-methoxy-acetophenone. The amount of *cis*-anethol, which is toxic, in commercially available anise oils is usually below about 0.2 to 0.5%. The anise oil also contains low quantities of mono- and sesquiterpene hydrocarbones, and γ-himachalene above all (Table 5.5). Other than mono- and sesquiterpenes, anise oil also contains derivatives of pseudoisoeugenol, up to 5%, as well as other phenolic compounds (Reichling et al., 1985, 1986; Kollmannsberger et al., 2000).

Field experiments on aniseed (fruits) of *P. anisum* were conducted in the winter seasons of 1983–1984 through 1985–1986 on medium clay-loam soil of average fertility in India. A higher than 100% yield of grain, essential oil, and *trans*-anethol was recorded by sowing anise on October 25 or November 5 in comparison with when grain was sown on November 30. A more than 20% higher yield in grain, essential oil, and *trans*-anethol was obtained by sowing aniseed broadcast or in 15-cm rows as compared to sowing it in 45-cm rows. The authors claim that a higher essential

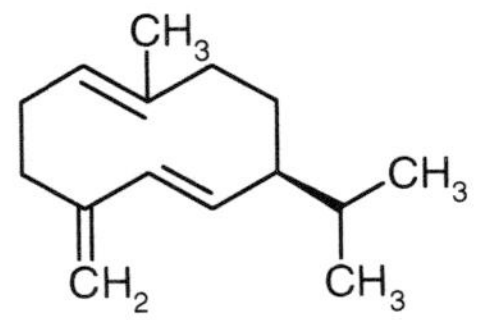

FIGURE 5.7 Germacrene-D

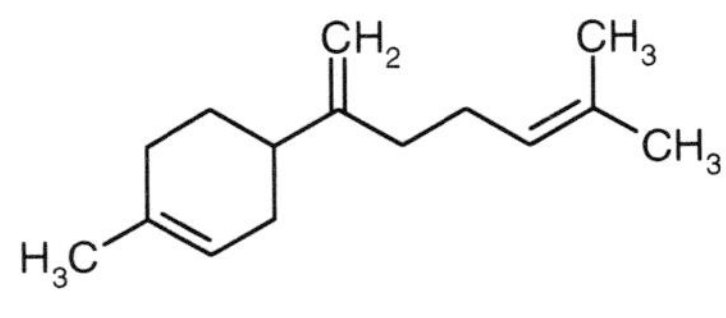

FIGURE 5.8 β-Bisabolene

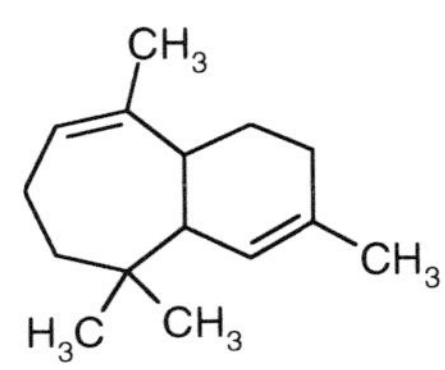

FIGURE 5.9 γ-Himachalene

oil content of 3.24 to 3.28% and anethol content of 92.5 to 93.5% along with a superior quality of the essential oil of light-yellow color was obtained throughout the studies (Maheshwari et al., 1989).

The essential oil of the herb of *P. anisum* contains considerable amounts of sesquiterpene hydrocarbons, such as germacrene-D (Figure 5.7), β-bisabolene (Figure 5.8), γ-himachalene (Figure 5.9), α-zingiberene, and ar-curcumene. The anise root oil (Table 5.2) is characterized by the presence of β-bisabolene, pregeijerene, and its Cope-rearrangement product geijerene, two terpenoic hydrocarbons with 12 C-atoms, as well as by *trans*-epoxypseudoisoeugenol-2-methylbutyrate. Pregeijerene and geijerene were not detected in the anise fruit oil (Kubeczka et al., 1986; Santos et al., 1998).

The antibacterial activity of *P. anisum* and its different dilutions in dimethyl sulfoxide against *Staphylococcus aureus, Streptococcus pyogenes, Escherichia coli,* and *Corynebacterium ovis* was evaluated. Pure penicillin and streptomycin were taken as standard inhibitors. Pure anise oil was more effective in inhibiting the growth of the bacteria than were dilutions in dimethyl sulfoxide (Gangrade et al., 1990). The essential oil of *P. anisum* L. (doses of 108 to 135 μL/L air) caused 95% mortality in *Tribolium confusum* Du Val adults, *Sitophilus oryzae* L. adults, and *Ephestia kuehniella* Zeller last instars within an exposure period of 24 hours in *S. oryzae* and 96 hours in *E. kuehniella* (Sarac and Tunc, 1995). In another investigation, anise oil was found to be toxic to two greenhouse pests, namely, the carmine spider mite, *Tetranychus cinnabarinus* Boisd, and the cotton aphid, *Aphis gossypii* Glov. In general, a minimum dose of 0.5 μL/L air and 2 to 3 days of exposure was required for 99% mortality (Tunl and Sahinkaya, 1998). The *in vitro* antifungal activity of essential oil distilled from seeds of *P. anisum* in pure state and four dilutions with dimethyl sulfoxide, 1:10, 1:100, 1:1000, and 1:10000, were tested against *Aspergillus niger, A. flavus, Fusarium oxysporum,* and *Penicillium* spp.; cycloheximide and hamycin were taken as standard inhibitors. The pure oil showed 70 to 80% growth inhibition of all tested pathogens as compared to cycloheximide and hamycin. The activity was reduced to below 50% inhibition after the 1:10 dilution of oil with dimethyl sulfoxide (Gangrade et al., 1991).

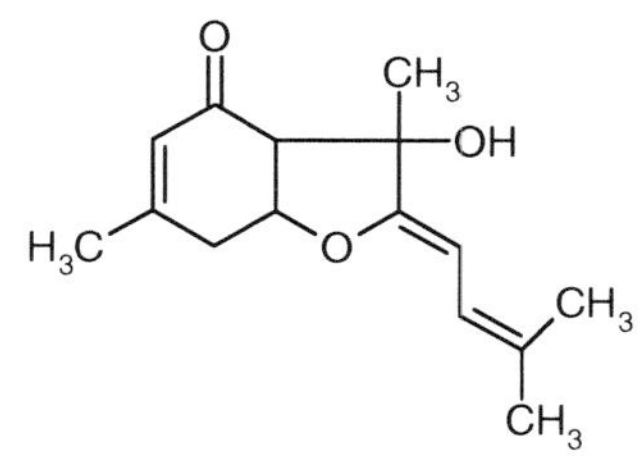

FIGURE 5.10 Bisabolangelone

The antiviral activity of the essential oil of *P. anisum* has been tested against potato virus X, tobacco mosaic virus, and tobacco ring spot virus on the hypersensitive host *Chenopodium amaranticolor*. At a concentration of 300 ppm, the essential oil totally inhibited the formation of local lesions (Shukla et al., 1989b).

Examination of the essential oil from fruits of *P. diversifolia* indicated the presence of carvone, geranyl acetate, limonene, myrcene, α-pinene, pulegone, santene, α-terpineol, and terpineolic acid (Bottini et al., 1986).

The essential oil of the root of *P. major* is a complex mixture of numerous components. The main component is epoxypseudoisoeugenol tiglate, followed by minor amounts of structurally related compounds; for example, pseudoisoeugenol tiglate and epoxypseudoisoeugenol-2-methylbutyrate. In addition to these characteristic compounds, a number of sesqui- and trinorsesquiterpenes occur in appreciable amounts. Monoterpenes were found only in traces. Among the C_{12}-hydrocarbons, pregeijerene is the most predominant compound, amounting to 5.1 to 12.1% of the root oil; smaller amounts of geijerene were present. Another norsesquiterpene found in the roots of *P. major* was trinoranastreptene (1,5-dimethyl-tricyclodeca-4,7-dien). The sesquiterpene pattern of the essential oil of plants deriving from different locations exhibited significant differences: the essential oil from the roots of plants collected in Germany contained about 12% δ-elemene and about 8% germacrene-C; in plants of Italian origin, the essential oil of roots was dominated by about 15% germacrone, about 6% germacrene-B, and about 10% γ-elemene (Bohn et al., 1989; Merkel and Reichling, 1989). The main constitutents of the essential oil of the fruit of *P. major* are 3,3a,7,7a-tetrahydro-3,6-dimethyl-2-(3-methyl-2-butenyliden)-4-benzofuran (bisabolangelone) (Figure 5.10), epoxyanol tiglate, and β-bisabolene. Because bisabolangelone is not present in the essential oil of the fruit of *P. saxifraga* and *P. peregrina,* these two species can be easily distinguished chemically from *P. major* (Stahl and Herting, 1976).

Epoxypseudoisoeugenol-2-methylbutyrate has been established as the main constituent of the essential oil from roots of *P. saxifraga* (Martin et al., 1985). Only very low levels of monoterpenes are present. In addition to the epoxy derivative, the presence of the corresponding pseudoisoeugenol-2-methylbutyrate has been confirmed. According to Kubeczka et al. (1986), the essential root oil of *P. saxifraga* contains no epoxypseudoisoeugenol isobutyrate. Kubeczka et al. postulated that this component acts as a useful indicator for a falsification with roots of *P. peregrina*. In addition, an anol ester and several terpene hydrocarbons, such as geijerene, pregeijerene, *trans*-β-farnesene, and germacrene-B, are also found in the root oil of *P. saxifraga* (Kubeczka et al., 1986).

In addition to epoxypseudoisoeugenol tiglate and epoxypseudoisoeugenol-2-methylbutyrate, the essential fruit oil of *P. saxifraga* contains β-bisabolene, but no bisabolangelone (Stahl and Herting, 1976).

The constituents of the essential oil from the roots of *P. peregrina* belong mainly to the group of sesquiterpenes and phenylpropene derivatives, which represent more than 80% of the total essential oil (Table 5.6). The hydrocarbons β-bisabolene and β-sesquiphellandrene are the main constituents of the sesquiterpene fraction, but minor amounts of α-bergamotene, *trans*-β-farnesene, and germacrene-D are also present. Epoxypseudoisoeugenol-2-methylbutyrate and epoxypseudoisoeugenol-2-methylpropionate are the prominent pseudoisoeugenol derivatives of the root oil (Kubeczka et al., 1986).

TABLE 5.6
Percentage Composition of Essential Oil of *P. peregrina* in Roots

Components	Root Oil (%)	Components	Root Oil (%)
α-Pinene	traces	α-Bergamotene	02.56
β-Pinene	traces	Carvacryl methylether	01.58
Sabinene	00.88	*trans*-β-Farnesene	03.20
Myrcene	00.07	Germacrene D	01.18
Limonene	traces	β-Bisabolene	10.00
β-Phellandrene	traces	Bicyclogermacrene	00.59
γ-Terpinene	01.30	β-Sesquiphellandrene	19.83
p-Cymene	00.45	Pseudoisoeugenol-2-methylbutyrate	01.68
Geijerene	00.96	Epoxypseudoisoeugenol-2-methylbutyrate	29.67
Pregeijerene	11.01	Epoxypseudoisoeugenol-2-methylpropionate	11.84

Source: Kubecza, K.H., Bohn, I., and Formacek, V., New constituents from essential oils of *Pimpinella* species, in *Progress in Essential Oil Research,* Brunke, E.J., Ed., Walter de Gruyter, Berlin, 1986, pp. 279–298.

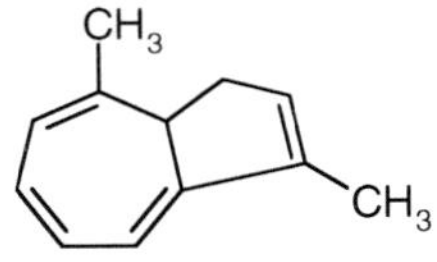

FIGURE 5.11 3,10-Dihydro-1,4-dimethylazulene

In the fruit of *P. peregrina*, α-bergamotene, β-bisabolene, and 2-methyl-5-methoxybenzofurane can be isolated, but no epoxypseudoisoeugenol derivatives are present (Stahl and Herting, 1976).

P. nigra Mill. is considered to be a subspecies of *P. saxifraga* by several authors. If the fresh root of the plant is cut, a blue oil emerges, indicating the formation of azulene. The azulene compound has been identified as 3,10-dihydro-1,4-dimethylazulene (Figure 5.11). The qualitative and quantitative composition of the dark blue essential oil of the plant roots differ considerably from those essential oils identified in fruits and roots of the *Pimpinella* species mentioned above. The essential oil of the plant roots (Table 5.7) contains only small amounts of phenylpropanoids, mainly epoxypseudoisoeugenol-2-methylbutyrate. In contrast, terpene hydrocarbons represent more than 70% of the essential oil. Other C_{12} hydrocarbones identified in the root oil are pregeijerene, its rearrangement product geijeren, and 1,5-dimethyl tricyclodeca-4,7-dien (Kubeczka et al., 1986).

5.6 FLAVONOIDS

Flavones and flavonols, such as apigenin, isoquercetin, kaempferol, luteolin, quercetin, and rutin, have been found in species of *Pimpinella.* The flavonoids are usually present in the O-glycoside form, and less often as C-glycosides, such as, for example, isorhamnetin and isovitexin.

In the fruits and leaves of *P. anisum*, several flavonoid glycosides were identified: luteolin, apigenin-7-O-glucoside, rutin, luteolin-7-O-glucoside, luteolin-7-O-xylosid, isoorientin, isovitexin, and quercetin-3-O-glucuronide (Kunzemann and Herrmann, 1977; Staesche et al., 1994). Quercetin and isorhamnetin glycoside were also found in the fruits of *P. peregrina* (Staesche et al., 1994). Up to 0.6% flavonoids, mainly composed of isoquercitrin, kaempferol-3-O-rutoside, and rutin, were identified in *P. saxifraga* when in blossom (Staesche et al., 1994).

TABLE 5.7
Percentage Composition of Essential Oil of *P. nigra* in Roots

Components	Root Oil (%)	Components	Root Oil (%)
Myrcene	0.02	*trans*-β-Farnesene	0.12
Limonene	0.04	Germacrene D	1.52
cis-Ocimene	0.28	β-Bisabolene	3.17
trans-Ocimene	0.83	β-Sesquiphellandrene	1.58
p-Cymene	0.02	3,10-Dihydro-1,4-dimethylazulene	16.10
Geijerene	3.60	Pseudoisoeugenol-2-methylbutyrate	0.20
Pregeijerene	6.01	Epoxypseudoisoeugenol-2-methylbutyrate	6.81
α-Bergamotene	0.16		

Source: Kubecza, K.H., Bohn, I., and Formacek, V., New constituents from essential oils of *Pimpinella* species, in *Progress in Essential Oil Research,* Brunke, E.J., Ed., Walter de Gruyter, Berlin, 1986, pp. 279–298.

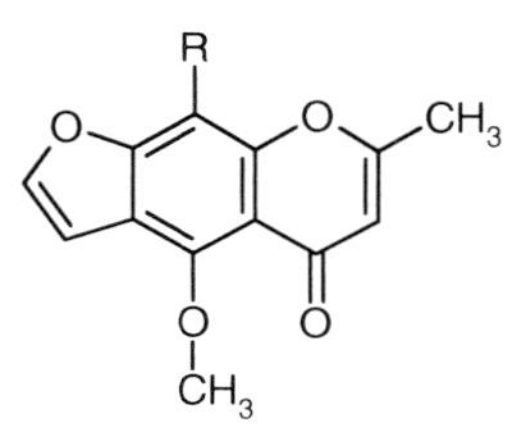

FIGURE 5.12 Visnagin (R = H) and khellin (R = OCH_3)

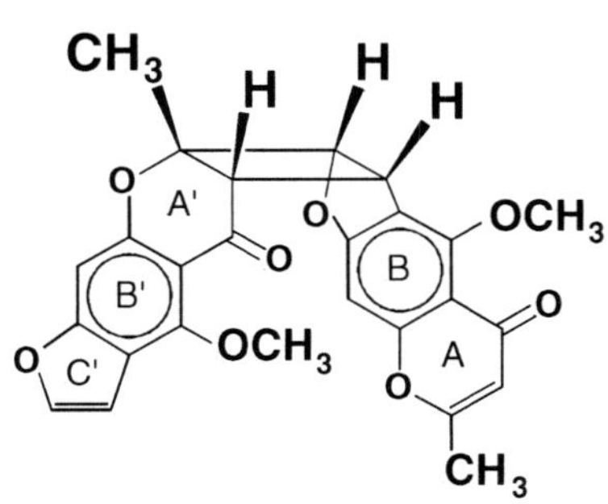

FIGURE 5.13 Pimolin

5.7 FURANOCHROMONES

Five biogenetically related furanochromones, khellin (Figure 5.12), visnagin (Figure 5.12), visamminol, amminol, and khellol, were found in *P. monoica* Dalz (Luthria and Banerji, 1994). Pradhan et al. (1994) detected pimolin (Figure 5.13), a novel dimeric furanochromone, in the roots of *P. monoica.* Novel oligomers di-, tri-, tetra-, and pentamers of the furanochromone visnagin were found in seeds of *P. monoica* in subsequent studies, and spectroscopy revealed a linear head-to-tail pyran–furan fusion in the oligomers (Pradhan and Banerji, 1998).

The hexane and methanol extract of *P. monoica* Dalz, commonly known as ladies lace, was shown to be a significant feeding deterrent for *Spodoptera litura* F. larvae. The biological activity involved was found to be the result of the furanocoumarins and furanochromones (Pradhan and Banerji, 1998).

5.8 FURANOCOUMARINS

Furanocoumarins occur in many plant species and are typical constituents of Apiaceae (Umbelliferae). They are well known for their mutagenic and photosensitizing activities (Schimmer et al., 1980; Schimmer, 1983). In the genus *Pimpinella,* only traces of furanocoumarins are found in the roots, leaves, and fruits.

In a search for bioactive compounds from indigenous plants, *P. monoica* was found to contain the furanocoumarin isopimpenellin (Luthria and Banerji, 1994). On the leaf surfaces of *P. major,* three furanocoumarins, psoralen, xanthotoxin, and bergapten, were found in amounts of 0.35 μg/g plant material (Zobel et al., 1990). Roots of *P. major* contain isopimpinellin (Schimmer et al., 1980). In Pimpinellae radix, eight furanocoumarins were isolated by semipreparative HPLC (high-performance liquid chromatography): iso-bergapten, angelicin, psoralen, bergapten, pimpinellin, sphondin, xanthotoxin, and isopimpinellin (Zogg et al., 1989). The fruits of *P. anisum* L. were analyzed for the presence of photoactive furanocoumarins using HPLC analysis and photobiological bioassay. No furanocoumarins could be detected in the fruits of anise by HPLC. However, using the ultrasensitive bioassay, traces of 5-methoxypsoralen and 8-methoxypsoralen were found in the fruits in amounts of less than 0.005 μg/g fruit (Ceska et al., 1987; Kollmannsberger et al., 2000).

5.9 LIPIDS/FATTY ACIDS

The seed oils of Apiaceae (Umbelliferae) species are characterized by a high content, over 50%, of petroselinic acid (*cis*-6-octadecenoic acid). This acid is the only naturally positioned isomer of significance of the most abundant octadecenoic acid, oleic acid (*cis*-9-octadecenoic acid), in plants (Placek, 1963; Nicolova-Damyanova et al., 1996). Petroselinic acid is qualified to produce a variety of valuable raw materials for industry, and plants of the Apiaceae family have been identified as promising sources (Kleiman and Spencer, 1982). For example, petroselinic acid, about 56%, was the main component of seed oils of *P. anisum.* Small amounts of other fatty acids, for example, oleic acid (18.5%) and *cis*-vaccenic acid (1.4%), were also present (Nikolova-Damyanova et al., 1996). In a further study, the composition of phospholipids in seed oil of *P. anisum* was analyzed. The phospholipid content amounted to 1.7 to 3.7% in the seed oil and 0.2 to 0.5% in the seeds. Phosphatidylcholine, 38 to 51%, phosphatidylinositol, 19 to 32%, and phosphatidylethanolamine, 9 to 19%, have been identified as major components. Minor components included phosphatidic acids, phosphatidylserine, phosphatidylglycerols, lysophosphatidylcholine, and lysophosphatidylethanolamine (Zlatanov, 1994). The herbs of *P. anisum* contain p-hydroxybenzoic acid glucoside, 18 to 21 mg/kg FG; salicylic acid glucoside, 16 mg/kg FG; vanillic acid glucoside, 6 mg/kg FG; and protocatechuic acid glucoside, 2 to 3 mg/kg FG (Staesche et al., 1994).

In the seed oil of *P. acuminata,* different classes of lipids were detected: hydrocarbons, 1.7%; sterol esters, traces; triglycerides, 74.1%; free fatty acids, 6.6%; 1,3-diglycerides, 1.6%; 1,2-diglycerides, 1.6%; sterols, 1.8%; monoglycerides, 2.0%; phosphatidyl ethanolamine, 0.8%; phosphatidyl choline, 1.2%; lysophosphatidyl ethanolamine, 0.4%; and phosphatidyl inositol, 1.2%. The fatty acids determined in seed oil were C_{10} to C_{22}, whereas $C_{16:0}$, $C_{18:0}$, $C_{18:1}$, and $C_{18:2}$ were predominantly present in all polar and nonpolar lipid classes (Raie et al., 1995).

5.10 MONO- AND SESQUITERPENES

Analysis of aniseed fruits of *P. anisum* L. extracted with diethylether revealed 22 monoterpenoids and 32 sesquiterpenoids (Kollmannsberger et al., 2000):

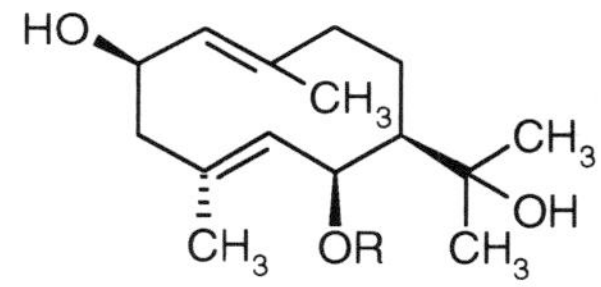

FIGURE 5.14 Germacradiene glycoside (R = β-glucopyranosyl)

- Monoterpenes: For example, β-citronellene, α,β-pinene, camphene, sabinene, myrcene, p-cymol, limonene, γ-terpinene, 1,8-cineol, linalool, fenchone, campher, menthone, borneol, terpinen-4-ol, α-terpineol, carvon, and geijerene
- Sesquiterpenes: For example, δ-elemene, α-longipinene, α-ylangene, α-copaene, β-elemene, α-bergamotene, β-farnesene, α-himachal-ene, ar-curcumene, α-zingiberene, α-/β-himachalene, β-bisabolene, β-sesquiphellandrene, δ-cadinene, α-calacorene, α-bisabolol.

The hydrocarbon fraction of the essential oil from anise fruits consisted predominantly of γ-himachlene. α-Himachlene, β-himachlene, and β-bisabolene were found in equal amounts. Minor constituents were δ-elemene, copaene, and ar-curcumene (Table 5.5) (Burkhardt et al., 1986).

β-Bisabolene and pregeijerene are also the major components of the essential oil isolated from the roots of *P. anisum*. Pregeijerene (1,5-dimethylcyclodeca-1,5,7-triene) has been identified in the root of 11 *Pimpinella* species, including *P. anisum* (Kubeczka and Ullmann, 1980). In the roots of *P. saxifraga*, a germacradiene glycoside (Figure 5.14) was found (Kisiel et al., 1998).

5.11 PHENOLS

In addition to flavonoids, other phenolic components can be found in species of *Pimpinella*, mainly esters and glycosides of hydroxybenzoic acid, hydroxy cinnamic acid, caffeic acid, and chlorogenic acid (Herrmann and Kunzemann, 1977; Dirks and Herrmann, 1984). In addition to the phenylpropanoids, a number of other phenolic and aromatic compounds were isolated from the fruits of anise, including guajacol, p-methoxyphenol, anisol, vinylanisol, methylchavicol, *cis-trans*-anethol, benzaldehyde, anisaldehyde, vanillin, phenylethanol, and anisalcohol, using diethylether extraction (Kollmannsberger et al., 2000).

5.12 POLYENES/POLYACETYLENES

Polyenes and polyacetylenes are present in the underground parts of many *Pimpinella* species. From the underground parts of *P. major* (syn.: *P. magna*), several polyacetylenes and one polyene could be isolated (Table 5.8) and the structures of five of these substances could be determined: 2,4,6,8-pentadecatetraene I, 2,8,10-tridecatriene-4,6-diyne II, 2,8-decadiene-4,6-diyn-1-al III, 2-tridecaene-4,6-diyn-8-ol-10-one IV, and 2,8-tridecadiene-4,6-diyn-10-ol V. Some of these substances could also be identified in the underground parts of *P. saxifraga* IV and *P. anisum* II (Schulte et al., 1970). In the genus *Pimpinella*, it is thought that the polyenes largely replace the furanocoumarins that are otherwise usually found in Apiaceae.

5.13 SAPONINS

Small amounts of triterpene saponin can be found in the roots and aerial parts of species of *Pimpinella* (Staesche et al., 1994). From the aerial parts of *P. magna*, eight triterpene saponines were isolated: β-sitosterol, ursolic acid, pomolic acid, tormentic acid, 19α-hydroxyasiatic acid, tormentic acid 28-O-glucosyl ester, and, 19α-hydroxyasiatic acid 28-O-glucosyl ester (Kojima et al., 1999).

TABLE 5.8
Occurrence of Various Polyenes/Polyacetylenes in Different Species of the Genus *Pimpinella*

	Pimpinella		
Compound	***major*** **(μg/100 g dry material)**	***saxifrage*** **(μg/100 g dry material)**	***anisum*** **(μg/100 g dry material)**
Pentadecatetraene-(2,4,6,8) (I)	40.0	—	—
Tridecatriene-(2,8,10)-diyne-(4,6) (II)	11.0	—	690.0
Decadiene-(2,8)-diyne-(4,6)-al-(1) (III)	55.0	—	—
Tridecaene-(2)-diyne-(4,6)-ol-(8)-on-(10) (IV)	300.0	133.0	—
Tridecadiene-(2,8)-diyne-(4,6)-ol-(10) (V)	110.0	—	—

Source: Schulte, K.E., Rücker, G., and Backe, W., Polyacetylene aus *Pimpinella*-Arten, *Arch. Pharm.*, 303, 912–919, 1970.

5.14 STEROIDS

A sterol content of 0.2 to 0.7% was detected in glyceride oil of *P. anisum*. Sitosterol (33 to 58%) and stigmasterol (27 to 48%) have been identified as major components. Minor components were cholesterol, brassicasterol, campesterol, 7-campesterol, 5-avenasterol, 7-stigmasterol, and 7-avenasterol (Zlatanov and Ivanov, 1995).

References

Andarwulan, N. and Shetty K., Phenolic content in differentiated tissue cultures of untransformed and Agrobacterium-transformed roots of anise *Pimpinella anisum* L., *J. Agric. Food Chem.*, 47, 1776–1780, 1999.

Andarwulan, N. and Shetty, K., Stimulation of novel phenolic metabolite, epoxy-pseudoisoeugenol-2-methylbutyrate EPB, in transformed anise *Pimpinella anisum* L. root cultures by fish protein hydrolysates, *Food Biotechnol.*, 14, 1–20, 2000.

Aoki, T., Takagi, K., Hirata, T., and Suga T., Two naturally occurring diterpene and norditerpene aldehydes from *Tetragonia tetragonoides*, *Phytochemistry*, 21, 1361–1363, 1982.

Becker, H., Vergleichende Untersuchung über die Zusammensetzung der ätherischen Öle verschiedener Handefrüchte von *Pimpinella anisum*, *Dtsch. Apoth. Ztg.*, 3, 41–43, 1971.

Blumer, M. and Thomas, D.W., Phytadienes in zooplankton, *Science*, 147, 1148–1149, 1965.

Bohn, I., Kubeczka, K., and Schultze, W., The essential root oil of *Pimpinella major*, *Planta Med.*, 55, 489–490, 1989.

Bottini, A.T., Dev, V., Garfragnoli, D.J., Mathela, CS., Melkani, A.B., Miller, A.A., and Sturm, N.S., Oxiranylphenyl esters from *Pimpinella diversifolia*, *Phytochemistry*, 25, 207–211, 1986.

Burkhardt, G., Reichling, J., Martin, R., and Becker, H., Terpene hydrocarbons in *Pimpinella anisum* L., *Pharmaceutisch Weekblad Sci. Ed.*, 8, 190–193, 1986.

Ceska, O., Chaudhary, S.K., Warrington, P.J., and Ashwood-Smith, M.J., Photoactive furocoumarins in fruits of some umbellifers, *Phytochemistry*, 26, 165–169, 1987.

Dev, V., Mathela, C., Melkani, A., Pope, N., Sturm, N., and Bottini, A., Diesters of 2-E-3-methyloxiranylhydroquinone from *Pimpinella diversifolia*, *Phytochemistry*, 28, 1531–1532, 1989.

Dirks, U. and Herrmann, K., 4-(S-glucopyranosyloxy)benzoic acid, a characteristic phenolic constituent of the Apiaceae, *Phytochemistry*, 23, 1811–1812, 1984.

Ernst, D., *Pimpinella anisum* L. Anise: Cell culture, somatic embryogenesis, and the production of anise oil, in *Biotechnology in Agriculture and Forestry*, Vol. 1, *Medicinal and Aromatic Plants II*, Bajaj, Y.P.S., Ed., Springer, Berlin, 1989, p. 381–397.

Gangrade, S., Shrivastava, R., Sharma, O., Jain, N., and Trivedi, K., In-vitro antifungal effect of the essential oils, *Indian Perfumer*, 35, 46–49, 1991.

Gangrade, S., Shrivastava, R., Sharma, O., Moghe, M., and Trivedi, K., Evaluation of some essential oils for antibacterial properties, *Indian Perfumer*, 34, 204–208, 1990.

Herrmann, K. and Kunzemann, J., A review with 12 references of flavonoids and phenolic acids of herbs and spices in relation to antioxidant and preservative effects in food, *Dtsch. Apoth. Ztg.*, 117, 918–920, 1977.

Karaali, A. and Basoglu, N., Essential oils of Turkish anise seeds and their use in the aromatization of raki, *Z. Lebens. Unters. Forsch.*, 200, 440–442, 1995.

Kemmerer, B. and Reichling, J., S-Adenosyl-L-methionine: Anol-O-methyltransferase activity in organ cultures of *Pimpinella anisum*, *Phytochemistry*, 42, 397–403, 1996.

Kisiel, W., Janeczko, Z., and Zgud-Walaszek, M., A germacradiene glycoside from roots of *Pimpinella saxifraga*, *Phytochemistry*, 49, 2031–2033, 1998.

Kleiman, R. and Spencer, G.F., Search for new industrial oils: XVI. Umbelliferae—seed oils rich in petroselinic acid, *J. Am. Oil Chem. Soc.*, 59, 29–39, 1982.

Kojima, K., Arai, T., and Ogihara, Y., Chemical constituents of *Pimpinella magna*, *Nat. Med.*, 53, 271, 1999.

Kollmannsberger, H., Fricke, G., Paulus, H., and Nitz, S., Über die Aromastoffzusammensetzung von Extrakten aus Umbelliferenfrüchten. 1. Anise *Pimpinella anisum*, *Adv. Food Sci.*, 22, 47–61, 2000.

Kubeczka, K.H., von Massow, F., Formacek, V., and Smith, M.A., A new type of phenylpropane from the essential fruit oil of *Pimpinella anisum* L., *Z. Naturforsch.*, 31b, 283–284, 1976.

Kubeczka, K.H. and Ullmann, J., Occurrence of 1,5-dimethylcyclodeca-1,5,7-triene in *Pimpinella* species and chemosystematic implications, *Biochem. Syst.*, 8, 39–41, 1980.

Kubecza, K.H. and Bohn, I., Pimpinella roots and its adulteration. Detection of adulteration by thin-layer and gas chromatography. Structure revision of the principal component of the essential oil, *Dtsch. Apoth. Ztg.*, 125, 399–405, 1985.

Kubecza, K.H., Bohn, I., and Formacek, V., New constituents from essential oils of *Pimpinella* species, in *Progress in Essential Oil Research*, Brunke, E.J., Ed., Walter de Gruyter, Berlin, 1986, pp. 279–298.

Kubeczka, K., Bohn, I., and Schultze, W., The composition of the essential root oils from *Pimpinella saxifraga* sensu lato and chemotaxonomic implications, *Z. Naturforsch. Sec. C, Biosci.*, 44, 177–182, 1989.

Kunzemann, J. and Herrmann, K., Isolation and identification of flavonol-O-glycosides in caraway *Carum carvi* L., fennel *Foeniculum vulgare* MILL., anise *Pimpinella anisum* L., and coriander *Coriandrum sativum* L., and of flavon-C-glycosides in anise, *Z. Lebensm. Unters. Forsch.*, 164, 194–200, 1977.

Lee, J.G., Known, Y.J., Jang, H.J., Kwag, J.J., Kim, O.C., and Choi, Y.H., A comparision of different extraction methods for the volatile components of Anis *Pimpinella anisum* L., *Agric. Chem. Biotechnol.*, 40, 144–147, 1997.

Li, F.X., Wang, C.D., Mi, C.F., Shi, H.L., and Qiao, B.L., Chemical constituents on the root of *Pimpinella thellungiana*, *Acta Bot. Sin.*, 40, 88–90, 1998.

Luthria, D.L. and Banerji, A., Biosynthesis of furanochromones in *Pimpinella monoica*, *Proc. Indian Acad. Sci. Chem. Sci.*, 106, 1149–1156, 1994.

Macias, M., Martin, V., Grande, M., and Kubeczka, K., Phenylpropanoids from *Pimpinella villosa*. *Phytochemistry*, 37, 539–542, 1994.

Maheshwari, S., Gangrade, S., and Trivedi, K., Effect of date and method of sowing on grain and oil yield and oil quality of anise, *Indian Perfumer*, 33, 169–173, 1989.

Martin, R., Reichling, J., and Becker, H., Reinvestigation of the phenylpropanoids from the roots of *Pimpinella* species, *Planta Med.*, 3, 198–202, 1985.

Martin, R. and Reichling, J., NIH-shift during biosynthesis of epoxy-pseudoisoeugenol-2-methylbutyrate in tissue cultures of *Pimpinella anisum*, *Phytochemistry*, 31, 511–514, 1992.

Merkel, B. and Reichling, J., Comparative investigation on formation and accumulation of rare phenylpropanoids in plants and *in vitro* cultures of *Pimpinella major*, *Z. Naturforsch.*, 45c, 602–606, 1990.

Nikolova-Damyanova, B., Momchilova, S., and Christie, W., Determination of petroselinic, *cis*-vaccenic and oleic acids in some seed oils of Umbelliferae by silver ion thin chromatography of their phenacyl esters, *Phytochem. Anal.*, 7, 136–139, 1996.

Placek, L.L., A review on petroselinic acid and its derivatives, *J. Am. Oil Chem. Soc.*, 40, 319–329, 1963.

Pradhan, P., Luthria, D., and Banerji, A., Pimolin, a new class of natural product from *Pimpinella monoica*: a novel dimeric furochromone, *Bioorganic Med. Chem. Lett.*, 4, 2425–2428, 1994.

Pradhan, P. and Banerji, A., Novel cyclobutane fused furochromone oligomers from the seeds of *Pimpinella monoica* Dalz, *Tetrahedron*, 54, 14541–14548, 1998.

Qiao, B.L., Wang, C.D., Mi, C.F., Li, F.X., Shi, H.L., and Takashima, J., Separation and identification of thellungianin A and B from the root of *Pimpinella thellungiana* Wolff, *Yaoxue Xuebao*, 32, 56–58, 1997.

Raie, M.Y., Ahmad, I., Javed, M.A., and Waheed, A., Lipid studies of *Pimpinella acuminata*, *Fett-Wissenschaft-Technol.*, 97, 455–457, 1995.

Reichling, J., Becker, H., Martin, R., and Burkhardt, G., Comparative studies on the production and accumulation of essential oil in the whole plant and in the cell-culture of *Pimpinella anisum* L., *Z. Naturforsch.*, 40c, 465–468, 1985.

Reichling, J., Martin, R., Burkhardt, G., and Becker, H., Comparative study on the production and accumulation of essential oil in the whole plant and in tissue cultures of *Pimpinella anisum*, in *Progress in Essential Oil Research*, Brunke, E.J., Ed., Walter de Gruyter, Berlin, 1986, p. 421–428.

Reichling, J., Kemmerer, B., and Sauer-Gürth, H., Biosynthesis of pseudoisoeugenols in tissue cultures of *Pimpinella anisum*, Pharmacy World Sci., 17, 113–119, 1995a.

Reichling, J., Martin, R., and Kemmerer, B., Biosynthesis of pseudoisoeugenol-derivatives in liquid tissue cultures of *Pimpinella anisum*, *Plant Cell Tissue Organ Cult.*, 43, 131–136, 1995b.

Santos, P.M., Figueiredo, A.C., Oliveira, M.M., Barroso, J.G., Pedro, L.G., Deans, S.G., Younus, A.K.M., and Scheffer, J.C., Essential oils from hairy root cultures and from fruits and roots of *Pimpinella anisum*, *Phytochemistry*, 48, 455–460, 1998.

Sarac, A. and Tunc, I., Toxicity of essential oil vapours to stored-product insects, *Z. Pflanzenkrankheiten und Pflanzenschutz*, 102, 69–74, 1995.

Schimmer, O., Beck, R., and Dietz, U., Phototoxzitität und Photomutagenität von Furocumarinen und Furocumarindrogen bei *Chlamydomonas reinhardii, Planta Med.*, 40, 68–76, 1980.

Schimmer, O., Bestimmung der phototoxischen und photomutagenen Potenz furocumarinhaltiger Drogenzubereitungen und Handelspräparate mit dem Testsystem Chlamydomonas, *Planta Med.*, 47, 79–82, 1983.

Schulte, K.E., Rücker, G., and Backe, W., Polyacetylene aus *Pimpinella*-Arten, *Arch. Pharm.*, 303, 912–919, 1970.

Shi, H., Li, S., Mi, C., Qiao, B., and Wang, C., Studies on chemical constituents of the root of *Pimpienlla thellungiana* WOLFF, *Zhongguo Zhongyao Zazhi*, 23, 421–422, 448, 1998.

Shukla, H., Upadhyay, P., and Tripathi, S., Insect repellent property of essential oils of *Foeniculum vulgare*, *Pimpinella anisum* and anethole, *Pesticides Bombay*, 23, 33–35, 1989a.

Shukla, H., Dubey, P., and Chaturvedi, R., Antiviral properties of essential oils of *Foeniculum vulgare* and *Pimpinella anisum* L., Agronomie Paris, 9, 277–280, 1989b.

Staesche, K., Schleinitz, H., and Burger, A., *Pimpinella*, in *Hagers Handbuch der Pharmazeutischen Praxis*, 5. Aufl., Bd. 6 Drogen, Hänsel, R., Keller, K., Rimpler, H., and Schneider, G., Eds., Springer, Berlin, 1994, p. 135–156.

Stahl, E. and Herting, D., Die Verteilung von Inhaltsstoffen in drei *Pimpinella*-Arten, *Phytochemistry*, 15, 997–1001, 1976.

Tunl, J. and Sahinkaya, S., Sensitivity of two greenhouse pests to vapours of essential oil, *Entomol. Exp. Appl.*, 86, 183–187, 1998.

Zlatanov, M., Studies upon phospholipid composition of some glyceride oils from family Apiaceae, *Fett Wissenschaft Technologie*, 96, 456–457, 1994.

Zlatanov, M. and Ivanov, S.A., Studies on sterol composition of some glyceride oils from family Apiaceae, *Fett-Wissenschaft-Technologie*, 97, 381–383, 1995.

Zobel, A., Brown, S., and Glowniak, K., Localization of furanocoumarins in leaves, fruits, and seeds of plants causing contact photodermatitis, *Planta Med.*, 56, 571, 1990.

Zogg, G., Nyiredy, S., and Sticher, O., Preparative medium pressure liquid chromatographic MPLC and semipreparative HPLC separation of furocoumarin isomers, *Chromatographia*, 27, 591–595, 1989.

6 Cultivation and Plant Raw Material of the Genus *Illicium*

Chun-shu Yang and Sheng-lin Li

CONTENTS

6.1 INTRODUCTION

The genus *Illicium* contains 34 species of evergreen trees or shrubs that are native to the Northern Hemisphere, most of them distributed in the east and southeast parts of Asia, with a minority existing in the southeast parts of North America and Middle South America; there are about 24

0-415-32246-4/04/$0.00+$1.50

species in southwestern and eastern China. The essential oil (anise oil) used for flavoring is obtained from the bark of *I. parviflorum* Michx.ex Vent. (yellow star anise) and the fruit, branch, and leaf of *I. verum* Hook. f. (Chinese star anise). Chinese star anise was first recorded in *Bencaopinhuijingyao* (1509), a book on Chinese herbal medicine, and was widely cultivated in the Guangxi and Yunnan provinces of China. The ripe fruit of this plant was named *Bajiaohuixiang* or *Daliao*; it is used as a carminative. Difengpi, the bark of *I. difengpi* B.N. Chang, is used as an antirheumatic. This latter species is distributed in the southwest part of Guangxi. These two drugs and anise oil were included in the Pharmacopoeia of the People's Republic of China (2000). The fruits of other species were also used locally as folk medicine. Because the morphology of the fruit of the genus was very similar, some fruits of toxic species appeared on the market as adulterants of that of Chinese star anise, which led to poisonings. All the species of *Illicium* contain shikimic acid, which has been shown to inhibit platelet aggregation and blood coagulation and to alleviate pain.

6.2 MORPHOLOGIC CHARACTER OF CHINESE STAR ANISE

The tree is 10 to 17 m high and 30 to 80 cm in diameter at chest height. The crown is tower-shaped, conical, or elliptic, and the bark is dark grey. The branches are dense. Leaves are alternate, or 3 to 6; subverticillate in the end of branch; coriaceous, thickly coriaceous, obovate-elliptic, oblanceolate, or elliptic; 5 to 15 cm long; 2 to 5 cm wide; acute or breviacuminate at apex; attenuate or cuneate at base; usually obscurely translucent-glandular beneath; the costa is slightly impressed above and prominent beneath. There are 4~6 lateral nerves per side, and the petioles are 20 mm long. Flowers are pink to deep red, axillary or subterminal, and solitary; pedicels are 15 to 40 mm long. Perianth segments are 7 to 12 mm long, usually being 9 to 11 mm long; the largest ones are broadly elliptic to obovate, 9 to 12 mm long, and 8 to 12 mm broad. Stamens are 11 to 20 mm long, usually being 13 or 14 mm long. There are one or two seriates, 1.8 to 3.5 mm long, the filaments are 0.5 to 1.6 mm long, the connective is truncated, and the thecae slightly protuberant (at 1 to 1.5 mm long). The pollen is trisyncolpate. Carpels are usually 8 mm long, though sometimes 7 or 9 mm long, rarely, at anthesis, being 2.5 to 4.5 mm long. The ovary is 1.2 to 2 mm long, the style subulate, slightly exceeding the ovary in length. The fruiting pedicel is 20 to 56 mm long, and aggregate fruits are 3.5 to 4 cm in diameter, plump, flattened-straight, usually with eight follicles, octagonal, 14 to 20 mm long, 7 to 12 mm broad, and 3 to 6 mm thick, gradually narrowed to a blunt beaked-acumen. Seeds are 7 to 10 mm long, 4 to 6 mm broad, 2.5 to 3 mm thick. Chromosome: $2n = 28$. Flowers March through May and August through October; fruits September to October and March to April the next year (Figure 6.1).

6.3 BIOLOGIC CHARACTER OF CHINESE STAR ANISE

6.3.1 Growing and Developing Period

The growing and developing period of the trees of Chinese star anise may fall into following three stages:

1. *Young forest stage:* This is the stage from sapling transplantation to presentation of the crown and flowering and fruitage. This stage takes about 13 to 15 years. The first 8 years are a growing stage, and the rest is a stage of growing and fruiting. The growing potential energy is stronger than the fruiting potential energy in the whole stage.
2. *The adult stage:* This is the stage of the flowering and fruiting peak. The development of fruit branches and the enlargement of crown are mainly completed in this stage.
3. *The senescent stage:* In this stage, the growing momentum of trees declines year by year. The fruit branches gradually wither so that the crown is destroyed. The fruit yield is down.

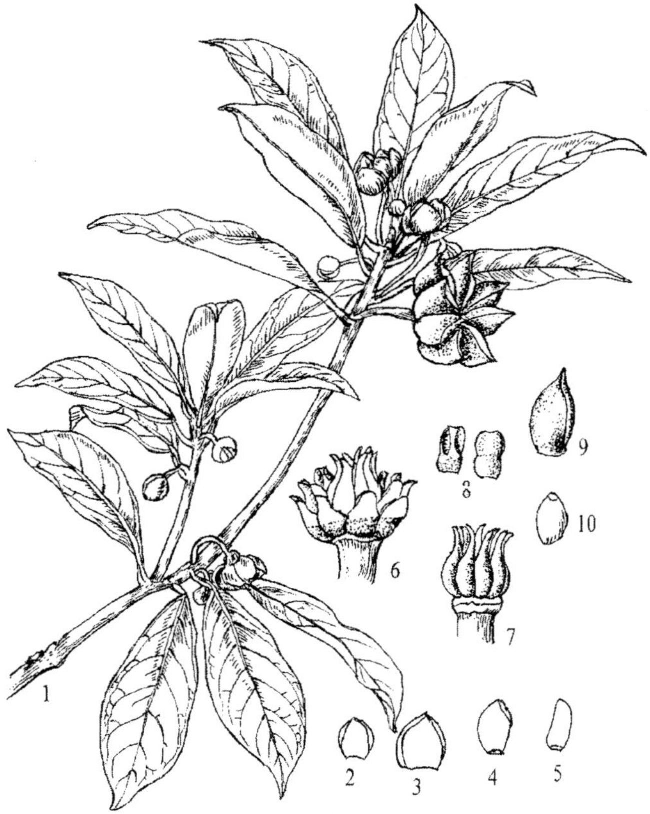

FIGURE 6.1 *I. verum* Hook. f.: (1) flowering and fruiting branch; (2–5) outer, middle, inner perianth segments; (6) androecium and gynoecium; (7) gynoecium; (8) stamen, dorsal and ventral view; (9) follicle; (10) seed

6.3.2 Shooting

The branches of the main trunk shoot out buds two to three times a year. The spring shooting leaf buds sprout in the second 10 days of February. The most luxuriant stage of leaf unfolding is the first 10 days of March. The shootings in summer sprout in the last 10 days of April. The shootings in autumn put out buds in the first days of August. The lateral branches (long fruiting branches) generally only shoot once in the first 10 days of March.

The terminal buds in the trunk and lateral branch are very developed. The new branches must have a period of 1 year to complete their development, so the flowers grow in the leaf axil of the branch from last year. It is possible for each leaf axil to sprout a flower bud or a leaf bud. Among these buds, the flower buds about make up 90% but the leaf buds, which comprise less than 10%, may develop into branches. The flower buds on branches do not totally blossom. Some of them successively blossom after the following year as a result of insufficient water, nutrients, and sunlight. The anthesis may be different between branches in the same tree, thus creating the phenomenon of successive blossoms and the appearance of flowers and fruit at the same time.

6.3.3 Blossoming and Fruiting

The trees blossom once and fruit two times every year. The fruits are developed from the flowers that bloom from July to November of the previous year. The fruits that mature in March to April are thin and small. The seeds that develop are unhealthful. They are not usually accustomed to

TABLE 6.1
Three Climate Zones

Climate Zone and Habitat	East Longitude	North Latitude	Average Temperature (°C)	Rainfall (mm)	Evaporation (mm)	Relative Humidity (%)	Extreme Minimum Temperature (°C)
			North Tropics				
Fangcheng	107°58	21°32	22.4	2904.2	1438.9	82	3.3
Longzhou	106°45	22°22	21.6	1397.4	1342.7	81	3.0 (1955)
Debao	106°38	23°20	19.5	1462.5	1437	79	2.6
			South Subtropics				
Lingyun	106°09	24°32	19.5	1762	1545.9	78	1.8
Shanglin	108°37	23°26	20.9	1750	1585.5	80	1.7 (1963)
Yulin	110°10	22°38	21.7	1647.8	1465.8	80	1
			Middle Subtropics				
Sanjiang	109°36	25°47	18	1527	1285.4	82	4.5
Rong'an	109°24	25°13	19	2049.1	1499.1	80	3.1
Quilin (Yanshan)	110°12	25°11	19.2	1655.6	1461	78.3	6 (1963)

reproduction. The fruits matured in the second to last 10 days of October are plump, with the seeds having high germinative rate and large quantities, making up 80 to 90% of the total yield.

6.4 ECOLOGICAL CHARACTERS

6.4.1 Temperature

The Chinese star anise tree is a species of tropical and south subtropical plant. It is mainly distributed in the mountainous area of the south, at 25°30 North latitude. The annual temperature in this region averages 19 to 23°F. The average temperature in the coldest month is not lower than 10°C. The absolute minimum temperature is above 6°C. Extreme low temperatures will affect the growth of the trees and yield of the fruit. The pedicels of spring fruiting, which lived through the winter, are injured, and some of the follicles fall off when the temperature drops to 0°C; fruits fruiting in spring are frozen, and those fruiting in autumn are injured and form incomplete fruit because some of the follicles drop when the temperature is 2°C. The autumn shootings of seedlings and young trees are injured, or the whole plant dies from cold, when the temperature is 3°C. The greater part of fruits fruiting in autumn are frozen when the temperature hits 4°C, and the branches of big trees are injured so that most plants die when it hits 6°C. It is unacceptable for the absolute minimum temperature of a new foresting area to be lower than 6°C; otherwise, cold injury is serious. If one wishes to transplant the seedlings to the north, one should consider the critical temperature values suitable for growth of the species (Table 6.1).

6.4.2 Water

Chinese star anise trees need not only a warm climate but also abundant rainfall. Table 6.1 shows that the annual rainfall and relative humidity of general producing areas are above 1300 mm and 78%, respectively. Although the rainfall of various areas is not well distributed, the trees are not injured by autumn and spring droughts because of shorter day length, increase in cloudy days, plenty of water in the soil, and great air humidity under the influence of the complex topography of mountain areas and thick vegetation.

6.4.3 Light

Although Chinese star anise is a kind of shade-tolerating tree, it wants different illumination in various growing and developing stages. The young trees prefer shade, and the adult trees prefer light. It is necessary that an awning is built in the stage of cultivating seedlings; otherwise, the leaves of the sapling change to yellow and all the plants die. In the young stage of the tree, it is beneficial for some other shrubs to be retained to act as shade trees, but cut them down after the saplings grow into trees because the differentiation of floral buds and the development of fruit branches of the tree need abundant sunlight to allow regular flowering and fruiting.

6.4.4 Soil

Chinese star anise grows best in the light acid soils with a deep layer, soil that is mellow and has good drainage. These trees will not grow well in alkaline soil. Chinese star anise is mainly distributed on acid soils formed from sandstone, sandy shale, shale, granite, and so forth. These kinds of soils are most suitable for cultivating this tree because of their deep layers, high content of humus, mellow and moist nature, and good ventilation (Table 6.2).

6.5 CULTIVATION

6.5.1 Cultivar

There are many cultivars of *I. verum* Hook. f. as a result of cultivation many years ago in China. Good cultivars and perfect management in cultivation have a direct bearing on the fruit yield. It is important to select an excellent tree and to build a nursery of scions, collecting an asexual series, and a seed-breeding garden of fine Chinese star anise trees. The cultivars were divided into four groups, including 17 cultivars according to the color of the flowers, the shape of the trees, and the habits and characteristics of the fruits, leaves, and branches. These cultivars are mainly distributed in Guangxi province of China.

Following is a list of the cultivars of red-flowered Chinese star anise:

- *Common red-flowered Chinese star anise:* A good kind of cultivar with a wide distribution and with an adaptable and high fruit yield. Flowers are red; sepals (three to four) are light green; petals (seven to nine) are imbricate; there are 12 to 22 stamens; the anther is pink and imbricate, in two to three whorls; there are eight pistils; and there are eight follicles in a radiate whorl, 2.5 to 3.4 cm in diameter. Pedicels are 2.2 to 3.2 cm long. Fresh fruits are about 200 per kg. Dried fruits yield 700 to 1000 per kg. This tree is distributed throughout the whole Guangxi.
- *Red-flowered Chinese star anise with drooping branches:* Its branchlets are slender and supple. The fruit is fat, and the pedicel is 3.0 to 4.8 cm long, with high oil content.
- *Red-calyx Chinese star anise:* The main distinction between the common red-flowered Chinese star anise and the red-calyx Chinese star anise is the latter's characteristics of a red pedicel, calyx, petal, and fruit-ridge. The yield of a single tree is low.
- *Large-fruited, red-flowered Chinese star anise:* Its fruit is more than 4 cm in diameter and over 1.1 cm in thickness. Fresh fruits count 100 to 140 per kg, and dried fruits total about 500 per kg. The oil content is high.
- *Many-angles red-flowered Chinese star anise:* The main distinction between the common red-flowered Chinese star anise and the many-angles red-flowered Chinese star anise is the latter's characteristics of about 21stamens, 9 to 13 pistils, and 9 to 13 follicles that are different in size and regularly arranged in a radiate whorl around the central axis or heaped on the axis.

TABLE 6.2
Nutritive Composition of Soil in Various Places of Origin

Collection Place	Layer (cm)	pH	Organic Matter (%)	Moisture Content	Total Amount (%)			Quick Effect (ppm)			Exchange (me/100g)		
					N	P_2O_5	K_2O	N	P	K	Ca	Mg	K
Conglomerate Red Soil													
Pingxiang No. 2	0–28	4.5	2.87	3.23	0.139	0.115	0.63	131	<0.5	90	—	0.40	
	28–105	4.6	0.94	3.18	0.071	0.075	1.27	56.2	<0.5	60	—	0.35	
Pingxiang No. 3	0–18	4.5	2.35	4.39	0.135	0.084	1.6	87.4	<0.5	90	—	0.63	
	18–181	4.7	0.82	3.67	0.093	0.093	1.59	45.2	<0.5	60	—	0.24	
Granite Red Soil													
Fangcheng Dalu	0–34	5.3	4.24	3.38	0.185	0.086	2.27	178.1	<0.5	120	—	0.77	
	34–100	5.1	2.16	3.36	0.102	0.088	1.9	114.1	<0.5	120		0.54	
Fangcheng Naliang	0–29	5.2	6.44	5	0.29	0.108	1.96	235.6	2.5	210	0.76	0.71	
	29–54	5	2.45	4.79	0.142	0.112	1.95	131.9	<0.5	105	—	0.85	
Purpule Sandstone Red Soil													
Ningming No.3	0–26	4.2	3.37	3.94	0.183	0.102	1.93	185	<0.5	120	0.09	1.08	
	26–71	4.5	1.07	3.96	0.071	0.085	1.92	65.6	<0.5	24	—	0.63	
Cangwu Tianghong-Ling No.1	0–20	4.4	3.74	3.01	0.134	0.076	2.73	102.3	0.8	52	0.34	0.07	
	20–70	4.7	1.28	2.25	0.058	0.07	2.65	46	<0.5	43	0.39	0.07	
Sandstone Red Soil													
Cangwu Gongqing-Lingchang No. 8339	0–15	4.5	10.22	4.29	0.268	0.075	2.39	183	1.2	90	0.56	0.05	
	15–53	4.2	1.34	2.17	0.047	0.055	3.35	38	<0.5	31	0.51	0.03	
	53–101	4.6	0.67	0.97	0.023	0.057	3.12	25	<0.5	26	0.45	0.03	

Cangwu Paiyang–Shan No. 2	0–14	4.3	2.86	2.94	0.08	0.068	0.48	142	5.7	37	0.04	0.08	0.10
	14–63	4.5	0.88	3.31	0.053	0.05	0.59	136	3.5	36	0.08	0.11	0.10
	63–90	4.6	0.8	3.7	0.053	0.046	0.73	50	0.5	13	trace	0.01	0.05
Debao Dongguan	0–37	4.7	8.22	4.16	0.33	0.164	0.93	251.2	5.0	105	0.48	0.07	
	37–176	5	3.93	3.89	0.172	0.168	0.92	120.3	<0.5	60	0.37	0.03	
Debao Zurong	0–45	4.8	2.57	3.2	0.122	0.131	2.56	88	<0.5	150	0.29	0.22	
	45–67	4.9	1.1	2.43	0.086	0.118	1.91	54.2	<0.5	135	0.25	0.49	
	67–168	4.6	0.37	2.14	0.07	0.14	1.91	28.5	<0.5	120	0.27	0.45	
Fangcheng Dalu	0–25	5.2	4.86	3.84	0.203	0.089	0.87	207.4	<0.5	90	—	0.83	
	25–102	5.4	1.5	3.36	0.084	0.112	0.05	82.5	<0.5	120	—	0.47	
Jinxiu No. 6	0–45	4.1	11.7	5.32	0.36	0.068	1.92	238	1.5	106	0.48	0.09	
	45–70	4.5	4.56	3.24	0.171	0.098	2.08	162	0.8	77	0.66	0.19	
	70–130	4.6	1.51	2.46	0.081	0.083	2.87	76	<0.5	33	0.48	0.08	
Lingyun Chengxiang No.1	0–15	4.9	4.68	5.28	0.205	0.118	2.58	156	<0.5	60	—	0.60	
	15–35	4.8	2.25	4.7	0.125	0.102	2.58	88	<0.5	45	—	0.48	
	35–70	4.9	1.36	4.63	0.119	0.105	2.90	74	<0.5	36	—	0.22	
Lingluo No.2	0–20	4.6	4.52	5.08	0.232	0.11	3.23	206	<0.5	75	0.29	0.33	
	20–36	4	1.08	1.81	0.114	0.111	3.18	67	<0.5	60	—	0.47	
	36–80	4.9	1.14	1.89	0.117	0.117	3.51	64	<0.5	30	—	0.34	
Ningming Nanan No.2	0–22	4.4	5.91	3.79	0.201	0.062	0.95	199.1	<0.5	120	—	1.05	
	22–81	4.5	1.16	3.21	0.063	0.042	1.58	61	<0.5	30	—	0.25	
	81–148	4.8	0.73	3.21	0.043	0.054	1.26	39.6	<0.5	30	—	0.13	
				Sandstone Yellow Soil									
Baise Yangwei No. 05	0–15	4.5	5.44	4.06	0.281	0.112	3.52	168	<0.5	105	—	0.83	
	15–36	4.5	1.75	3.17	0.154	0.142	3.80	76	<0.5	18	—	0.32	
	36–120	4.8	0.47	3.19	0.14	0.129	3.49	58	<0.5	27	—	0.35	

- *Hawk's beak-like, red-flowered Chinese star anise*: This plant has eight follicles, with the summit acuminated and incurvated as hawk's beak.
- *Small-fruited, red-flowered Chinese star anise:* Of low economic value; only eight fruits, less than 2.5 cm in diameter; the color is bright green, and the odor is strong aromatic. Fresh fruits number about 400 per kg.
- *Thick-leaved, red-flowered Chinese star anise:* Of low economic value; leaves are more than twice as thick as those of the common red-flowered Chinese star anise, and they are leathery and dark green; there is little fruit.
- *Dwarf-typed, red-flowered Chinese star anise:* This plant exhibits natural dwarfing: the tree is less than 8 m high, and it branches low with a large crown diameter, long lateral branched, dense branchlet, and thin and leathery leaves. This cultivar is a variety of little economic value.

Following is a list of the cultivars of light-red-flowered Chinese star anise:

- *Common light-red-flowered Chinese star anise:* A good cultivar variety. There are six to nine sepals, which are pale red or have a white white with a red center, imbricately arranged in two to three whorls. There are about 10 to 12 stamens, and the anther is pale red to light yellow. There are eight pistils and eight follicles 8, stellately arranged with the pedicel unequal in length. The tender leaves are dull red, and adult leaves are green to deep green.
- *Many-angles light-red-flowered Chinese star anise:* A good cultivar variety. The flower is light red, and there are 9 to 13 follicles, 3.5 to 4.5 cm in diameter, stellately arranged or heaped on the top of pedicel. Other characteristics are similar to those of the light-red-flowered Chinese star anise.
- *Light-red-flowered Chinese star anise with drooping branches:* Also a good cultivar. The flowers of this tree are light red. The characteristics of the branch are similar to those of red-flowered Chinese star anise with drooping branches; other characteristics are similar to the light-red-flowered Chinese star anise.
- *Thick-leaved, light-red-flowered Chinese star anise:* A cultivar with some economic value. The leaf is elliptic to obovate and is over twice as thick as that of the common red-flowered Chinese star anise.

Following is a list of the cultivars of white-flowered Chinese star anise:

- *Common white-flowered Chinese star anise:* This cultivar has 2 to 4 pale green sepals. The 7 to 10 petals are white and imbricately arranged in three to four whorls, and the 12 to 17 stamens are light yellow, with a few pink once. The eight follicles are octagonal.
- *White-flowered Chinese star anise with drooping branches:* A good cultivar. The flower is white, and the characteristics of the branch are similar to those of the red-flowered Chinese star anise with drooping branches; other characteristics are similar to those of the common white-flowered Chinese star anise.
- *Many-angles white-flowered Chinese star anise:* The flower is white. The characteristics of the fruit are similar to that of the many-angles red-flowered Chinese star anise, and other characteristics are similar to those of the common white-flowered Chinese star anise.

Yellow-flowered Chinese star anise. Although it is not advisable to extend cultivation because of the poor fruit yield of a single tree and the low oil content, this tree may act as a decorative plant, with its bright, yellow flower. This plant has three to four sepals, which are light green, and 10 to 11 petals. There are 14 to 16 stamens, 7 to 10 pistils, and 7 to 10 follicles (Table 6.3).

TABLE 6.3
Morphologic Character of Fruit of Different Cultivars

Cultivar	Follicles (number)	Fresh Fruits (number/kg)	Rate of Dried Fruits (%)
Common red-flowered Chinese star anise	8	160~240	21~28
Red-flowered Chinese star anise with drooping branches	8	160~240	23~26
Red-calyx Chinese star anise	8	116~130	19~25
Large-fruited, red-flowered Chinese star anise	8	360~440	17~28
Many-angles red-flowered star anise	8	170~196	25~28
Hawk's beak-like, red-flowered Chinese star anise	8~13	140~180	24~27
Small-fruited, red-flowered Chinese star anise	7~10	160~202	24~26
Thick-leaved, red-flowered Chinese star anise	7~8	190~230	25~28
Short-typed, red-flowered Chinese star anise	8~10	100~150	24~26
Common light-red-flowered Chinese star anise	8	160~240	23~26
Many-angles light-red-flowered Chinese star anise	8~13	160~184	25~27
Light-red-flowered Chinese star anise	8	120~180	20~26
Thick-leaved, light-red-flowered Chinese star anise	8	160~240	26~28
Common white-flowered Chinese star anise	8	200~240	23~27
White-flowered Chinese star anise with drooping branch	8~11	180~240	25~27
Many-angles white-flowered Chinese star anise	8	100~120	26~28
Yellow-flowered Chinese star anise	8	170~230	24~28

6.5.2 Reproduction

6.5.2.1 Seed Collecting

Adult trees with developing crowns, numerous fruits, excellent character, no plant diseases, and no insects should be selected as seed-reserving mothers. The fruits are picked with a wooden hook, cloth bag, small basket, or a special collector when they turn from green to yellowish-brown, and the pericarps have not split, in October. The collector is an obconiform bamboo basket, with ropes tied on the ends of it. The fruits are pulled down when the collector is put on the ground by the ropes (Figure 6.2).

Spread the fresh fruits out on the floor, turning several times a day to avoid mildew and rot. Collect the seed dropped by the pericarp's splitting after 3 to 4 days, or dry fresh fruits in the sun until the pericarp splits, and then move them indoors and collect the seeds. The thousand-grain weight of fine seed should be 105 to 120 g, or 8,400 to 10,000 grains per kilogram.

6.5.2.2 Preservation of Seed

First, uniformly scatter the seeds on the loess, immediately sprinkling a small quantity of water on them. Mix up the seeds and loess until the seeds have been wrapped to forming a granule by loess, then store in a cool and moist house. Second, put the seeds into a fresh-keeping bag, and store at 0 to 2°C in a warehouse.

6.5.2.3 Sowing and Cultivating Seedlings

The seed plot should be chosen in a place with sufficient water resources, fertile soil, good draining, and a shady natural environment. Turn up the soil and apply abundant base fertilizer and make a bed. The surface of the bed should be 1 to 1.2 m in width and 20 to 25 cm in height, and the distance between beds should be about 40 cm.

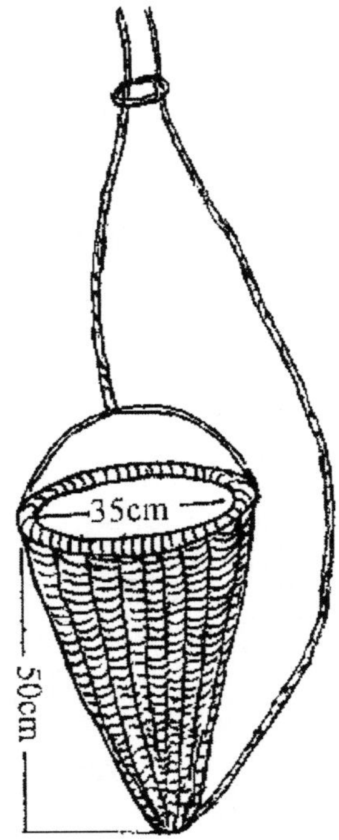

FIGURE 6.2 Collecter of fruit

Autumn sowing is suitable for areas with no winter or a short winter season, but spring sowing is best for areas with frost. The seeds almost do not germinate if the sowing season is delayed to March or April.

In general, adopt drilling to nurture and field management: row spacing should be 15 to 20 cm, and the sowing furrow should be 4 cm deep. Dibble one to two seeds per furrow, spacing them 3 to 4 cm apart, and then cover the seed with plant ash and soil, with the thickness of the cover being about 3 cm. It is necessary to cover the bed with straw or cogon grass and sprinkle sufficient water to quicken germination of the seeds and to inhibit growth of the weeds after sowing.

Field management is vital to cultivating healthy and strong seedlings. Frequently sprinkle the field with water to promote the germination of seeds, and then remove the cover on the furrows so that the seedlings can grow normally. When half the number of seeds has germinated and the young seedlings come out of the ground, immediately put up a shed to keep out the sun. The shed should be 1 to 1.5 in height. The shed should transmit about 30% of the sunlight. Then remove the shed until October to November, weed grass from the bed, thin out the seedlings, apply fertilizer, and prevent and eliminate plant diseases and insect pests.

When the seedlings are 5 to 8 cm in height, begin to top-dress the field with fertilizers: Apply mainly nitrogenous fertilizer in first stage, and phosphate and potash fertilizer in latter stage. However it is necessary that the annual seedling be 30 to 45 cm in height, the biennial seedling be 45 to 60 cm in height, and the triennial seedling be above 1.3 m in height before transplantation.

6.5.3 Selection of an Afforested Place and Soil

The light acid soil formed from the granite, sandstone, shale, breccias, and so forth is suitable for a forest of Chinese star anise. Among these options, granite is best and breccias are the worst. The thickness of the soil layer should be more than 100 cm. A mountain slope facing east-northeast is best for afforestation; one facing west is unsuitable because of the strong sunshine.

Soil preparation is performed on a gentle slope with cohesive soil or thick shrubs and weeds, and strip or hole reclamation should be applied to steep slopes with loose soil or sparse shrubs and weeds. Earth ridges should be formed for the preservation of water. A drainage ditch should be created in overall reclamation, terraced ridges in strip reclamation, and open holes in hole reclamation. Soil preparation and the excavation of the tree pit should be completed before afforestation. The tree pit should be 0.5 m in both length and width, and 0.4 m in depth.

6.5.4 Seedling Transplant

The density of the seedlings reproduced from seed differs according to the cultivated objective, the row spacing, and plant spacing. The distance between plants should be 5 ∞ 5, 4.5 ∞ 6, or 4 ∞ 5 m if it is a fruit-bearing forest, or for a leaf-bearing forest, the distances should be 1.33 ∞ 1.33 or 1.5 ∞ 1.5 m.

The density of planting is determined by cultivated objective, site condition, and management level. It is generally believed that the density should be slightly lower in the mountain area, at the foot of the mountain, and in fertile soil, and slightly higher in hilly regions, mountainside, and in poor soil. Afforesting work usually begins in February, before the seedlings have sprouts; plant annual seedlings for a fruit-bearing forest and triennial seedlings for a leaf-bearing forest. In the transplantation of large seedling, one should cut away three-quarters of each lamina and most lateral branches to reduce leaf transpiration. It is beneficial to wrap the roots of seedlings with loess, treating them with slurry of loess, and to use timely transplant on rainy days. To improve the survival rate of afforestation, one may also cut off the seedling 10 cm from the root for a fruit-bearing forest and 130 cm for a leaf-bearing forest 2 weeks before transplantation if the nursery is far from the afforested area and it is dry. The seedling may shoot out its new buds after about 20 days.

6.5.5 Management of the Field

It is good to intercrop green manure crops or other crops in yang forest; it cannot only increase the fertility of the soil but also expand the ground coverage to preserve the soil moisture and to prevent sunscald. In general, one may intercrop for 3 years, but one should turn up the soil completely every 3 to 5 years and cut the grass two times every day after 10 years, when the trees came to the fruiting peak stage. In February through March, and again in August through September, one can collect the spring fruits and autumn fruits, respectively.

For the leaf-bearing forest, one may intercrop the crops during the first 2 years after afforestation, then intertill and apply manure; otherwise, the trees will lose their ability to sprout and will die.

6.5.6 Prevention and Elimination of Plant Diseases and Insect Pests

6.5.6.1 Anthracnose

Anthracnose is a plant disease caused by infection by *Colletotrichum* sp. (family Melancoiaceae). It can harm the leaves, fruits, and stems of nursery stock. Infected organs show brown to black spots. To prevent and eliminate this pest, the nursery should be built in a place far from the pathogens; the seeds should be sterilized with bactericide before sowing; injured branches, leaves, and fruit should be removed in a timely fashion to reduce the pathogens; and locations that have exhibited signs of disease should be sprayed with a Bordeaux Mixture of 1:1:100 or a solution of 5% cabendazin in a 1:1000 ratio.

6.5.6.2 Sooty Mold

Sooty mold results from an infection by *Capnodium* sp. (family Capnodiaceae). It mainly harms the leaves of the plants, which, when injured, show a soot-like substance on both surfaces. To

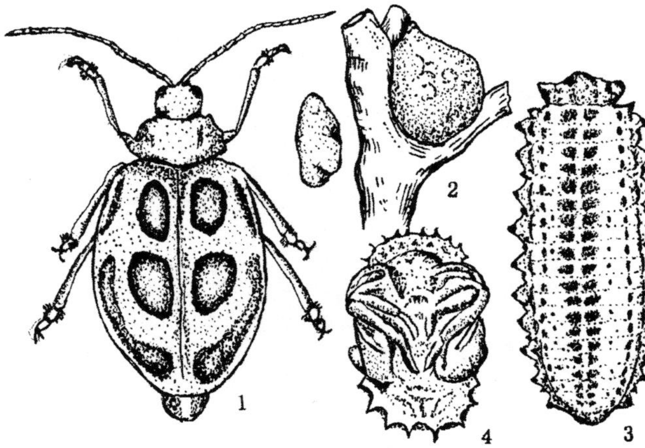

FIGURE 6.3 *Oides leucomelaena* Weise: (1) imago; (2) ova and enlargement; (3) larva; (4) pupa

prevent and eliminate this pathogen, remove seedlings infested with insects before afforestation, because the pathogenic germs of sooty mold draw nutrients from the excretion of insects. To eliminate as much as possible the activity of the insects, cut away the branches and leaves that have insects in winter or early spring, and in the hatching stage of the insect nymphs, spray the plants with an emulsion of 50% DDV in a ratio of 1:500 to 1:1000, or with a solution of 40% dimethoate or 50% naled.

6.5.6.3 *Oides leucomelaena* Weise (Family Chrysomelidae)

Oides leucomelaena is a very dangerous, harmful insect that infests Chinese star anise (Figure 6.3). All the imago and larva bite the new branches and tender leaves. In a serious attack, all the leaves and tender branches are bitten, leading to the death of the whole plant. To prevent and eliminate the pest, catch the imago, remove the ova of the insect, eliminate the pupas in pupation of the larva, or kill the imago by spraying an emulsion of 90% trichlorphon or 80% DDV or malathion in a ratio of 1:500 to 600. In addition, spraying powdered *Beauveria bassiana* (Bals.) Vuill. when the insects are in the larva stage is very effective; the larvae die as a result of infection by the fungus.

6.5.6.4 *Dilophodes elegans sinica* Prout (Family Geometridae)

Dilophodes elegans can occur in three to five generations every year, and its larvae have a liking for the leaves of Chinese star anise (Figure 6.4). In a serious infestation, all the leaves of the tree are eaten, causing the death of the plant. To prevent and eliminate the pest, catch the larvae, trap and kill the geometrids with black light lamp, and in the young stadium of the insect, spray them with an emulsion of 90% trichlorphon in a ratio of 1:800, or with 80% DDV in a ratio of 1:1500, or with malathion in a 1:1000 ratio.

6.5.6.5 *Agrotis ipsilon* (Rottemberg) (Family Noctuidae)

Agrotis ipsilon can occur in five to seven generations every year; its larvae have six instars (Figure 6.5). They gather together between stem and leaf of young plant and eat the tender leaves before the second instar. The larvae of the fourth instar hide in soil, but emerge from the ground at night and bite the young trees. To prevent and eliminate the pest, trap and kill imago with a black light lamp or a mixture of sugar, vinegar, and wine. Remove weeds, as they are a place for the ovipositioning of imago and the feeding of the larva, and spray with an emulsion of 50% trichlorphon in a ratio of 1:800.

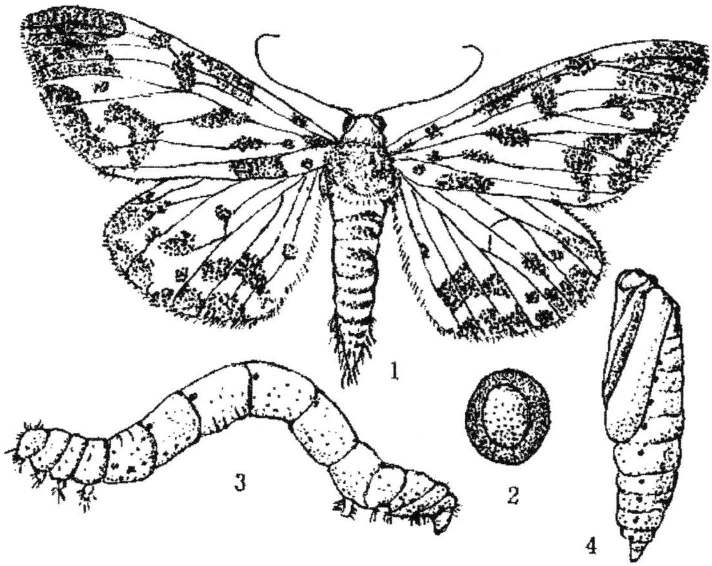

FIGURE 6.4 *Dilophodes elegans sinica* Prout: (1) imago; (2) ovum; (3) larva; (4) pupa

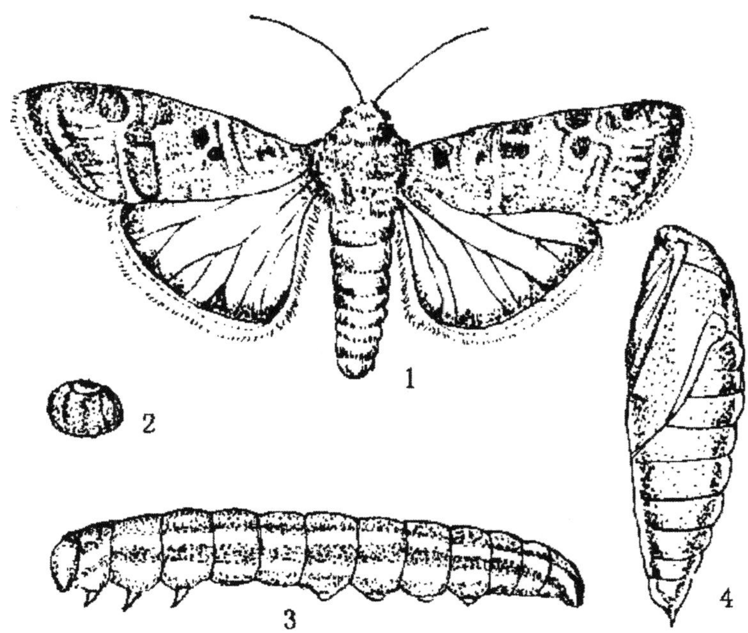

FIGURE 6.5 *Agrotis ipsilon* (Rottemberg): (1) imago; (2) ovum; (3) larva; (4) pupa

6.5.7 Gathering and Processing

The fruits are picked in spring and autumn, when they turn from green to yellow, and are dried in the sun or by gentle heat after treating them with boiling water for a moment to deactivate the enzymes. Anise oil is obtained via steam distillation from fresh branches and leaves or from the ripe fruits.

FIGURE 6.6 *I. verum* Hook. f.; place of collection: Wuzhou, Guangxi, China

6.5.8 Morphological Identification of Fruits from 16 Species of *Illicium*

Because the morphology of the fruits from various species of *Illicium* appears to be very similar, the fruit of some toxic species has appeared in the market as an adulterant of *Illicium verum* Hook. f., and has caused poisonings.

The morphologic characters of the fruits of 16 species of *Illicium* are described as follows:

1. *I. verum* Hook. f. Place of collection: Wuzhou of Guangxi, China. An aggregate fruit, mostly consisting of eight (rarely nine or 10) follicles arranged in a radiate whorl around the central axis, about 25 to 40 mm in diameter. Pedicel about 20 to 40 mm long, 1.5 to 2 mm thick, curved in the end near the fruit. Follicle: plump, about 15 to 20 mm long, 7 to 11 mm broad, 0.3 to 0.5 mm thick; outer surface: reddish-brown, irregularly wrinkled, summit beaked; upper lateral part: mostly dehiscent; inner surface: light brown, smooth, shiny, hard, and brittle. Each follicle contains a pressed ovoid seed, about 7 to 8 mm long, 5 to 6 mm broad, reddish-brown or yellow-brown, shiny, and with a hilum at the pointed end; endosperm is white and rich in oil. Odor: strongly aromatic; taste: acrid and sweet (Figure 6.6).
2. *I. anisatum* Linn. Place of collection: Nagasaki, Japan. Aggregate fruit mostly consists of eight follicles and is similar to that of *I. verum* but is readily distinguished from the latter by its fruit, which is about 20 to 30 mm in diameter; follicle: 6 to 7 mm broad; pedicel: 10 to 20 mm long; odor: weakly aromatic; taste: bland, with tongue-numbing sensation when chewed (Figure 6.7).
3. *I. henryi* Diels. Place of collection: Wuxi of Sichuan, China. Aggregate fruit mostly consists of eight (rarely nine) follicles, about 15 to 25 mm in diameter. Pedicel: straight or curved, 10 to 35 mm long, about 1 mm thick. Follicle: small, plump, about 7 to 14 mm long, 4 to 7 mm broad, 2 to 3 mm thick; summit: acute and warped; outer surface:

FIGURE 6.7 *I. anisatum* Linn.; place of collection: Nagasaki, Japan

brown. Seed: 5 to 6 mm long and about 4 mm broad. Odor: aromatic; taste: bland, with tongue-numbing sensation when chewed (Figure 6.8).

4. *I. dunnianum* Tutch. Place of collection: Longsheng of Guangxi, China. Aggregate fruit consists of eight follicles and is similar to that of *I. henryi*, but it is readily distinguished from the latter by its pedicel, which is about 1 to 2 mm long; follicle: thin and small, 1 to 4 mm broad; seed: 4 to 5 mm long, 3 mm broad (Figure 6.9).
5. *I. micranthum* Dunn. Place of collection: Chongqing, Shizhu of Sichuan, Wenshan, and Pingbian, Yunnan, China. Aggregate fruit consists of about 5 to 8 follicles, 15 to 20 mm in diameter. Pedicel: about 15 to 62 mm long. Some follicles are often smaller and thinner than others in the same fruit. Odor: slight; taste: bland (Figure 6.10 through Figure 6.13).
6. *I. merrillianum* A.C. Smith. Place of collection: Wenshan, Yunnan, China. Aggregate fruit consists of eight follicles and is similar to that of *I. verum*, but is readily distinguished from the latter by its follicle: about 20 to 30 mm long, 8 to 13 broad, with a long acuteness 6 to 10 mm long; seed: about 9 to 10 mm long, 6 to 7 mm broad; taste: slightly bitter, with tongue-numbing sensation when chewed (Figure 6.14).
7. *I. simonsii* Maxim. Place of collection: Nujiang, Yunnan, China. Aggregate fruit consists of about 8 to 12 follicles, approximately 25 to 35 mm in diameter. Pedicel: mostly straight, rarely curved in the end near fruit, about 8 to 20 mm long, 1 to 2 mm thick. Follicle: about 12 to 18 mm long, with an acuteness, 2 to 3 mm long; outer surface: brown. Seed: 6 to 8 mm long, 3 to 5 mm broad. Odor: aromatic; taste: bland, with tongue-numbing sensation when chewed (Figure 6.15).
8. *I. pachyphyllum* A.C. Smith. Place of collection: Longsheng, Guangxi, China. Aggregate fruit mostly consisting of eight follicles is similar to that of *I. simonsii*, but is easily distinguished from the latter by its fruit of 20 to 28 mm in diameter; 3 to 7 carpels are usually incompletely developed in same fruit; follicle: about 4 to 6 mm broad (Figure 6.16).

FIGURE 6.8 *I. henryi* Diels.; place of collection: Wuxi, Sichuan, China

9. *I. fargesii* Finet et Gagnep. Place of collection: Nanchuan, Sichuan, China. Aggregate fruit mostly consists of 12 follicles (rarely 10 to 13), 25 to 43 mm in diameter. Pedicel: straight or curved in the end near the fruit, about 10 to 15 mm long, 2 to 2.5 mm thick. Follicle: about 12 to 20 mm long, 6 to 6.5 mm broad, with a slender acuteness, about 5 mm long. Odor: aromatic; taste: bland, with slightly tongue-numbing sensation when chewed (Figure 6.17).
10. *I. burmanicum* Wils. Place of collection: Jingdong, Yunnan, China. Aggregate fruit consisting of 13 follicles is similar to that of *I. fargesii*, but it is easy distinguished from the latter by its fruit of about 34 to 48 mm in diameter; follicle: about 8 to 13 mm broad, with a slightly curved acuteness (Figure 6.18).
11. *I. majus* Hook. f. et Thoms. Place of collection: Quanzhou, Guangxi and Pingbian, Yunnan; China. Aggregate fruit consists of 11 to 13 follicles, 25 to 45 mm in diameter. Pedicel: usually curved in the end near the fruit, about 30 to 70 mm long, 1.5 to 2 mm thick. Follicle: plump, 12 to 21 mm long, 8 to 11 mm broad, with an acute, slightly warped summit, 2 to 3 mm long, outer surface: brown. Seed: about 8 to 9 mm long, 6 to 7 mm broad. Odor: aromatic; taste: bland, slightly bitter, with tongue-numbing sensation when chewed (Figure 6.19 and Figure 6.20).
12. *I. lanceolatum* A. C. Smith. Place of collection: Longquan, Zhejiang, and Lushan, Jiangxi, China. Aggregate fruit consists of about 11 to 13 follicles, similar to that of *I. majus*, but it is easy distinguished from the latter by its fruit of 25 to 30 mm in diameter; follicle: 6 to 8 mm broad, seed: about 5 to 6 mm long, 4 to 5 mm broad (Figure 6.21 and Figure 6.22).
13. *I. spathulatum* Wu. Place of collection: Jinxiu, Guangxi, China. Aggregate fruit consists of about 12 to 14, and mostly 13, follicles, 26 to 40 mm in diameter. Pedicel: thick,

FIGURE 6.9 *I. dunnianum* Tutch.; place of collection: Longsheng, Guangxi, China

FIGURE 6.10 *I. micranthum* Dunn.; place of collection: Chongqing, Sichuan, China

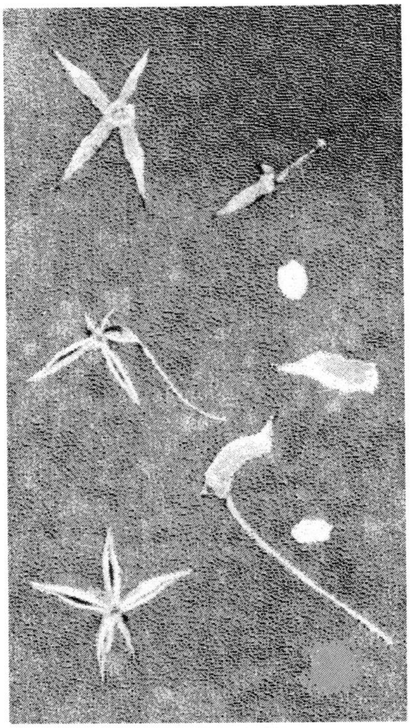

FIGURE 6.11 *I. micranthum* Dunn.; place of collection: Shizhu, Sichuan, China

FIGURE 6.12 *I. micranthum* Dunn.; place of collection: Wenshan, Yunnan, China

FIGURE 6.13 *I. micranthum* Dunn.; place of collection: Pingbian, Yunnan, China

FIGURE 6.14 *I. merrillianum* A.C. Smith; place of collection: Wenshan, Yunnan, China

FIGURE 6.15 *I. simonsii* Maxim.; place of collection: Nujiang, Yunnan, China

FIGURE 6.16 *I. pachyphyllum* A.C. Smith; place of collection: Longsheng, Guangxi, China

FIGURE 6.17 *I. fargesii* Finet et Gagnep.; place of collection: Nanchuan, Sichuan, China

FIGURE 6.18 *I. burmanicum* Wils.; place of collection: Jingdong, Yunnan, China

FIGURE 6.19 *I. majus* Hook. f. et Thoms.; place of collection: Quanzhou, Guangxi, China

FIGURE 6.20 *I. majus* Hook. f. et Thoms.; place of collection: Pingbian, Yunnan, China

FIGURE 6.21 *I. lanceolatum* A.C. Smith; place of collection: Longquan, Zhejiang, China

FIGURE 6.22 *I. lanceolatum* A.C. Smith; place of collection: Lushan, Jiangxi, China

FIGURE 6.23 *I. spathulatum* Wu; place of collection: Jinxiu, Guangxi, China

straight, rarely curved in the end near fruit, about 23 to 40 mm long, 22.5 mm thick. Follicle: plump, 13 to 20 mm long, 6 to 13 mm broad, with an obtuse summit; outer surface: brown. Seed: about 8 to 9 mm long, 6 to 7 mm broad. Odor: aromatic; taste: bland, with tongue-numbing sensation when chewed (Figure 6.23).

14. *I. terstroemioides* A. C. Smith. Place of collection: Changjiang, Hainan, China. Aggregate fruit consists of 12 to 13 follicles and is similar to that of *I. spathulatum*, but is readily distinguished from the latter by its slender pedicel of 1 mm in diameter, curved in the end near the fruit; follicle: about 5 to 8 mm broad; seed: about 6 to 7 mm long, 4 to 5 mm broad (Figure 6.24).
15. *I. angustisepalum* A. C. Smith. Place of collection: Longsheng, Guangxi; Xiuning, Anhui; and Qingyuan, Zhejiang, China. Aggregate fruit consists of about 11 to 14 follicles. Pedicel: 10 to 20 mm long, 1 to 2 mm thick, straight or curved in the end near the fruit. Follicle: 12 to 18 mm long, 5 to 8 mm broad, with an acute summit, about 2 mm long, warped; outer surface: brown. Pericarp extends to form wing-like extensions on two sides of the dorsal surface. Seed: 6 to 9 mm long, 4 to 6 broad. Odor: aromatic; taste: bitter, with little tongue-numbing sensation when chewed (Figure 6.25 through Figure 6.27).
16. *I. difengpi* K.I.B. et K.I.M. ex B.N. Chang. Place of collection: Hechi, Guangxi, China. Pericarp with wing-like extension is similar to that of *I. angustisepalum*, but is easy distinguished from the latter by its fruit consisting of about 9 to 13, but mostly nine, follicles, with an obtuse or acute summit; pedicel: 3 mm thick; taste: slightly sour (Figure 6.28).

FIGURE 6.24 *I. terstroemioides* A.C. Smith; place of collection: Changjiang, Hainan, China

FIGURE 6.25 *I. angustisepalum* A.C. Smith; place of collection: Longsheng, Guangxi, China

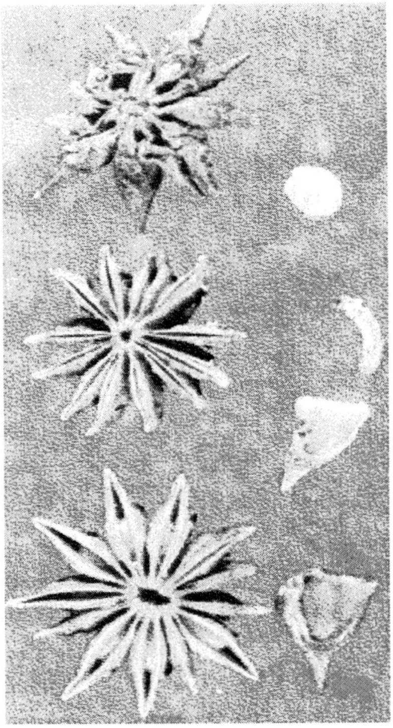

FIGURE 6.26 *I. angustisepalum* A.C. Smith; place of collection: Xiuning, Anhui, China

FIGURE 6.27 *I. angustisepalum* A.C. Smith; place of collection: Qingyuan, Zhejiang, China

FIGURE 6.28 *I. difengpi* K.I.B. et K.I.M. ex B.N. Chang; place of collection: Hechi, Guangxi, China

References

Anonymous, *The Flora of China,* Tome 30 (1), Science Press, Beijing, 1996.

Hui, L. and Chunshu, Y., Pollen morphology of Illiciaceae and its significance in systematics, *Chinese J. Bot.*, 1, 104–115, 1989.

Institute of Medicinal Plants, Chinese Academy of Medical Sciences, *The Cultural Sciences of Chinese Medicinal Plants*, Agriculture Press, Beijing, 1991.

Jianmei, H., *Studies on Morphology, Chemical Constituents and Pharmacodynamics of Fruits of Illiciaceae of China*, Doctoral thesis, School of Chinese Pharmacy, Beijing University of Chinese Medicine, 1997.

Jianmei, H., Hui, L., Chunshu, Y., et al., Morphological Identification of fruits from 16 *Illicium* Species, *Chinese Trad. Herbal Drugs*, 31, 54–58, 2000.

Liu, J.S. and Zhou, Q.R., The toxic principle of *Illicium henryi* Diels and structure of 6-deoxypseudoanisatin, *Acta Pharmaceut. Sin.*, 23, 221–223, 1988.

Qi, L., *Studies on Systematic Taxonomy and Evolution of Illiciaceae*, Doctoral thesis, School of Forestry, South China Agricultural University, Guangzhou, 1997.

Yang, C.S. and Wang, J.L., et al., Studies on the toxic constituents of *Illicium simonsii* Maxim, 26, 128–131, 1991.

Yi, M., Jianning, S., et al., Inhibitory effects of shikimic acid on platelet aggregation and blood coagulation, *Acta Pharmaceut. Sin.*, 35, 1–3, 2000.

Zhuomin, H., et al., *Bajiao*, Chinese Forestry Press, Beijing, 1984.

7 Cultivation and Plant Raw Material of the Genus *Pimpinella*

Jürgen Reichling and Ulrich Bomme

CONTENTS

0-415-32246-4/04/$0.00+$1.50

7.1 *P. ANISUM* L.

7.1.1 Cultivation

7.1.1.1 Soil and Climate Conditions

For cultivating anise, a warm, sunny, and dry climate with long and dry autumns is ideal. Therefore, as a rule, cultivation in the northern regions of the earth does not pay, as the fruits do not usually ripen, and harvests are often poor. The plants grow best in light- to medium-weight, loose, chalky humus soil that is free of weeds but rich in nutrients and not too dry. Cold, argillaceous, and moist ground is not suitable. The cultivation plain should be protected from wind (Ebert, 1982; Heeger, 1956; Noack, 1996; Poss, 1991).

7.1.1.2 Crop Rotation

To avoid losses in harvest yield, *P. anisum* should only be planted in the same field every 4 years; during this 4-year period, no other Apiaceae, such as carrots, celery, dill, or parsley, should be cultivated in this field. Using root crops as the preceding crop and grain as the following crop has proven to be effective (Heeger, 1956).

7.1.1.3 Sowing

The 1000-grain weight of the fruits lies between 1.07 and 1.53 g. The purity of the seeds for drilling should be at least 90% and the germination rate at least 70%. Because the ability to germinate decreases rapidly when using inadequate conditions of storage, the best seeds will come from the previous year's harvest (Ebert, 1982; Heeger, 1956).

Anise is an annual plant. In Central Europe, the fruits are sown in open fields between the middle of April and the beginning of May, where using 15 to 20 kg/ha of seeds and a distance of

20 to 30 cm between rows has proven to be successful. Because anise germinates in the dark, the fruits are sown 1.0 to 1.5 cm into the ground, pressed lightly with a roller, and then covered with soil. After 2 to 3 weeks, depending on the weather, the anise germinates. In subtropical cultivation areas, other times may be more suitable for the planting. In Turkey, for example, the fruits are sown at the end of January or beginning of February. Even sowing at the end of November has produced positive yields (Boshart, 1942; Fazecas et al., 1985; Heeger, 1956; Noack, 1996; Taysi et al., 1977).

7.1.1.4 Fertilization

The soil should be fertilized before sowing. The amount of fertilizer depends on the nutrient uptake by the plants and the mineral supply of the soil. In practice, 80 to 100 kg K_2O/ha and 50 to 70 kg P_2O_5/ha have proven to be favorable. Care should be taken when adding nitrogen to the ground, as it can result in a too-exuberant growth of weeds as well as poor fruit setting and storing conditions. About 20 to 30 kg N/ha given some weeks after emergence has proven to be successful (Ebert, 1982; Heeger, 1956; Noack, 1996).

7.1.1.5 Cultivation Measure

At the beginning of the vegetation period, the anise grows slowly. Therefore, it is necessary to hoe the ground several times to fight the weeds. Because anise is extraordinarily sensitive to weeds, heavy weed growth can lead to a severe loss in yield (Ebert, 1982; Heeger, 1956; Taysi et al., 1977).

7.1.1.6 Country of Origin

The actual country of origin of *P. anisum* is not exactly known. Anise is indigenous in Cyprus, Egypt, Greece, Syria, and the Aegean islands. Today, it is also cultivated in European countries with a mild climate, in Argentina, Chile, India, Japan, Mexico, and the United States.

The bulk of the drug *Anisi fructus*, which is used frequently in Europe, comes mostly from Bulgaria, the Czech Republic, Egypt, Greece, Italy, Russia, Spain, and Turkey. Small amounts of anise are also cultivated in Belgium, France, and Germany (Czygan, 1997; Heeger, 1956; Staesche et al., 1994).

7.1.1.7 Plant-Line Selection

There are no special cultivars of anise used for cultivation. In the various countries, mostly regional plants are cultivated, selected according to certain characteristics (i.e., appearance, growth, size of fruit, content of oil). Thus, the fruits of German and Russian origin, for example, are supposed to be shorter, thicker, and have a brownish yellow color, whereas those of Spain and Italian origin are thinner and are gray in color. In Germany, for example, "Thüringer Anis" is cultivated. This cultivar can be stored, and the fruits are firmly attached to the plant. So as not to be dependent on a commercial product whose quality may fluctuate greatly, it is best to be in a position to have your own seed production (Ebert, 1982; Heeger, 1956; Staesche et al., 1994).

7.1.1.8 Pests, Diseases, and Their Control

Anise is very sensitive to weeds. The use of herbicides in cultivation of anise depends on the actual concessions of the individual countries. In Germany, there are no concessions for herbicides that could be used in cultivation of anise. Therefore, it is advisable to contact the respective official advisory board for plant protection.

Fungus diseases and pests are likely to develop under disadvantageous climatic conditions and heavy weed growth. Such conditions hamper and weaken the development of anise. In well-cultivated and healthy anise stocks, diseases and pests only seldom appear (Heeger, 1956). *Plasmopara nivea* and *Puccinia pimpinella* infest the leaves of anise.

Spring tails cause serious damage at the root collar. Caterpillars of various butterflies damage the leaves, and the caterpillar of the moth *Depressaria depresella* destroys the umbels. Sometimes plant louses appear (Heeger, 1956).

7.1.2 Plant Raw Material

7.1.2.1 Collection

Anise is not gathered from the fields anymore. The raw plant material (anise fruits) that is commercially available originates exclusively from cultivation.

7.1.2.2 Harvesting Methods

The fruits are harvested from the end of July until the beginning of September, depending on the crop area and the course of the summer, when most of the fruits have a brownish color. In wet and cold European summers, the fruit ripens very late. The whole plant is harvested with a combine shortly below the ripe umbels (Ebert, 1982; Heeger, 1956; Noack, 1996; Staesche et al., 1994).

7.1.2.3 Drying Conditions/Storage Conditions

Immediately after harvesting the ripe umbels have to be dried at 45°C. Thereafter the dried umbels are threshed and the anise fruits purified mechanically in a seed cleaner (Ebert, 1982; Heeger, 1956; Noack, 1996).

To prevent a substantial loss of the essential oils, the anise fruits should always be kept in paper sacks, jute sacks, or tin cans—never in plastic boxes. In addition, the anise fruits should be protected from moisture, strong light, and pests.

Class I goods may contain a minimum of 0.2% (v/v) essential oil, a maximum of 2% foreign constituents, a maximum of 7% moisture, and 12% ashes.

7.1.2.4 Yield

The fruit yield fluctuates greatly. In good years, yields of up to 5 to 10 dt/ha and, although rarely, sometimes even 20 dt/ha can be obtained. The greatest amount of essential oil is gained from the ripe central umbels. The annual world production of anise oil amounts to as much as 500 tons (Ebert, 1982; Heeger, 1956; Noack, 1996).

7.1.2.5 Description and Illustration of Raw Material

The whole fruit is a two-part schizocarp, with a thin, stiff, and slightly bent fruit-stalk (Figure 7.1 and Figure 7.2). It is grayish green to yellow green, egg- to pearshaped, about 3 to 6 mm long, up to 3 mm wide, delicately ribbed, slightly laterally compressed, and covered with bristle hairs. On the upper end, there is a pistil cushion with two short, bent pistils. Both mericarps are grown together, with the carpophore at the top. When mature, the two mericarps separate along the commissural face but remain suspended to the carpophore. The dorsal side of each fruit part is convex. It has five primary ridges, which are not very prominent. Two of them are very close to the commissural face, and three are close on the dorsal side.

On cross section, the two mericarps appear to be oval, pentagonal, almost kidney shaped, and connected along the commissural face (Figure 7.3). A highly developed endosperm is visible, surrounded by a narrow pericarp. Numerous epidermal cells of the pericarp are grown out to short, mostly monocelled hairs, which are covered by a papillary cuticle. The mesocarp is parenchymatous. Up to 30 excretory canals are arranged in an uninterrupted row at the dorsal side of the mesocarp. Along the

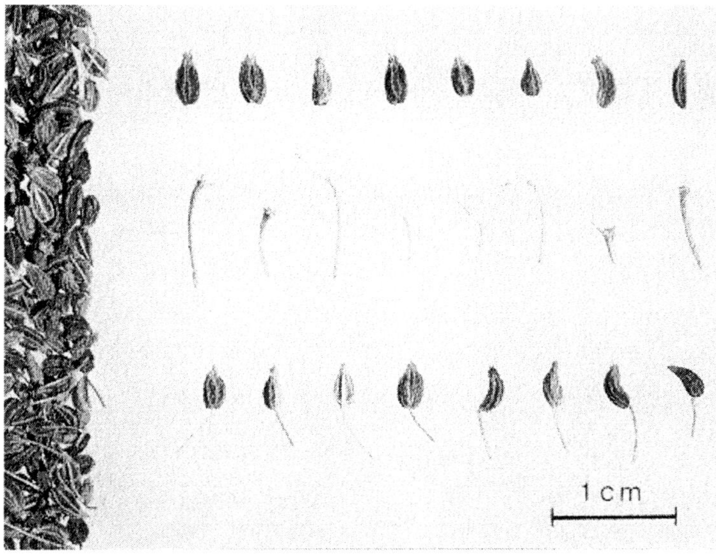

FIGURE 7.1 Anise fruits (*Anisi fructus*): In the central row, you see the rest of the stem and carpophore (Photo: M. Wichtl, Wissenschtl. Verlagsgesellschaft, Stuttgart)

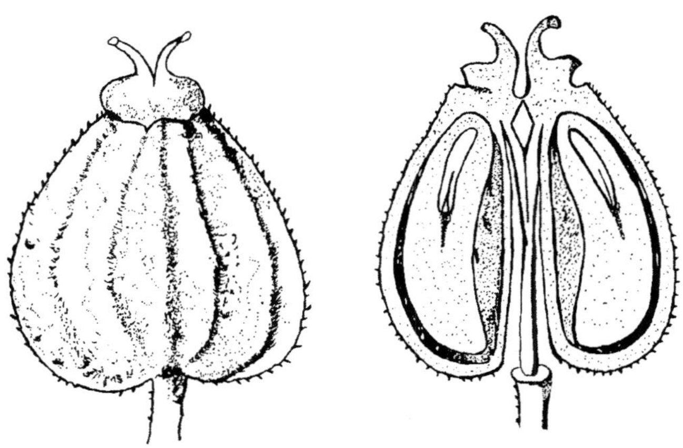

FIGURE 7.2 Left: Outer shape of a whole anise fruit; right: a longitudinal section through the same fruit

commissural face on each side of the carpophore, a wide excretory canal is visible. Narrow vascular bundles run through the ridges. The fruit has a pleasant and strongly aromatic smell and tastes sweet to spicy (Berger, 1952; Czygan, 1992, 1997; Heeger, 1956; Pahlow, 1986; Staesche et al., 1994).

Because anise originates mainly from cultivation, adulterations are rather seldom. In the past, poisonous fruits of *Conium maculatum* were occasionally found in Italian and Russian drugs. From time to time, adulterations with fruits of *Coriandrum sativum* are also reported (Berger, 1952; Czygan, 1997).

7.2 *P. MAJOR* (L.) HUDS.

7.2.1 CULTIVATION

P. major is not cultivated. The drug, Pimpinellae radix, derives exclusively from wild growing plants. In Germany, for example, it is hardly ever commercially available, and sometimes it is not available at all.

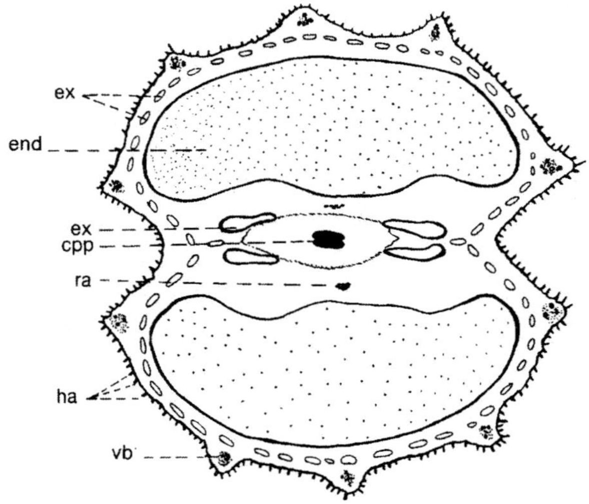

FIGURE 7.3 Cross-section of whole anise fruit: ex., excretory canal; end., endosperm; cpp., carpophore; ra., raphe; ha., hair; vb., vascular bundle

7.2.1.1 Country of Origin

P. major (great burnet saxifrage) is widely spread in Europe, except in Greece, Portugal, Scandinavia, and in the southern part of the Balkans. Great burnet saxifrage is often found in Australia and Switzerland. The plant has been naturalized in North America (e.g., in Pennsylvania). It grows best in chalky soil, mainly in meadows, on the edge of bushes and fields, on fallow land, and at the edge of thick woods and sparse woods.

The herbal drug, Pimpinellae radix, mainly originates in regional wild collections. The former Yugoslavia used to be the main exporting country of Pimpinellae radix (Berger, 1960; Staesche et al., 1994; Willuhn, 1997).

7.2.2 Plant Raw Material

7.2.2.1 Collection

The roots are dug out in spring or in September and October. The fresh roots have an unpleasant, rancid-tallowy, carrot-like odor (Berger, 1960; Staesche et al., 1994).

Because the roots of *P. major* grow wild, adulterations with roots of other Apiaceae are common. The authentic herbal drug, Pimpinellae radix, is occasionally not obtainable at all. Roots of *P. saxifraga*, *Pastinaca sativa*, and *Heracleum sphondylium* are most often mistaken for Pimpinellae radix. Adulterations with roots of *Peucedanum oreoselinum*, *Daucus carota*, and *Angelica archangelica* have been seen, although rarely (Staesche et al., 1994; Willuhn, 1997).

Adulterations can be identified by comparing the coumarin fingerprinting (Wagner et al., 1983; Zogg et al., 1989) or the composition of the essential oils (Kubeczka and Bohn, 1985) chromatographically (TLC, GC, HPLC).

7.2.2.2 Drying Conditions

When the roots are taken out of the earth, the soil is shaken off and the roots are cleaned with a brush. The roots are washed but not chopped, so as to avoid loss of essential oils. Then they are spread out on a cloth and air dried. For the actual drying, the temperature should be about 40°C (Staesche et al., 1994).

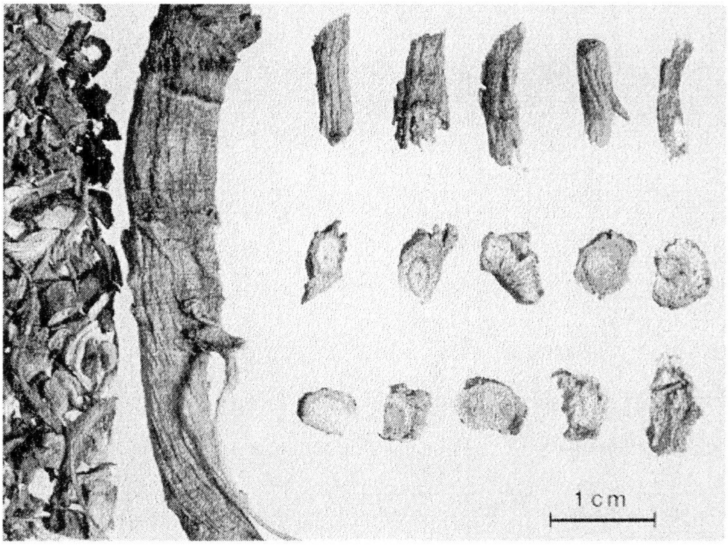

FIGURE 7.4 Burnet-saxifrage root (Pimpinellae radix). Upper row, left side: Pieces of the whole root; below there are two rows of cross-sections through the root (Photo: M. Wichtl, Wissenschaftl. Verlagsgesellschaft, Stuttgart)

7.2.2.3 Description and Illustration of Raw Material

The herbal drug, Pimpinellae radix, is composed of the dried roots and rootstocks (Figure 7.4). The rootstock is short and 0.5 to 3 cm thick. Downward, the short rootstock turns into a 50-cm-long to 1.5-cm-thick spindle-shaped tap-root. The surface is yellow-gray to dark yellow, never black. The roots have coarse vertical ridges and are covered with horizontally running papillae and lateral root stigmas.

The freshly cut drug smells sweet and spicy. It is initially aromatic and then becomes very spicy and produces a burning sensation.

In a cross section of the root, the periderm is light to dark yellow. The cortex is whitish and wide, and the outer half is torn greatly in a radial direction. It contains numerous radially arranged small yellow-brown schizogenic excretory canals that extend up to the phellem. They decrease in width from the phellem inward to the cambium. Close to the phellem, the excretory canals are usually 80 to 120 µm; near the cambium, in contrast, the canals are only 40 to 50 µm wide. The centrally located woody body is relatively small and lemon yellow in color and constitutes only about half of cross section. It does not separate from the cortex when cut. The root stock has large central pith. All the parenchymatous cells contain roundish, 2 to 4 µm, large, simple, or composed starch grains (Bohn, 1991; Staesche et al., 1994; Willuhn, 1997).

7.3 *P. PEREGRINA* L.

7.3.1 Cultivation

7.3.1.1 Soil and Climate Conditions

P. peregrina is frost resistant. Seeds should be sown in deep soil with no stones, only few weeds, and no stagnant water. Young plants are sensitive to high concentrations of mineral salt (Bomme, 1984).

7.3.1.2 Crop Rotation

To avoid losses in harvest yield, *P. peregrina* should be cultivated in the same field every 4 years. During this 4-year period, no other Apiaceae, such as carrots, celery, dill, anise, or parsley, should be cultivated in this field either. Root crops as previous crop and grain as following crop have proven to be effective (Bomme, 1984).

7.3.1.3 Sowing

P. peregrina is cultivated as an annual plant. The 1000-grain weight of the fruits is 1 g. In central Europe, sowing of the fruits (0.5 to 1.0 kg/ha) is carried out from the end of March to mid-April directly on the field at a depth of 1 cm. An autumn sowing is not recommended. It is important that the fruits are sown in rows that are between 30 and 50 cm apart. Germination takes between 2 and 4 weeks. Rain promotes steady germination of the fruits and development of the young plants. After germination of the fruits, the soil should be hoed and weeded if necessary (Bomme, 1984).

7.3.1.4 Fertilization

The amount of fertilizer depends on the nutrient uptake of the plants and soil analysis. Guiding values: 120 to 150 kg N/ha, 40 to 60 kg P_2O_5/ha, and 220 to 240 kg K_2O/ha. Because the young plants are sensitive to mineral salt, the field should be fertilized with P_2O_5 and K_2O in the fall. Nitrogen can be given at two or three different times, both weeks after emergence and during cultivation (Bomme, 1984).

7.3.1.5 Cultivation Measure

Usually, at the beginning of cultivation it is necessary to hoe and weed the soil two or three times. In June/July the rows have closed together, and no more weeds are likely to develop (Bomme, 1984).

7.3.1.6 Country of Origin

P. peregrina is found mainly in southern Europe from southern France to Greece, Turkey, Crimea, Caucasia, Crete, Asia Minor, Syria, Egypt, and Ethiopia. The plants grow best in sparse bushy areas. In Germany, *P. peregrina* is grown mainly in the southern part of the country (Staesche et al., 1994).

7.3.1.7 Plant-Line Selection

There are no particular planting cultivars. In southern Germany, plants of various origins are cultivated. It is best to be in a position to have your own seed production, as it is not always commercially available. To do this, the fruit must be harvested, dried at 35°C, stored in airtight containers, and kept dry and cool (Bomme, 1984).

7.3.1.8 Pests, Diseases, and Their Control

Genuine mildew is found in *P. peregrina* (white coating), and brown spots are found on the leaves from *Septoria. Rhizoctonia* infestation often causes root rot. The roots are also often eaten by voles (Bomme, 1984).

To reduce diseases during germination, the fruits can be disinfected; for example, with a metiram-containing solution (wet staining in 0.2% solution).

The use of herbicides in cultivation of *P. peregrina* depends on the actual concessions of the individual countries. In Germany, there are no concessions for herbicides that could be used in cultivation of *P. peregrina*. Therefore, it is advisable to contact the respective official advisory board for plant protection.

7.3.2 Plant Raw Material

7.3.2.1 Collection

P. peregrina cannot be gathered in the wild. The herbal drug P. peregrina root is obtained exclusively from cultivation (Staesche et al., 1994).

7.3.2.2 Harvesting Methods

The roots and rootstocks of the plants are harvested from the mid-October to the beginning of November. First the leaves are cut off mechanically and removed. Then the roots are machine harvested; for example, using a shaking sieve digger (Bomme, 1984).

7.3.2.3 Drying Conditions/Storage Conditions

After harvesting, the roots are washed carefully in washing machines, cut into large pieces, and dried at 45°C. To avoid any possible loss of essential oils, the roots are coarsely cut and should not be dried at higher temperatures. The drying process only goes on until the roots can be broken easily (about 20 hours). The dried roots must be protected from moisture. They have a very peculiar, unpleasant smell and a very strong taste, which is spicy at first and then bitingly hot. This is said to be the result of the pimpinellin that the herbal drug contains (Bomme, 1984).

7.3.2.4 Yield

The average yield of fresh roots in central Europe is about 60 to 100 dt/ha; of dry roots, it is approximately 15 to 25 dt/ha (Bomme, 1984).

7.3.2.5 Description and Illustration of Raw Material

The herbal drug P. peregrina root consists of dried roots with rootstocks (Figure 7.5). The rootstock is up to 3 cm thick, short, and brownish gray, and it often contains short residual stems and the bases of the rosette-like leaves. Downward, the short rootstock turns into a carrot-like main root. The main root is about 2 cm thick and up to 20 cm long and is covered by several extended, 2- to 8-mm-thick side roots in the lower area. The surface of the root is light grayish yellow to brownish yellow, has coarse vertical ridges, and is covered with horizontally running papillae and lateral root stigmas. The odor of the root is weak and strange. Initially, the root tastes sweet and slimy, and later, spicy and slightly hot.

In cross section, a small, pale-yellow-colored, woody body is visible, which only makes up about one-fourth to one-third of the cross-section. The cortex seems to be almost white and spongy and is often full of gaps. The schizogenic excretory canals are colorless, poorly developed, and only 20 to 40 μm wide. Sclerenchymatous fibbers are completely absent. There are usually single starch grains 4 to 12 μm in size. Composed starch grains, which can be as large as 15 μm, are much rarer (Staesche et al., 1994).

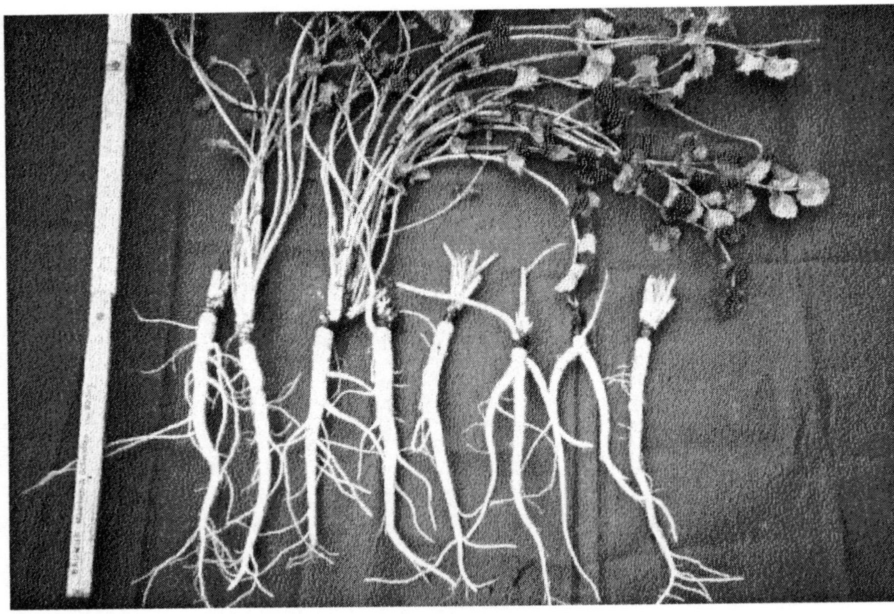

FIGURE 7.5 Roots of *Pimpinella peregrina* L. (Photo: U. Bomme, Freising)

7.4 *P. SAXIFRAGA* L.

7.4.1 Cultivation

7.4.1.1 Soil and Climate Conditions

P. saxifraga (saxifrage) prefers dry, sunny locations. Humus-rich soil is especially suitable for its cultivation (Franz, 1981; Schmid and Imhof, 1988; Schulz et al., 1983).

7.4.1.2 Crop Rotation

P. saxifraga should not be planted after umbellifers, such as dill, carrots, parsley, or celery. Cultivation after other root crops, however, is recommended. In summer, sowing cereals after saxifrage cultivation is favorable.

7.4.1.3 Sowing

P. saxifraga is an annual plant. Because seeds are not, or are hardly ever, commercially obtainable, it is best to be in a position to have your own seed production. The umbels ripen at very different rates, whereby the fruits have to be harvested more often. The fruits are harvested as soon as the umbels start to become brown.

Studies have shown that sowing *P. saxifraga* in the spring is better than doing so in the fall. The drilling is carried out from the end of March until the middle of April, and using amounts of between 3 and 5 kg/ha is favorable. The fruits are seeded in the open field with a distance of 30 to 40 cm between rows. After 3 to 4 weeks, the fruits germinate. A sprinkler system facilitates the even germination and prevents extreme dryness during the vegetation period (Franz, 1981; Schmid and Imhof, 1988; Schulz et al., 1983).

Seedlings can also be grown in a greenhouse and then planted in the open field by the end of April or the beginning of May. This cultivation method, however, has certain drawbacks: more shoots develop, the stock density is lower, and the roots branch out more, which has disadvantages for the harvest later on. In addition, the yield is lower (Franz, 1981).

7.4.1.4 Fertilization

Seedlings are very sensitive to high amounts of fertilizer and high concentrations of mineral salt. Therefore, it is better to fertilize the ground in the fall, long before the sowing. As a base, fertilization by 170 to 200 kg K_2O/ha and by about 70 kg P_2O_5/ha has proven to be effective. The nitrogen can be given in several doses during the vegetation period; 100 to 120 kg N/ha has been used successfully. No more than 50 kg N/ha should be used. Fertilizer containing magnesium is preferable (Franz, 1981; Schmid and Imhof, 1988).

7.4.1.5 Cultivation Measure

The plants prefer loose, chalky soil that is free of weeds. Cultivation is only possible in sandy ground with additional sprinkling. According to Franz (1981), *P. saxifraga* can be cultivated with carrots and parsley. Cultivation after root crops fertilized with stable manure, for example, seems to be especially suitable. Hoeing two to three times during the vegetation period is enough to keep the plants free of weeds and to loosen the ground. Between June and July, the rows close together, and no more weeds are likely to develop (Franz, 1981; Schulz and Imhof, 1983).

7.4.1.6 Country of Origin

P. saxifraga can be found in nearly the whole of Europe with the exception of Portugal, southern Spain, some Italian isles, Greece, parts of Turkey, and northern Russia. The plant is found in Asia Minor, the Caucasus, Armenia, and western Siberia, as well as, supposedly, in East Asia. Saxifrage was exported to North America and naturalized there.

The plants grow best in dry, meager meadows; pastures; hills; open pinewoods; dry riverbanks; dry, warm taluses; on banks; and on roadsides.

The herbal drug, Pimpinellae radix, is hardly even commercially obtainable. The raw material (roots and rootstocks) is mainly supplied by eastern European countries (Berger, 1960; Bohn, 1991; Schulz et al., 1983; Staesche et al., 1994).

7.4.1.7 Plant-Line Selection

There is no particular planting cultivar. Plants of regional origin are cultivated locally. Seeds for the cultivation have to be reproduced (Schulz et al., 1983; Seidel, 1986).

7.4.1.8 Pests, Diseases, and Their Control

Saxifrage is very sensitive to weeds, and high weed growth affects the yield. Mechanical weed control through hoeing has proven to be very effective. The use of herbicides in cultivation of saxifrage depends on the actual concessions of the individual countries. In Germany, there are no concessions for herbicides that could be used in cultivation of saxifrage. Therefore, it is advisable to contact the respective official advisory board for plant protection.

In culture, the plants are susceptible for mildew and other fungus infection diseases. Attacks by *Erysiphe heracli*, *Plasmospora nivea*, *Puccinia pimpinella*, *Septoria pimpinella*, *Rhizoctonia* sp., and others have been observed. Animal pests such as *Smithurus luteus* and *Phytomyza albiceps* destroy the plants as well. On seedlings, plant louses can be observed from time to time, too (Franz, 1981).

7.4.2 Plant Raw Material

7.4.2.1 Collection

The herbal drug, Pimpinellae radix, originates in part from cultivation and in another part from collecting plants that grow wild. In the latter case, rootstocks and roots are dug out in spring or fall. The danger of adulteration exists in the collection of wild plants. Roots from *H. sphondylium* or *P. sativa* frequently have been observed (Berger, 1960; Bohn, 1991).

Adulterations can be identified by comparing the coumarin fingerprinting (Wagner et al., 1983; Zogg et al., 1989) or the composition of the essential oils (Kubeczka and Bohn, 1985) chromatographically (TLC, GC, HPLC).

7.4.2.2 Harvesting Methods

The roots are harvested, as a rule, at the end of October or the beginning of November. First the leaves are removed and then the roots are rooted out mechanically; for example, using a shaking sieve digger. Then the roots are washed thoroughly. Finally, they are coarsely chopped to prevent an unnecessary loss of essential oil (Franz, 1981; Schmid and Imhof, 1988; Schulz et al., 1983).

7.4.2.3 Drying Conditions/Storage Conditions

The harvested roots are dried in well-ventilated drying plants at 40 to 45°C. The drying ratio of fresh and dried roots is 3(4):1. As a rule, there should be about 10% residual moisture. The dried roots should be stored in dry, well-ventilated rooms. They should be stored in paper sacks or jute sacks (Schmid and Imhof, 1988; Schulz et al., 1983).

7.4.2.4 Yield

The annual root yields fluctuate. Yields from 40 to 50 dt/ha can be reached under good weather conditions (summer with high precipitation). Otherwise, the average yield is more like 20 to 30 dt/ha (Schmid and Imhof, 1988; Schulz et al., 1983).

7.4.2.5 Description and Illustration of Raw Material

The herbal drug, Pimpinellae radix, consists of dried rootstocks and roots. The roots of *P. saxifraga* are similar to those of *P. major*, with the exception that the roots of *P. saxifraga* are usually thinner and darker on the outside. Their surface is brown or blackish.

On *P. saxifraga*, the cortex is as wide as the woody body. The brown, schizogenic excretory canals in the cortex are narrower than they are in *P. major*. Excretory canals located in the periphery are 50 to 80 μm wide, and near the cambium they are only 20 to 50 μm wide (Berger, 1960; Staesche et al., 1994; Walther, 1938).

References

Berger, F., *Handbuch der Drogenkunde, Bd. 3*, Wilhelm Maudrich, Vienna, 1952, p. 63–71.

Berger, F., *Handbuch der Drogenkunde, Bd. 5*, Wilhelm Maudrich, Vienna, 1960, p. 332–336.

Bohn, I.U., *Pimpinella saxifraga* und *Pimpinella major*: Die Kleine und Große Bibernelle, *Zeitschrift für Phytotherapie*, 12, 98–104, 1991.

Bomme, U., Kleine Bibernelle, Frucht für Praktiker, die neue Wege gehen wollen? *Deutsche Landwirtschafts Zeitung*, 35, 871–873, 1984.

Boshart, K., Über Anbau und Düngung aromatischer Pflanzen, *Heil- und Gewürzpflanzen*, 21, 84–91, 1942.

Czygan, F.C., Anis (Anisi fructus DAB 10): *Pimpinella anisum* L., *Zeitschrift für Phytotherapie*, 13, 101–106, 1992.

Czygan, F.C., Anisi fructus, in *Teedrogen und Phytopharmaka, 3. Aufl.*, Wichtl, M., Ed., Wissenschaftliche Verlagsgesellschaft, Stuttgart, 1997, p. 66–68.

Ebert, K., *Arznei- und Gewürzpflanzen*, Wissenschaftliche Verlagsgesellschaft, Stuttgart, 1982, p. 53–54.

Fazecas, I., Borcean, I., Tabara, V., Lazar, S., Samaila, M., and Nistoran, I. Studies on the effects of fertilizers and sowing date on the yield and essential oil content in *Pimpinella anisum* in the year 1978–1980, *Horticulturae Abstracts*, 55, 387, 1985.

Franz, C., Bibernellwurzeln—eine gärtnerische Freilandkultur, *Deutscher Gartenbau*, 47, 1993–1994, 1981.

Heeger, E.F., *Handbuch des Arznei- und Gewürzpflanzenanbaues*, Deutscher Bauernverlag, Leipzig, 1956, p. 578–583.

Kubeczka, K.H. and Bohn, I., Radix Pimpinellae und ihre aktuellen Verfälschungen, *Deutsche Apotheker Zeitung*, 125, 399–402, 1985.

Noack, P., Seminarvortrag an der Technischen Universität München-Weihenstephan, July 1996 (unpublished).

Pahlow, M., Anis—Anisi fructus, *Deutsche Apothekerzeitung*, 126, 367–368, 1986.

Poss, E., Untersuchungen zum Einfluß einer Saatgutbeizung auf das Keimverhalten ausgewählter Heil- und Gewürzpflanzenarten, Diplomarbeit, Technische Universität München-Weihenstephan, 1991.

Schmid, O., and Imhof, P., *Naturgemäßer Anbau von Heilkräutern,* Forschungsinstitut für biologischen Landbau, Eigenverlag, Bernhardsberg, 1988.

Schulz, C., Schießl, E., and Voerkelius, S., *Die Bibernelle*, Diplomarbeit, Technische Universität München-Weihenstephan, 1983.

Staesche, K., Schleinitz, H., and Burger, A., *Pimpinella*, in *Hagers Handbuch der Pharmazeutischen Praxis,* 5. Aufl., Bd. 6, Hänsel, R., Keller, K., Rimpler, H., and Schneider, G., Eds., Springer, Berlin, 1994, p. 135–156.

Taysi, V., Vömel, A., and Ceylan, A., Neue Anbauversuche mit Anis (*Pimpinella anisum* L.) im Ege-Gebiet der Türkei, *Zeitschrift für Acker- und Pflanzenbau*, 145, 6–21, 1977.

Wagner, H., Bladt, S., and Zgainski, E.M., *Drogenanalsyse. Dünnschicht-chromatographische Analyse von Arzneidrogen*, Springer, Berlin, 1983.

Walther, K., Biologie und Systematik der kleinen Bibernelle, *Die Deutsche Heilpflanze*, 4, 33–36, 1938.

Willuhn, G., Pimpinellae radix, in *Teedrogen und Phytopharmaka,* 3. Aufl., Wichtl, M., Ed., Wissenschaftliche Verlagsgesellschaft, Stuttgart, 1997, p. 440–442.

Zogg, G.C., Nyiredy, S., and Sticher, O., Apiaceenwurzeln. Qualitative und quantitative Furanocumarinbestimmung in Apiaceenwurzeln, *Deutsche Apotheker Zeitung*, 129, 717–722, 1989.

8 Therapeutic and Pharmacological Properties of *Illicium*

Emi Okuyama

CONTENTS

8.1 OVERVIEW

The genus *Illicium* is small, comprising about 50 species of which therapeutic use is not very extensive, with the exception of star anise (the ripe fruit of *I. verum*). Star anise is known as a spice as well as for its medicinal use. Star anise oil distilled from the fruits of *I. verum* has been widely used as a substitute for anise oil (*Pimpinella anisum*) or fennel oil (*Foeniculum vulgare*) because of the aromatic resemblance caused by anethole. *Illicium* species are found in South and Southeast Asia and North America, where some of them have been used as a traditional medicine. Most *Illicium* species, however, are poisonous, and special care is necessary for their use. Scientific investigations of the bioactive components including toxic properties have been carried out on *I. verum*, *I. anisatum*, and some other *Illicium* species.

0-415-32246-4/04/$0.00+$1.50

8.2 THERAPEUTIC USES OF *ILLICIUM*

8.2.1 Traditional Use of *Illicium* Plants

The medicinal use of star anise has a long history in Asian countries, especially in China. The dried ripe fruit (anisi stellati fructus; star anise or Chinese anise in English) had been already written about in a Chinese herbal classic, the *Compendium of Materia Medica,* in the sixteenth century (Zhen, 1590). The *Pharmacopoeia of the People's Republic of China*, Vol. 1 (Pharmacopoeia Commission of the Ministry of Public Health, 1995), stated the properties of Anisi Stellati Fructus (bajioahuixiang in Chinese) as warming the viscera, expelling "cold," and ensuring normal flow of *qi* to relieve pain. Doses of 3 to 6 g are used to treat abdominal and lower back pain and to treat emesis. The *Pharmacopoeia* also includes the dried stem bark of *I. difengpi* K.I.M. (cortex illicii; *difengpi* in Chinese) for the treatment of pain, explaining that the drug dispels "wind," removes "damp," and ensures normal flow of *qi*.

Because of the growth of these plants in the southwest area of China, the *Guangxi Medicinal Herb Index* (Guangxi Institute of Traditional Medical and Pharmaceutical Sciences, 1984) and the *Flora of Guangxi* (Guangxi Institute of Botany Academia Guangxiana, 1991) also describe the use of other *Illicium* species such as *I. brevistylum* A.C. Smith, *I. dunnianum* Tutcher, *I. henryi* Diels, *I. jiadifengpi* B. N. Chang, *I. majus* Hook f. et Thoms., *I. micranthum* Dunn, *I. oligandrum* Merr. et Chun, and *I. pachyphyllum* A.C. Smith. The fruits of these plants are mostly very toxic and are not used. The parts used are the root or stem bark, for the relief of joint and lower back pain. Some of these species, such as *I. brevistylum, I. dunnianum, I. jiadifengpi*, and *I. majus*, are applied only externally for joint pain and injury because of their high toxicity. *I. henryi* can be taken orally for the same purpose as *I. difengpi*. The root bark is generally used at a dose of 1 g/day and not more than 5 g should be taken, whereas *I. difengpi* is used at a dose of 6 to 9 g/day. The *Yun Nan Medicinal Plants List* (Yun Nan Drug Material Co., 1993) includes the medicinal usage of some fruits of *Illicium*; that is, the fruits and the leaves of *I. macranthum* A.C. Smith, *I. majus*, and *I. simonsii* Maxim. for the treatment of emesis and pain, and the fruit of *I. ternstroemioides* A.C. Smith as a stomachic and an antiemetic. Because most *Illicium* species are more or less toxic, special care should be taken in their application.

In Japan, the medicinal use of *Illicium*, including star anise (*dai-uikyo* or *hakkaku-uikyo* in Japanese), is not popular in the traditional medical system, Kampo. The *Japanese Pharmacopoeia* (Committee of the Japanese Pharmacopoeia Evaluation and Licensing Division, 1996) only includes star anise oil in Oleum Foeniculi. Japanese star anise (*shikimi* in Japanese) is the fruit of *I. anisatum* L. (*I. religiosum* Sieb. et Zucc.), which has long been cultivated in Japan. It is known to be highly toxic, and cases of poisoning have often been reported. Tsao (1929) gave a detailed bibliography of this plant. The volatile oil obtained from the fruits by distillation has little resemblance to star anise oil (Burkill, 1966). *I. anisatum* is not generally used for medicinal purposes, although Chin and Keng (1990) described its use as a fish and rat poison. They also mentioned that the plant is used to treat skin problems.

Ilyas (1980) reported on the use of star anise in India: It is obtained in the markets of Hong Kong, Malaysia, Singapore, and Indonesia under the name "badyan" or "badian" seed. The fruit, called "sonf" and "anasphal" because of the similarity of its aroma to that of fennel, is chewed for a stomachic and carminative. *I. griffithii* H. f. and Thoms is used as a substitute for *I. verum* as a vernacular of bâdayân (Kirtikar and Basu, 1918).

A *Dictionary of the Economic Products of the Malay Peninsula* (Burkill, 1966) described how the fruits of *I. verum* that were exported from China were retailed in Malaya as "bunga lawang" (clove flowers) and as "adas china" (Chinese anise). The plant is an ingredient in a compound emmenagogue. The Chinese use it to make it a medicinal tea and to flavor foods and confectionery, and they regard it as good for colic and constipation. The herb is also used for the treatment of insomnia, and in external applications after childbirth. The *Medicinal Herbs Index in Indonesia* (Kasahara and Hemmi, 1995) described similar medicinal uses.

In a survey of medicinal herbs in markets of central and northern Mexico and the southwestern United States, Linares and Bye (1987) reported on a plant complex, "hierba anis," which is commonly associated with "anis de estrella" (*I. verum*). The fruits (star anise) are used in traditional teas to treat nervousness and sleeplessness and as a sedative. The most common use in Mexico and adjacent United States is to alleviate colic of babies and stomach aches.

8.2.2 Usage of Essential Oils

Star anise oil distilled from the fruits of *I. verum* is more popular than the fruit itself, as it is often used instead of the oil from aniseed (anise; *Pimpinella anisum* L.) (Stary, 1996). Both oils are officially recognized as anise oil in the United States (Duke, 1985), and in Europe, its anise-like smell is reflected by its names in various languages. In India, it is also used in lieu of anise oil for the same properties (Ilyas, 1980). Star anise oil contains 85 to 90% of *trans*-anethole, which is also the major component of fennel (*Foeniculum vulgare* Miller). This is the reason that the *Japanese Pharmacopoeia* (Committee of the Japanese Pharmacopoeia Evaluation and Licensing Division, 1996) recommends both oils as oleum foeniculi (fennel oil). It is used for carminative and stomachic remedies (Namba and Tuda, 1993). In China, star anise oil is mainly used for flavoring and treatment of stomach aches at 0.06 to 0.6 mL/day, having the Chinese name of bajiao huixiangyou (oleum anisi stellati). Chin and Keng (1990) suggested other uses in their book; namely, that the oil can be an antidote for a number of poisons and used to treat rheumatism. You may refer to *Pimpinella*, *Foeniculum*, or anethole for the usage and pharmacological properties of the oil.

8.3 PHARMACOLOGICAL PROPERTIES OF *ILLICIUM*

8.3.1 Pharmacological Profile of Star Anise

8.3.1.1 Mutagenicity

In a mutagenic screening of herbs and spices, star anise showed weak mutagenicity toward *Salmonella typhimurium* strain TA100, but not TA98, when it was fed to rats and the metabolites that were extracted from the 24-hour urine samples had been assayed (Rockwell and Raw, 1979). The general method of using the extract, however, means that the mutagenic potential could not be analyzed, as the use of the extract was accompanied by a growth inhibition of the bacterial tester strain. To examine the effect of the oil on the carcinogen metabolism system of benzo[*a*]pyrene and aflatoxin B_1 in mouse liver, the ethanol extract of *I. verum* was added to a diet fed to male and female mice (Hendrich and Bjeldanes, 1983, 1986). After 14 and 10 days, respectively, their liver microsomal fractions were assayed for cytochrome P-450 content, aryl hydrocarbon hydroxylase, 7-ethoxycoumarin-*O*-deethylase, epoxide hydratase, and glutathione *S*-transferase. Liver weight of the animals fed on *I. verum* was significantly higher than that of the basal and chow diet groups. In male mice, epoxide hydratase activity increased, whereas glutathione *S*-transferase activity decreased, compared with the activity in animals of the basal group. In benzo[*a*]pyrene and aflatoxin B_1 mutagenicities determined with the Ames assay, *I. verum* did not differ significantly from the basal group. The 7-ethoxycoumarin-*O*-deethylase and glutathione *S*-transferase activities of the *I. verum* group were significantly elevated over the basal level in female mice, and the DNA binding with microsomes used for aflatoxin B_1 activation was significantly higher than in the basal group.

Trans-anethole (Figure 8.1), the major component of star anise oil, indicated mutagenic activity only against strain TA100 in the presence of S9 via Ames Salmonella reversion assay, using several bacterial strains (Sekizawa and Shibamoto, 1982). In the assay using growing yeast cells without S9, however, *trans*-anethole was shown to be negative (Nestmann and Lee, 1983), and it showed no hepatocarcinogenic activity at 31 mmol/kg of diet after administration to mice for 12 months (Miller et al., 1983). Safrole is known as a weak carcinogen in rodents, and it was detected in star anise at 9325 ppm, although the mutagenicity resulting from the presence of safrole can be decreased

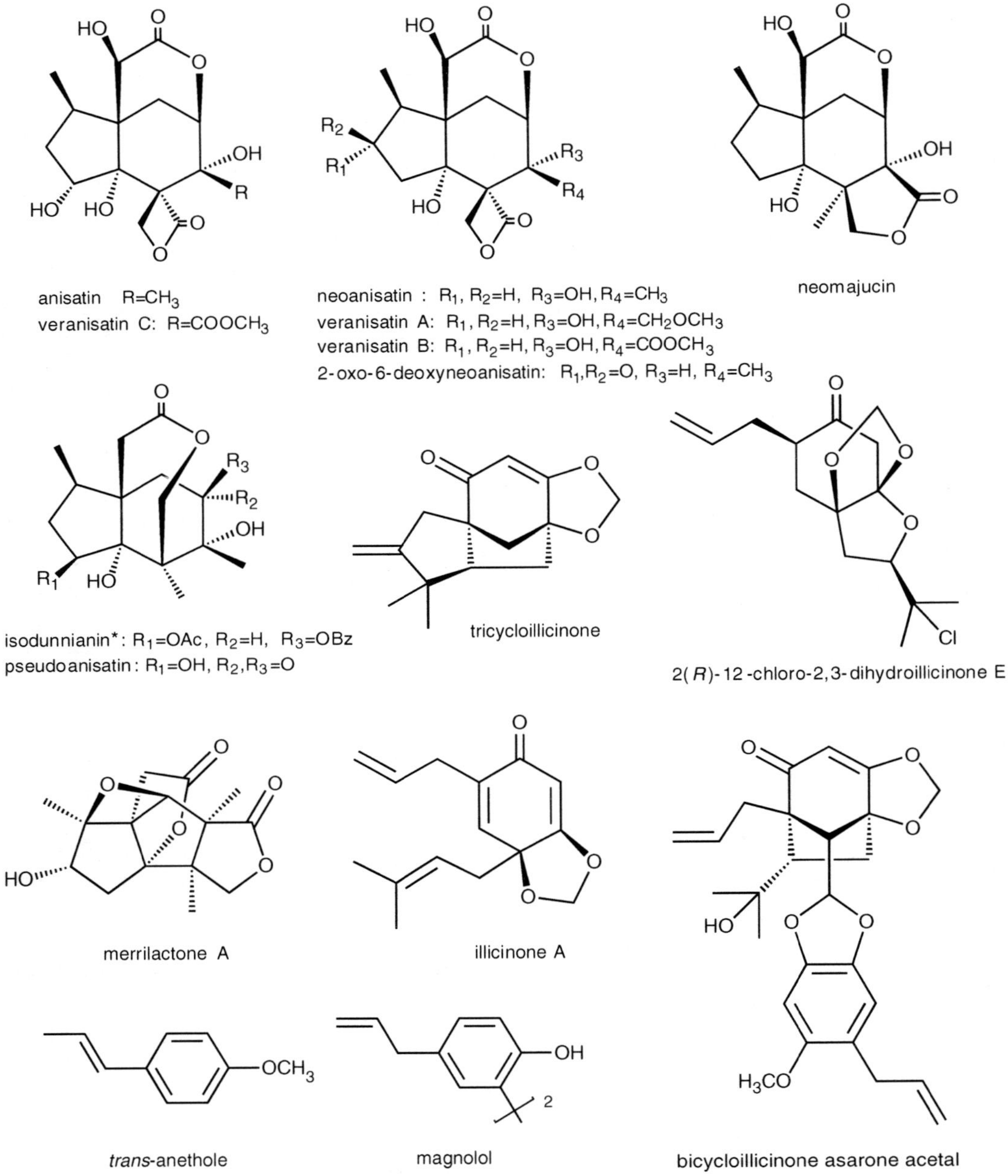

FIGURE 8.1 Chemical structures of *Illicium* components; * = a report by Schmidt and Peters (1997) suggested revision of isodunnianin.

by processing methods (Farag and Abo-Zeid, 1996). For more information about the toxicity of these compounds, you may refer to a review written by Buchanan (1978).

8.3.1.2 Antimicrobial and Other *in vitro* Activity

Star anise exhibited antifungal activity against the mycotoxin-producing fungi such as *Aspergillus flavus*, *A. ochraceus*, *A. versicolor*, and *Fusarium sporotrichioides* (Morozumi et al., 1989). Addition of the ground material at concentration of 10% to the media inhibited growth of these fungi completely. Anethole from star anise was not a potent inhibitor; it caused inhibition at 2000 μg/mL or more. De et al. (2002), however, identified anethole as the antimicrobial component of *I. verum*, which showed MIC 5~50 μg/mL against 12 bacteria and 15 fungi/yeast. Antioxidative activity (Shang et al., 1994) and the inhibition of some digestive enzymes (Yamasaki et al., 1986) of *I. verum* were also reported.

8.3.1.3 Insecticidal and Molluscicidal Activity

Insecticidal activity of star anise against fruit flies (*Drosophila melanogaster*) was observed, and the activity-oriented separation of the methanol extract yielded *trans*-anethole as the active component. *Trans*-anethole's 5-day LD_{50} on *D. melanogaster* was 0.2 mg/mL diet (Miyazawa et al., 1993). The insecticidal activity of *trans*-anethole also presented against two stored product insects, *Tribolium castaneum* and *Sitophilus zeamais* (Ho et al., 1997), and against adults of *Blattella germanica* (Chan and Ahn, 2002). In a search for herbicides, extracts of star anise were found to inhibit velvetleaf germination (Wolf, 1986). Followed by bioassay, the hexane extract was separated to yield *p*-methoxycinnamaldehyde as the responsible component. After 4 days of exposure to a 2.5-mM dose of this compound, essentially no velvetleaf seeds germinated, whereas anethole did not inhibit even at 5 mM. Toxicity of star anise oil to golden snail was reported by Maini and Morallo-Rejesus (1992), showing 100% mortality at 10 to 20 ppm against young snails.

8.3.1.4 Irritation

Sensitizing and irritating properties of star anise oil were disclosed when testing consecutive dermatitis patients with 36 essential oils (Rudzki and Grzywa, 1976). The oil of 2 and 1% concentrations produced a positive patch test in 36 and 34%, respectively, of the patients. Patients positive to this oil are frequently positive to anethole and to other constituents of this oil such as α-pinene, limonene, and safrole. A 1% concentration is strongly irritant, and 0.5% reveals sensitivity in only one fifth of the actively sensitized subjects.

8.3.1.5 Analgesic and Sedative Effects and Convulsive Components

Because star anise has been used for the treatment of pain, and sometimes for sedation, we tested both activities of the methanol extract in mice by acetic acid-induced writhing inhibition and hypothermic effect, respectively. The extract indicated 23% writhing inhibition ($P < .001$) at an oral dose of 500 mg/kg and a significant hypothermic effect at 3 g/kg (ΔT_{max} 5.2°C, $P < .001$). Using the bioassay guides then carried out the isolation of the active component. During the isolation processing in the preliminary experiment, we observed convulsive effect and lethal toxicity in mice, depending on the doses, in the same fractions that caused hypothermia. The convulsion and lethal toxicity in mice were therefore taken as isolation guides. The ethyl acetate extract of star anise indicated hypothermia (ΔT_{max} 5.6°C, $P < .001$) at an oral dose of 100 mg/kg, and 500 mg/kg administration caused severe convulsions and lethal toxicity (in four out of four cases) in mice. From the extract, the active components were finally isolated and named as veranisatins A, B, and C (Fig. 8.1) (Okuyama et al. 1993; Nakamura et al. 1996). Oral administration of these compounds at 3 mg/kg produced convulsions and lethal toxicity (three out of three cases) in mice, whereas they also showed a hypothermic effect without convulsions at lower doses such as 0.5 or 1 mg/kg. Anisatin (Fig. 8.1), a toxic component of *I. anisatum*, also presented similar behavior at a dose of 0.5 mg/kg. Veranisatin A and anisatin were further tested for their pharmacological effects. Analgesic effects were detected by acetic acid-induced writhing and by tail-pressure methods in mice. Both compounds exhibited analgesia at 0.1 mg/kg by these methods. To analyze their sedative effects, locomotor activity enhanced by methamphetamine was tested, in addition to the hypothermic effect. Veranisatin A and anisatin decreased the locomotion at 0.1 and 0.03 mg/kg, respectively. These results indicate that toxic components such as veranisatins and anisatin may greatly contribute to the pharmacological properties and clinical usage of *Illicium* as well.

8.3.2 Toxic Components of *I. anisatum* and the Other *Illicium* Species

8.3.2.1 Toxic Components of *Illicium species*

The convulsive effect of Japanese star anise (Shikimi in Japanese, the fruit of *I. anisatum*) had been known for several centuries. Contamination of Japanese star anise with star anise (*I. verum*) as a result of their similarity has been very serious. Montoya-Cabrera (1990) described the application of star anise to babies for the treatment of colic and flatulence in Mexico, but warned that the contamination of *I. anisatum* caused tonic and chronic convulsions, alteration of consciousness, and epilepsy. Recently, a series of serious adverse health effects following consumption of an herbal tea were observed in Spain (Garzo Fernandez et al. 2002) and in the Netherlands (Johanns et al. 2002). The investigations indicated that Japanese star anis (*I. anisatum*) might have been inadvertently mixed into the herbal tea. The persons reported symptoms of general malaise, nausea, and vomiting about 2 to 4 hours following the consumption of the herbal tea, and some people required hospitalization as a result of tonic-clonic seizures.

Among the respective parts of this plant, the pericarps are the most toxic (LD_{50} 510 mg/kg in mice), whereas the seeds and other parts are less so (LD_{50} 2170 mg/kg and more doses in mice) (Kawano and Matsuo 1958). Attempts at isolation were undertaken during many years from the nineteenth century, and Lane and colleagues (1952) succeeded in purifying the toxic component, named anisatin (Fig. 8.1). They described the behavior in detail as follows: When anisatin was injected intraperitoneally (ip) into mice at dosages on the order of the minimum lethal dose. Anisatin exhibited partial paralysis of the hind limbs shortly after injection, followed by generalized body trembling with convulsive seizures from 15 to 30 minutes after injection. These seizures increased rapidly in severity and duration, with death usually occurring within 90 to 120 minutes. Mice that survived this period, however, almost always recovered. They estimated the LD_{50} of 1.1 ± 0.1 mg/kg for anisatin and 1.1 ± 0.1 g/kg for the powdered crude drug. Later, Kouno et al. (1989) and Yang et al. (1991) independently reported the LD_{50} of anisatin as 1.03 ± 0.1 mg/kg (95% confidence limits) and 0.76 ± 0.07 mg/kg, respectively, in mice by intraperitoneal injection. Oral administration to mice at a dose of 1 mg/kg also caused convulsion (in all of them) and lethal toxicity (from three of four to three of three) in mice from our data (Nakamura et al., 1996, and additional unpublished data). Another convulsant of *I. anisatum* is neoanisatin, with LD_{50} being 1.62 mg/kg (ip) in mice (Kouno et al., 1989). Anisatin content was also reported in some other species such as *I. henryi* (Liu and Zhou, 1988), *I. maxus* (Kouno et al., 1989), *I. simonsii* (Yang et al., 1991), and *I. minwanense* (Wang et al., 1994). The toxicity (and possibly clinical advantage as well) of these plants seems to be caused mainly or partly by anisatin.

Other toxic compounds related to anisatin were isolated from *I. majus* by Kouno et al. (1989). Neomajucin (Figure 8.1) displayed a LD_{50} of 12.2 mg/kg via intraperitoneal injection for mice, whereas majucin did not indicate toxicity at a dose of more than 40 mg/kg. At doses of 20 to 40 mg/kg (ip), other majucin-related compounds are not toxic in mice in 72 hours (Kouno et al., 1990). The plant also contains 2-oxy-6-deoxyanisatin, causing convulsion, and its LD_{50} (ip) in mice was reported as 1.46 mg/kg (Kouno et al., 1991a). The same compound was isolated from *I. simonsii* by Yang et al. (1991), and the LD_{50} (ip) was described as 0.94 mg/kg in mice. Veranisatins, which are the analgesic and sedative components of *I. verum*, also indicated convulsion and lethal toxicity, depending on the doses (Okuyama et al., 1993; Nakamura et al., 1996), as described in Section 3.1.

8.3.2.2 Action Mechanism of Anisatin and the Related Compounds

After the isolation of anisatin from *I. anisatum*, biochemical studies on anisatin were carried out, although some pharmacological studies on the crude toxic material of *I. anisatum* had already been attempted. The effect of anisatin on the frog spinal cord and on the crude synaptic membrane from rat brain was compared with that of picrotoxin and bicuculline on the same tissues, presenting its picrotoxin-like, noncompetitive GABA-antagonist (Kudo et al., 1981). Using crayfish neuromuscular junction, the presynaptic inhibition caused by GABA was irreversibly blocked by anisatin, whereas the effect of anisatin on postsynaptic inhibition subsided after prolonged washing of the preparation (Shinozaki et al., 1981). The effects of anisatin were investigated on the diazepam binding enhanced by GABA or by pentobarbital (Matsumoto and Fukuda, 1982). Anisatin inhibited the enhanced bindings without affecting the basal-specific binding to rat brain membranes. It temperature-dependently modified the effect of chloride ion on muscimol binding to rat brain membranes (Matsumoto and Fukuda, 1983). Anisatin also inhibited with IC_{50} 0.10 μM *t*-butylbicyclophosphorothionate, a ligand that binds to the picrotoxinin site (Ramanjaneyulu and Ticku, 1984). In addition to the similarity of anisatin's behavior to picrotoxin-convulsion, these findings indicated that there was anisatin modulation of the GABA receptor through the picrotoxinin-sensitive sites.

Recent reports on the mechanisms of anisatin interaction with the GABA system include the interacting region in rat brain $GABA_A$ receptors that is allosterically linked to the binding site of competitive antagonists (Kakemoto et al., 1999), as well as the modulation of GABA receptor-channel in rat dorsal root ganglion neurons by whole-cell and single-channel patch clamp experiments (Ikeda et al., 1999). A common pharmacophore structure for convulsant sesquiterpenes of the *seco*-prezizaane such as anisatin and veranisatins and picrotoxane types was proposed following comparison of the three-dimensional molecular shape and electrostatic properties of active and inactive compounds based on computer molecular models (Schmidt et al., 1999). Ozoe and colleagues investigated the structure–activity relationships of *seco*-prezizaanes in GABA receptors of houseflies and rats, including three-dimensional QSAR analysis (Kuriyama et al., 2002). For its ability to inhibit the specific binding of [^{3}H]4-ethynyl-4-*n*-propylbicycloorthobenzoate, a noncompetitive antagonist of GABA receptors, veranisatin A was the most potent inhibitor in both housefly-head (IC 78.5 nM) and rat-brain membranes (IC 271 nM), followed by anisatin (IC 123 nM, IC 282 nM). Interestingly, pseudoanisatin displayed a high (more than 26-fold) selectivity for housefly vs. rat GABA receptors (IC 376 nM, IC > 10,000 nM).

8.3.3 Other Biological Activities of the Components from *Illicium*

Fukuyama et al. (1993) reported that some components of *I. tashiroi* Maxim showed *in vitro* activity. Isodunnianin enhanced neurite sprouting during the development of neurons in a primary culture of fetal rat cerebral hemisphere at 10 μM and increased choline acetyltransferase activity to 162% at the same concentration (Fukuyama et al., 1993). Tricycloillicinone and 2(*R*)-12-chloro-2,3-dihydroillicinone E (Figure 8.1) increased the latter activity to 143 and 228%, respectively, at 30 μM in a culture of P10 rat septal neurons (Fukuyama et al., 1994, 1995). Bicycloillicinone asarone acetal was also reported as adding to the enhancement of the choline acetyltransferase activity (Fukuyama et al., 1997). A neurotrophic activity in the cultures of fetal rat cortical neurons at 0.1 to 10 μM was observed in merrilactone A (Figure 8.1) and 11-*O*-debenzoyltashironin from *I. merrillianum* A.C. Smith (Huang et al., 2000), and jiadifenin and (2*S*)-hydroxy-3,4-dehydroneomajucin from *I. jiadifengpi* (Yokoyama et al., 2002). Yakushijin et al., (1980) reported that illicinones A (Figure

8.1) and B, phytoquinoids isolated from the same plant, indicated repellent effect against *Monodonata neritoides* with EC_{50} of 6.8 and 7.9 μg/cm^2, respectively.

Many phenylpropanoids in Illicium were reported; among them, magnolol (Figure 8.1) was isolated from the barks of *I. difengpi* (Huang et al., 1996) and *I. dunnianum* (Kouno et al., 1991b). Magnolol is a bioactive principle of *Magnolia officinalis* Rehd et Wils. (Sarker and Maruyama, 2002), and the pharmacological properties of it and its related compounds were reported for their centrally acting muscle relaxant effect (Watanabe, 1985) and microbiological activity against Gram-positive and acid-fast bacteria and fungi (Clark, 1981). Alkenyloxybenzene derivatives including 1-[(3-methyl-2-butenyl)oxy]-4-*trans*-propenylbenzene from star anise seeds have also been investigated for their melanin-formation inhibitors (Tada et al., 1999).

8.4 CONCLUSION

Among *Illicium* species, which are mostly toxic, star anise (*I. verum*) has been extensively used worldwide as a spice and an essential oil because of its resemblance to anise oil and fennel oil. Star anise, however, has a long history as a traditional medicine, especially in China and other Southeast Asian countries. Although the pharmacological study is still far from a complete understanding of the herb, the activity indicating clinical usage may extend not only to *trans*-anethole, the major component of the oil but also to the other components, like veranisatins, having an anisatin-related structure. Anisatin was first isolated as a picrotoxin-like convulsive component from *I. anisatum*, and recently it was also indicated to have analgesic and sedative properties at lower doses. Further study on the details of the pharmacological action and mechanism, including the GABA-antagonistic property of anisatin and its related compounds, may help us to understand the pharmacological aspect of Illicium, because these compounds probably have a role in the medicinal use, together with the toxicity of *Illicium*. Biochemical findings on other constituents such as phenylpropanoids and C6-C3 components may also contribute to therapeutic use.

References

Buchanan, R.L., Toxicity of spices containing methylenedioxybenzene derivatives: a review, *J. Food Safety*, 1, 275–293, 1978.

Burkill, I.H., *A Dictionary of the Economic Products of the Malay Peninsula Vol. 2*, Government of Malaysia and Singapore, Ministry of Agriculture and Co-operatives, Kuala Lumpur, 1966, pp. 1244–1246.

Chan, K.S. and Ahn, Y.J., Fumigant activity of (E)-anethole identified in *Illicium verum* fruit against *Blattella germanica*, *Pest. Manag. Sci.*, 58(2), 161–166, 2002.

Chin, W.Y. and Keng, H., *An Illustrated Dictionary of Chinese Medicinal Herbs*, Times Editions, Singapore, and Eu Yan Sang Holdings, Malaysia, 1990, p. 96.

Clark, A.M., El-Feraly, F.S., and Li, W.-S., Antimicrobial activity of phenolic constituents of *Magnolia grandiflora* L., *J. Pharm. Sci.*, 70(8), 951–952, 1981.

Committee of the Japanese Pharmacopoeia Evaluation and Licensing Division, Pharmaceuticals and Medical Safety Bureau, Ministry of Health, Labor, and Welfare, *The Japanese Pharmacopoeia*, 13th ed., Ministry of Health and Welfare, Tokyo, 1996, p. 2000 (in Japanese).

De, M., De, A.K., Sen, P., and Banerjee, A.B., Antimicrobial properties of star anise (*Illicium verum* Hook f), *Phytother. Res.*, 16(1), 94–95, 2002.

Duke, J.A., *CRC Handbook of Medicinal Herbs*, CRC Press, Boca Raton, FL, 1985.

Farag, S.E.A. and Abo-Zeid, M., Mutagenicity and degradation of natural carcinogenic compound-safrole in spices under different processing methods, *Al-Azhar J. Pharm. Sci.*, 17, 149–158, 1996.

Fukuyama, Y., Hata, Y., and Kodama, M., A novel prenylated C6-C3 compounds increasing choline acetyltransferase (ChAT) activity from *Illicium tashiroi*, *Planta Med.*, 63, 275–277, 1997.

Fukuyama, Y., Okamoto, K., Kubo, Y., Shida, N., and Kodama, M., New chlorine-containing prenylated C6-C3 compounds increasing choline acetyltransferase (ChAT) activity in culture of postnatal rat septal neurons from *Illicium tashiroi*, *Chem. Pharm. Bull*, 42(10), 2199–2201, 1994.

Fukuyama, Y., Shida, N., and Kodama, M., Isodunnianin: a new sesquiterpene enhancing neurite outgrowth in primary culture of fetal rat cerebral hemisphere from *Illicium tashiroi*, *Planta Med.*, 59, 181–182, 1993.

Fukuyama, Y., Shida, N., Kodama, M., Chaki, H., and Yugami, T., Tricycloillicinone, a novel prenylated C6-C3 compound increasing choline acetyltransferase (ChAT) activity, isolated from *Illicium tashiroi*, *Chem. Pharm. Bull.*, 43(12), 2270–2272, 1995.

Garzo Fernandez, C., Gomez Pintado, P., Barrasa Blanco, A., Martinez Arrieta, R., Ramirez Fernandez, R., and Ramon Rosa, F., Cases of neurological symptoms associated with star anis consumption used as a carminative, *An. Esp. Pediatr.,* 57(4), 290–294, 2002.

Guangxi Institute of Botany Academia Guangxiana, Eds., *Flora of Guangxi Vol. 1*, Guangxi Science and Technology Publishing House, Nanning, 1991, p. 98–106 (in Chinese).

Guangxi Institute of Traditional Medical and Pharmaceutical Sciences, Eds., *Guangxi Medicinal Herb Index*, Guangxi People's Publishing House, Nanning, 1984, pp. 50–51 (in Chinese).

Hendrich, S. and Bjeldanes, L.F., Effects of dietary cabbage, brussels sprouts, *Illicium verum*, *Schizandra chinensis* and alfalfa on the benzo[*a*]pyrene metabolic system in mouse liver, *Food. Chem. Toxicol.*, 21(4), 479–486, 1983.

Hendrich, S. and Bjeldanes, L.F., Effects of dietary *Schizandra chinensis*, brussels sprouts and *Illicium verum* extracts on carcinogen metabolism systems in mouse liver, *Food. Chem. Toxicol.*, 24(9), 903–912, 1986.

Ho, S.H., Ma, Y., and Huang, Y., Anethole, a patent insecticide from *Illicium verum* Hook F., against two stored product insects, *Int. Pest Control,* 39(2), 50–51, 1997.

Huang, J.-M., Yokoyama, R., Yang, C.-S., and Fukiyama, Y., Merrilactone A, a novel neurotrophic sesquiterpene dilactone from Illicium merrillianum, *Tetrahedron Lett.*, 41, 6111–6114, 2000.

Huang, J.-M., Yokoyama, R., Yang, C.-S., and Fukiyama, Y., Structure and neurotrophic activity of *seco*-prezizaane-type sesquiterpenes from *Illicium merrillianum*, *J. Nat. Prod.*, 64, 428–431, 2001.

Huang, P., Yang, M., Lai, M.X., Zheng, X.Z., Nisimasa, S., and Nakanisi, C., Studies on the chemical constituents of the traditional Chinese medicine "difengpi," *Yaoxue Xuebao*, 31(4), 278–281, 1996 (in Chinese).

Ikeda, T., Ozoe, Y., Okuyama, E., Nagata, K., Honda, H., Shono, T., and Narahashi, T., Anisatin modulation of the γ-aminobutyric acid receptor-channel in rat dorsal root ganglion neurons, *Br. J. Pharmacol.*, 127(7), 1567–1576, 1999.

Ilyas, M., Spice in India: III, *Econ. Bot.*, 34(3), 236–259, 1980.

Johanns, E.S., van der Kolk, L.E., van Gemert, H.M., Sijben, A.E., Peters, P.W., and de Vries, I., An epidemic of epileptic seizures after consumption of herbal tea, *Ned. Tijdschr. Geneeskd.*, 146(17), 808–811, 2002 (in Dutch).

Kakemoto, E., Okuyama, E., Nagata, K., and Ozoe, Y., Interaction of anisatin with rat brain γ-aminobutyric acid A receptors: allosteric modulation by competitive antagonists, *Biochem. Pharmacol.*, 58(4), 617–621, 1999.

Kasahara, S., and Hemmi, S., Eds., *Medicinal Herb Index in Indonesia*, 2nd ed., P.T. Eisai, Indonesia, 1995, p. 10.

Kawano, N. and Matsuo, A., Studies on the toxic components of *Illicium Anisatum* L. I., *Yakugaku-Zasshi*, 78(11), 1220–1223, 1958 (in Japanese).

Kirtikar, K.R. and Basu, B.D., *Indian Medicinal Plants I*, Sudhindra Nath Basu, Bahadurganj, 1918, p. 43.

Kouno, I., Baba, N., Hashimoto, M., Kawano, N., Takahashi, M., Kaneto, H., Yang, C.S., and Sato, S., Isolation of three new sesquiterpene lactones from the pericarps of *Illicium majus*, *Chem. Pharm. Bull.*, 37(9), 2448–2451, 1989.

Kouno, I., Baba, N., Hashimoto, M., Kawano, N., Takahashi, M, Kaneto, H., and Yang, C.S., Sesquiterpene lactones from the pericarps of *Illicium majus*; 2-oxy derivatives of neomajucin and 3,4-dehydroxy-neomajucin, *Chem. Pharm. Bull.*, 38(2), 422–425, 1990.

Kouno, I., Hashimoto, M., Enjoji, S., Takahashi, M., Kaneto, H., and Yang, C.S., Isolation of neoanisatin derivatives from the pericarps of *Illicium majus* with other constituents, *Chem. Pharm. Bull.*, 39(7), 1773–1778, 1991a.

Kouno, I., Morisaki, T., Hara, Y., and Yang, C.S., Two new sesquineolignans from the bark of *Illicium dunnianum*, *Chem. Pharm. Bull.*, 39(10), 2606–2608, 1991b.

Kudo, Y., Oka, J., and Yamada, K., Anisatin, a potent GABA antagonist, isolated from *Illicium anisatum*, *Neurosci. Lett.*, 25, 83–88, 1981.

Kuriyama, T., Schmidt, T.J., Okuyama, E., and Ozoe, Y., Structure-activity relationships of seco-prezizaane terpenoids in γ-aminobutyric acid receptors of houseflies and rats, *Bioorg. Med. Chem.*, 10(6), 1873–1881, 2002.

Lane, J.F., Koch, W.T., Leeds, N.S., and Gorin, G., On the toxin of *Illicium anisatum*. I. the isolation and characterization of a convulsant principle: anisatin, *J. Am. Chem. Soc.*, 74(13), 3211–3215, 1952.

Linares, E. and Bye Jr., R. A., A study of four medicinal plant complexes of Mexico and adjacent United States, *J. Ethnopharmacology*, 19(2), 153–183, 1987.

Liu, J.S. and Zhou, Q.R., The toxic principle of *Illicium henryi* Diels and structure of 6-deoxypseudoanisatin, *Yao Hsueh Hsueh Pao*, 23(3), 221–223, 1988 (in Chinese).

Maini, P.N. and Morallo-Rejesus, B., Toxicity of some volatile oils against golden snail (*Pomacea* spp.), *Philippine J. Sci.*, 121(4), 391–397, 1992.

Matsumoto, K. and Fukuda, H., Anisatin modulation of GABA- and pentobarbital-induced enhancement of diazepam binding in rat brain, *Neurosci. Lett.*, 32, 175–179, 1982.

Matsumoto, K. and Fukuda, H., Anisatin modulation of temperature-dependent inhibition of [3H]muscimol binding by chloride ion, *Brain Res.*, 270, 103–108, 1983.

Miller, E.C., Swanson, A.B., Phillips, D.H., Fletcher, T.L., Liem, A., and Miller, J.A., Structure-activity studies of the carcinogenicities in the mouse and rat of some naturally occurring and synthetic alkenylbenzene derivatives related to safrole and estragole, *Cancer Res.*, 43, 1124–1134, 1983.

Miyazawa, M., Ota, H., Ishikawa, Y., and Kameoka, H., An insecticidal compound from *Illicium verum*, *Chem. Express*, 8(7), 437–440, 1993.

Montoya-Cabrera, M.A., Poisoning by star anise (*Illicium verum*) tea, *Gaceta Med. Mexico*, 126(4), 341–342, 1990 (in Spanish).

Morozumi, S., Wauke, T., Kudoh, Y., and Hitokoto, H., Antifungal effects of commercial foods and spices, and their components, *Bioactive Mol.*, 10, 155–160, 1989.

Nakamura, T., Okuyama, E., and Yamazaki, M., Neurotropic components from star anise (*Illicium verum* Hook. fil.), *Chem. Pharm. Bull.*, 44 (10), 1908–1914, 1996.

Namba, T. and Tuda Y., Eds., *Outline of Pharmacognosy, a Textbook*, Nankodo, Tokyo, 1993, pp. 257, 309 (in Japanese).

Nestmann, E.R. and Lee, G.-H., Mutagenicity of constituents of pulp and paper mill effluent in growing cells of *Saccharomces cerevisiae*, *Mutation Res.*, 119, 73–280, 1983.

Okuyama, E., Nakamura, T., and Yamazaki, M., Convulsants from star anise (*Illicium verum* Hook.f.), *Chem. Pharm. Bull.*, 41(9), 1670–1671, 1993.

Pharmacopoeia Commission of the Ministry of Public Health, *Pharmacopoeia of the People's Republic of China*, Vol. 1, Guangdong Science and Technology Press, Guangzhou, Chemical Industry Press, Beijing, 1995, pp. 3–4, 99 (in Chinese).

Ramanjaneyulu, T. and Tcku, M.K., Binding characteristics and interactions of depressant drugs with [35S]*t*-butylbicyclophosphorothionate, a ligand that binds to the picrotoxinin site, *J. Neurochem.*, 42(2), 221–229, 1984.

Rudzki, E. and Grzywa, Z., Sensitizing and irritating properties of star anise oil, *Contact Dermatitis*, 2, 305–308, 1976.

Rockwell, P. and Raw, I., A mutagenic screening of various herbs, spices, and food additives, *Nutr. Cancer*, 1(4), 10–15, 1979.

Sarker, S.D. and Maruyama, Y., Eds., *Magnolia; The genus* Magnolia, Taylor & Francis, London, 2002.

Schmidt, T.J., Okuyama, E., and Fronczek, F.R., The molecular structure of 2α-hydroxyneoanisatin and structure-activity relationships among convulsant sesquiterpenes of the *seco*-prezizaane and picrotoxane types, *Bioorg. Med. Chem.*, 7(12), 2857–2865, 1999.

Schmidt, T.J. and Peters, W., Revision of the structure of dunnianin and related sesquiterpene lactones from *Illicium* species, *J. Nat. Prod.*, 60, 783–787, 1997.

Sekizawa, J. and Shibamoto, T., Genotoxicity of safrole related chemicals in microbial test systems, *Mutation Res.*, 101, 127–140, 1982.

Shang, S., Wu, X., Yang, L., and Lan, S., Screening antioxidants from Chinese herb medicines, *Tianran Chanwu Yanjiu Yu Kaifa*, 6(1), 36–39, 1994 (in Chinese).

Shinozaki, H., Ishida, M., and Kudo, Y., Effects of anisatin on the GABA action in the crayfish neuromuscular junction, *Brain Res.*, 222, 401–405, 1981.

Starý, F., *The Natural Guide to Medicinal Herbs and Plants*, Barnes & Noble, New York, 1996, p. 152–153.

Tsao, M., *Illicium religiosum* Siebold, *Am. J. Pharm.*, 101(10), 687–715, 1929.

Tada, A., Kato, T., and Katagiri, T., Skin-lightening cosmetics containing alkenyloxybenzene derivatives, *Jpn. Kokai Tokkyo Koho*, JP 11322528 A2, 1999, p. 6 (in Japanese).

Wang, J.L., Yang, C.S., Yan, R.N., Yao, B., and Yang, X.B., Sesquiterpene lactones from *Illicium minwanense*, *Yaoxue Xuebao*, 29(9), 693–696, 1994.

Watanabe, K., Kouboku, *Biomed. Therapeutics*, 14(5), 697–703, 1985 (in Japanese).

Wolf, B.R., Effects of *p*-methoxycinnamaldehyde from star anise and related cinnamic acid derivatives on velvetleaf germination, *J. Nat. Prod.*, 49, 156–158, 1986.

Yakushijin, K., Suzuki, R., Tohshima, T., Murata, H., Furukawa, H., Nakabayashi, T., and Kurotaki, M., The constituents of *Illicium tashiroi* Maxim. leaves. Structure and biological activities of novel phytoquinoids, *Tennen Yukikagobutsu Toronkai Koen Yoshishu*, 23, 648–655, 1980 (in Japanese).

Yamasaki, K., Yokoyama, H., Nunoura, Y., Umezawa, C., and Yoneda, K., Studies on the effect of crude drugs on enzyme activities (IV) Influence of stomachic crude drugs on digestive enzymes, *Shoyakugaku Zasshi*, 40(3), 289–294, 1986 (in Japanese).

Yang, C.S., Wang, J.L., Zhang, Z.L., and Kouno, I., Toxic constituents of *Illicium simonsii* Maxim, *Yaoxue Xuebao*, 26(2), 128–131, 1991 (in Chinese).

Yokoyama, R., Huang, J.-M, Yang C.-M., and Fukuyama, Y., New *seco*-prezizaane-type sesquiterpenes, jiadifenin with neurotrophic activity and 1,2-dehydroneomajucin from *Illicium jiandifengpi*, *J. Nat. Prod.*, 65, 527–531, 2002.

Yun Nan Drug Material Co., Eds., *Yun Nan Medicinal Plants List*, Science Publishing House, Beijing, 1993, pp. 150–151 (in Chinese).

Zhen, L.S., *Compendium of Materia Medica* (1590), reproduced by People's Hygiene Publication Co., Beijing, 1979, pp. 1636–1639 (in Chinese).

9 Therapeutic and Pharmacologic Properties of the Genus *Pimpinella*

Anna De Pasquale

CONTENTS

9.1 TRADITIONAL USES

Numerous species belong to the genus *Pimpinella* L., only some of which have been used in medicine. The most widely used species, as a flavor and as a medicament, is *P. anisum* L., of which the fruits and the essential oil are used. However, most of the biological investigations carried out so far essentially deal with the pharmacological activity and toxicology of the essential oil and of its main component, *trans*-anethol.

P. anisum L., the original distribution of which is not known, was indigenous to Egypt, Asia Minor, and the island of Crete; from there, in ancient times, it became diffused in the countries of the Mediterranean basin, where it became naturalized. Its cultivation extended also to Asia and successively to other continents, especially to the countries of Latin America, where it became acclimatized and naturalized.

The use of *P. anisum* in traditional medicine is closely linked to the history of the civilizations that have followed one another in the course of centuries in the countries of the Mediterranean basin, through which the herb became diffused in the areas in which it became acclimatized.

The medicinal use of anise fruit goes back to a far-remote time. It seems that, even some millennia before Christ, the plant was already represented in Egyptian papyri and in the mural paintings of Luxor, and the fruit was already used as a carminative and digestive. In these ancient times, however, drugs originating from different plants were undoubtedly called with the same name. For example, in Babylonian times, anise and fennel were exchanged one for the other.

However, in some authorized versions of the Bible, there is the following translation of the Gospel according to Matthew (23:23):

> Woe unto you, scribes and Pharisees, hypocrites! for you pay tithe of mint and anise and cummin, and have omitted the weightier matters of the law.

0-415-32246-4/04/$0.00+$1.50

P. anisum is also now cultivated in the Holy Land, but the plant to which Jesus referred in the biblical text is dill, *Anethum graveolens*, another umbellifer, cultivated in Palestine for its aromatic fruits and medicinal use (*La Sacra Bibbia,* 1940; Moldenke and Moldenke, 1952).

There is no mention of the use of anise in the most ancient treatises of Indian medicine, but in the Susruta (fifth century A.D.), anise is indicated among other galactogogue drugs and is given the name *atichatra* (Tschirch, 1910). Even practitioners of ancient Chinese medicine widely used anise.

More certain reports of the use of anise come from Greece at the time when medical practices started to be stripped of mythological beliefs. Hippocrates of Coo (fifth century B.C.), the father of Greek and rational medicine, formed the basis of pathology and therapy. In *Corpus Hyppocraticum*, a compendium of the teachings of Hippocrates and his disciples, over 300 plants are indicated that were used for the preparation of medicaments; anise was classified among the plants to be used in the maladies of women and for fumigations. It is in Dioscorides, however, that we find the description of the plant and more copious news on the use of anise.

Pedanius Dioscorides, born (50 A.D.) at Anazarba in Cilicia, was a Greek physician who, in Nero's era (during the first century A.D.), followed the Roman legions to Egypt, other African countries, Spain, and Italy. Dioscorides compiled an important treatise titled *De materia*, to which the copyists added the adjective *medica*. All medical and botanical knowledge on the medicaments used in the countries he visited were gathered in this work.

De materia medica, translated and annotated, was for centuries the book on which generations of physicians were formed. In the *Liber III*, in which, together with the *Liber IV*, the medicinal plants are described, he writes about *Anison:*

> Anisum (some call it Sion, ye Romans Anisum) hath generally a warming, drying, sweet breath-making, paine easing, dissolving, ureticall, discussing faculty. Being drank it takes away thirst from ye hydropicall. It is good also both for outcasting ye poison of venemous beasts and for inflations, and it stops ye belly and ye white flux and draws down milk and incites conjunctions and being suffited to ye nostrils, it quieteth ye headach, and being beaten small, and droped in with Rosaceum it heals the chaps of ye ears. But ye best of it is ye new and fuller, and not branny, strong in ye scent. But ye Creticall claims ye first place, and the next ye Egyptian.
>
> **—Gunther, 1968**

Dioscorides's book had various editions with comments and additions. One of the most famous editions is that of Pietro Andrea Mattioli (1504–1577), a Sienese doctor who amplified and illustrated the descriptions made by Dioscorides (Mattioli, 1557). Very many reprints of this book were made, and up to the eighteenth century, generations of physicians and pharmacists were trained on it.

Because it was of distinctly Greek origin, Roman medicine largely used vegetal remedies. There are no specific documents on the use of anise in the first centuries, but some information can be gained from the works of writers of that era (Celsus, Marcellus Empiricus, Trallianus, Lucius Columella, etc.).

In the *Historia naturalis* (Plinii Caii Secundi, 1831), an encyclopedia of the time, Caius Plinius Secundus (23–79 A.D.), known as Pliny the Elder, gathered information on the most diverse subjects: cosmetology, astronomy, geography, physics, agriculture, medicine, etc. Of the 37 books that made up the work, at least 15 refer to the *Materia medica*; of these, 8 books deal with vegetable drugs. It seems that Pliny described the plants that a famous Roman physician of the time, called Castor, cultivated in his garden. Pliny gives diverse information about the plants, much of which is similar to that referred by Dioscorides, his contemporary, so it would seem that they gleaned some of their information from the same sources. Pliny reports that Pythagoras greatly appreciated anise both raw and cooked, and to the already mentioned therapeutic uses of anise as a carminative and expectorant, Pliny adds pieces of information that he gathered, some of which is clearly unlikely; for example, that anise rejuvenates the face and that, hung on the pillow, it alleviates insomnia to those who smell it.

In 47 A.D., Scribonius Largus, in the *Compositiones medicamentorum seu compositiones medicae*, reports on the use of *Anesum*.

The emperor Nero asked Andromachus, a physician of the Roman court who was born in Crete (about 54 A.D.), for a remedy that could render him invulnerable; Andromachus thus created the famous *Theriaca*, the prescription of which we can find in the formularies and in the pharmacopoeias up to the last century and in which, among 61 components that also include snake venom, we can find anise. The term *Theriaca* is the title of a chapter of a poem, *Georgica*, about agriculture, which was written by the poet Nicandrus of Colophonia, who lived during the first century B.C. The poisons originating from the animal kingdom were included in *Theriaca*, whereas the chapter titled *Alexifarmaca* dealt with poisons of vegetable origin. Andromachus called the preparation *Theriaca* in the sense of an antidote against all poisons; for its preparation Andromachus based his text on an electuary, the *Mithridaticum*, compounded by Cratevas, who was the physician to Mithridates IV, the King of Ponto. Mithridates IV, for fear of being poisoned, consumed ever increasing doses of arsenic daily, to acquire a form of resistance that, after the king's name, is called mithridatism. *Mithridaticum*, made up of 54 substances of vegetable and mineral origins and snake venom, was also still in use in medicine up to the last century.

During his studies in Greece, Claudius Galenus (138–201 A.D.), born in Pergamum in Asia Minor, came under the influence of Dioscorides; subsequently, living in Rome from the age of 32, Galenus elaborated and prepared medicaments for the Emperors Marcus Aurelius, Lucius Versus, and Commodus. The so-called galenic pharmacy has its origins in his work. In his formulas, Galenus made wide use of the anise fruit — which he considered the most important part of the plant — as a diuretic, aphrodisiac, and antidote for poison (Galenus, ed. 1600).

In addition to being used as a spice, anise fruit was also widely used in medicine in Europe during the early and late Middle Ages, to the extent that the Emperor Charlemagne (812 A.D.) ordered it cultivated on the imperial farms.

The *Patent of Pontage*, issued by Edward I of England in 1305 to procure funds to repair London Bridge, enumerated anise among the goods to be taxed when they crossed the bridge (Thomson, 1827). The consumption of anise was so large at that time that the Grocer's Company of London was given the task of supervising its commerce and weight, and as with other foreign drugs, the government resorted to increasing the tax on anise every time the treasury needed to be enriched. This happened, for example, in 1635, under the reign of Charles V (Hebert, 1834).

The *Scuola Medica Salernitana*, the origin of which is unknown but that was already considered old in 846 A.D., included medical knowledge of the early and late Middle Ages; it was initially based on Greco-Roman medicine and made use of the contribution of the experience of Arabs and Spanish Jews. Of this text has remained the *Regimen Sanitatis Salerni* (1963) or *Regola Sanitaria Salernitana*, of which more than 300 codices have been preserved with various titles: *De Conservandi Bona Valetudine*, *Medicina Salernitana*, *Flos Medicinae*, and so forth. Among the aphorisms used to express the rules to keep in good health there is one that refers to anise:

> *Caput XLIX*
> *Emendat visum, stomachum confortat anisum.*
> *Copia dulcoris aniso sit melioris.*
>
> Item XLIX
> Anise clears the sight and restores the stomach.
> Of all the varieties the sweetest ones are to be appreciated.

The date of publication of the *Theatrum Sanitatis*, an illuminated codex of the Casanatense Library of Rome, is uncertain but might be at the end of the fourteenth century. It may belong either to the Veronese or the Tuscan School, and as far as the contents are concerned, it might have been derived from the *Tacuini Sanitatis* or *Tabulae Sanitatis*, works written in Arabic and translated into Mediaeval Latin to which were added the medicaments and food more commonly used in

Italy. A miniature of the anise plant completes the description of the activity and toxicity of anise fruits (*Theatrum Sanitatis,* 1940):

LXXV. Anice
Natura calda e secca in 3°
meglio quella recente avente grossa grana
giovamento. espelle le ventosità provoca il latte ed eccita la libidine
danno. quando stringe il ventre
danno remoto. se consumato intero o tritato

LXXV. Anice
Of a warm and dry nature in 3°
better the recent one having a coarse grain
relief. eliminates flatulence, increases the milk and excites the libido
harm. when it squeezes the stomach
long-term harm. if eaten whole and not minced.

The Arabs gathered and revised the medical knowledge of the Greek and Roman Schools, and anise occupied a place of honor in their therapeutic equipment.

Among the authors of innumerable works, Abu Yusuf Ya'qub ibn Ishaq al-Kindi (about 800–870 A.D.), born at Kufa (now in Iraq), an eclectic philosopher, wrote 24 pieces on medicine. Most of these works have not yet been found, but a medical formulary, known as *Aqrabadhin of Al-Kindi,* has been discovered and has also been translated into English (Levey, 1966). In this formulary, al-Kindi gathered the botanical and pharmacological knowledge of the Islamic, Greek, and Latin ancient masters such as Hippocrates, Pythagoras, Dioscorides, Galen, and so forth. *Aqrabadhin of Al-Kindi* is a list or register of prescriptions. In each formula, the dosage of each single component, the method of preparation, and its therapeutic indications are listed, and each formula ends with the sentence, "It is effective, with God's help." The prescriptions that contain anise are the following (Levey, 1966):

- Electuary of fennel: Prepared from fennel root, celery root, fennel seed, and anise, it was good for clearing obstruction of the liver; increasing urine; treating pain in the sides, the kidneys, and the bladder; and treating fever heat penetrating to the bones
- Drugs for the eyes: Anise, yellow arsenic, verdigris, and burnt copper were pounded, sieved, mixed, and made viscous by warming over a low fire for these prescriptions
- Stomachic of Chinese cinnamon according to al-Kindi: This was a strengthener for the stomach, for digestion of food, and for driving away the rheums, the composition of which includes anise among the 16 components

Ibn Sînâ, who was known as Avicenna (978?–1037) and was born in Afshana, in a village near Boukara, wrote the *Kitab al-Quanum fi at-Tibb* — the laws of medicine — which was known under the pseudo-Latin name of the *Canon of Avicenna*, or the *Antidotary*, in which Greco-Latin medical knowledge mixed with Arabic and Indian medicine. Avicenna added various Asian drugs to the prescription for Theriaca, which was very popular among the Arabs, and used anise as an ingredient in many prescriptions.

Toward the eleventh and twelfth centuries, Ibn Sarabi, better known as Serapione the Young, published the *Liber Serapionis aggregatus in medicinis simplicibus*, in which the medico-pharmacological doctrines of Dioscorides and Galen integrate with Arabic medicine; in this book, the galactogogue, carminative, and diuretic properties of anise are exalted, but it is then added that when the seed is eaten alone, it diminishes the sperm (Ineichen, 1962).

In asthma and pulmonary illnesses, Moshe ben Maymon or Maimonides (1135–1204) counseled the use of drinks and, in severer cases, fumigations of mixtures of vegetable drugs, among which was found anise, which permitted the expulsion of mucous and the cleaning and fortification of the head.

As in other European countries, in Italy anise was included in numerous official preparations. The very famous *Ricettario Fiorentino* (1550), written in Italian and printed in Florence in 1498, which had 10 reprints before 1789, and which served as a model for many Italian and European pharmacopoeiae, contains as many as 49 preparations in which anise is present; among these there is the *Theriaca di Andromacho vecchio secondo Galeno*, enriched with as many as 63 components. For example, we meet the *Theriaca Andromachi senioris ex Galeno* in the *Pharmacopoeia Augustana* of 1734.

In his *Dizionario*, that is, the *Trattato universale delle droghe semplici*, Nicolò Lemery (1766) included the *Anisum:*

> Il seme di Anice ha in se molto olio esaltato e sale volatile. E' cordiale, stomacale, pettorale, carminativo, e digestivo: promove il latte alle balie; ferma i dolori colici. (The seed of anise has in itself a large quantity of exalted oil and volatile salt. It is cordial, stomachic, pectoral, carminative and digestive: it increases milk in wet-nurses; stops colic pains.)

A *Tavola delle infermità* (table of diseases) is added to the dictionary, in which remedies are indicated for every infirmity. According to this table, the anise seeds can cure

> ...la colica ventosa, il cuore debole, la digestione — da aiutare, flati o ventosità, latte — da promuovere, stomaco (the windy colic, the weak heart, the digestion — to help, flatulence or wind, the milk — to promote, the stomach)

In the *Herbario nuovo* (Durante 1684), by Castore Durante, a physician and a Roman citizen, there is a description of the virtues of anise that makes it seem as if it were, in that period, a panacea for all ills.

In the *Catanense Dispensatorium* by Nicolò Catanuto (1666), it is written in Latin, next to *Theriaca Andromaci senioris, ex Galeno, primo de Antidotis*, that

> Flatus dissipat, humores in ventriculo, et aliis visceribus putres absumit, obstructionem tollit, colorem commendat, cordis palpitationem multum opitulatur, mulieribus menstrua mouet, fetus mortus eijcit, et canis rapidi morsum opportune sanat. Quae non datur ante sex menses. (Dissipates flatulence and the bad humor of the stomach, and eliminates what is putrid in the other viscera, liberates obstructions, betters the color, greatly helps palpitations of the heart, facilitates menstruation in women, expels dead fetuses, and cures dog bites rapidly. Not to be used before 6 months [of age].)
>
> *Oleum anisi*, which is listed apart,
>
> Valet ad colicum dolorem, a flatibus, e frigore natum: ventriculi infactionem, cruditatem, acidum rectum, et intestinorum rugitum (Is useful against colics provoked by effluvia and cold: corruption of the stomach, difficulty in digestion, acid belching, and borborygmus in the intestine).

Among the many preparations that contain anise, the *Antidotarium Messanense*, compiled by Cortesi (1629), included *Oleum anisi*, which is useful for

> Ad vertiginem, pectoris angustiam ex capitis defluxu natam, praefocationem, vomitionem, cruditatem stomachi ex flatibus confert. Aque inter cute opitulatur, uteri fluorem album exiccat, oris halitusque faetorem emendat (Dizziness, difficulty in respiration caused by catarrh, vomit, poor digestion caused by effluvia. Reduces edemas, dries up white leaks of the uterus, corrects bad breath and a foul mouth).

At the end of the *Antidotarium*, the *Semina Anisi* are included in the "Catalogus rerum omnium quae asservari debent Messanae a Pharmacopolis in suis officinis ad usum medicum" ("List of all the things Pharmacists must keep in their apothecaries for medical use").

Numerous preparations containing anise are included also in the *Farmacopea Messinese* of P. Alojsio (1815), written in Italian, among which are

- Alcool aromatico di aniso composto: Spiritus carminativi (anise spirit)
- Elettuario di sena composto: Electuarium diatartari (compound senna electuary)
- Elettuario di salsa composto: Electuarium di salsa compositum; electuarium ad gonorrea gallicam Josephi Galeani ph. Gerv. (*Gervasius* 1772) (compound sarsaparilla electuary)
- Elettuario di teriaca: Theriaca andromachi (Theriaca electuary)
- Elettuario teriacale: Theriaca Brug. Farm Gen. (Brugnatelli 1808) (Theriaca according to Brugnatelli)
- Sciroppo di erismo composto: Syrupus de erysmo lobellii (*Reus* 1686) (compound elfdock root syrup)
- Sciroppo di salsa pariglia composto: Syrupus salsaparillae compositus; syrupus magistralis de sarsaparilla; sciroppo del cuoco (*Gervasius,* 1772) (compound sarsaparilla syrup)
- Olio volatile d'Aniso: Oleum essentiale anisi (anise essential oil)

In his book on theoretical and practical pharmacology, J. Friedericius Cartheuser (1756) included both the fruit and the essential oil of anise in various chapters that deal with the most disparate series of therapeutic activity such as the carminative, aperitif, diuretic, diaphoretic, pectoral purifier, emmenagogue, febrifuge, balsamic, etc.

In the *Codex Medicamentarius seu Pharmacopoea Parisiensis* (1748), the steam distillation of the seeds of anise to obtain the essential oil is reported. Again in the eigtheenth century, among *Acido-austera Vegetabilia*, Boerhaave (1772) classifies *Pimpinella* in Sect. 28 n.4; its species is not indicated and its leaves are included in the *Infusum* adstringens and served to prepare a *Vinum medicamentum*. The *anisum* was indicated for acidity ("Ad morbos ex acido spontaneo") and listed among the essential oils, leaves, and *seeds* considered aromatic stimulants.

Among the carminative seeds, De Rochefort (1789) recommended those of various umbelliferae, among which anise belongs; however, De Rochefort thought that it was preferable to use the anise oil, at a dose of 12 to 15 drops in association with mint water, ether, and Hoffman's liquor. The anise fruits, as a macerate in wine (*enolita*), were also recommended as sudorifics, tonics, and stimulants.

In the chapter on the preparation of *Theriaca* in the *Antidotario Romano* (1678) translated by Ippolito Ceccarelli, the "Anisi di Puglia, seme fresco, pieno, grasso, odourifero, non semoloso, al gusto acuto, è dolce" ("anise from Puglia, a fresh seed, full, fat, fragrant, not branny, with a strong taste, it is sweet") is listed among the 76 ingredients for the *Theriaca*. In the "Sommario dell'elettione de' semplici e del modo da tenersi in hauergli à metter'in opera" ("summary of the choice of simples and the way to choose them for use"), anise is also included.

Up to the beginning of the 1900s, anise occupied a place of honor in all the pharmacopoeias, antidotaries, and treatises of pharmacology, *Materia Medica*, and Therapy (Odier, 1813; Alibert, 1823; Edwards and Vavasseur, 1829; Chevallier and Richard, 1830; Semmola, 1853; Cantani, 1887; Binz, 1888; Giacosa, 1901). In a general repertory of practical pharmacy (Dorvault, 1880), anise was considered one of the fundamental medicaments of the doctor's therapeutic equipment; it was used as a fruit, essential oil, aromatic water, spirits, a macerate in the wine, an anisated ammonia liquor, and a balsam of anise oil and sulfur. It also formed part of the composition of the *elettuario lenitivo* and of *cattolicon doppio* (Chevallier and Richard, 1830). Anise was one of the medicaments that by law had to be kept in every apothecary's shop (Semmola, 1853).

In Middle America, an infusion of anise fruits is commonly used as a carminative. In the Yucatan, a tincture, made with equal parts anise fruits and the flowers of *Matricaria courrantiana*, at doses of 20 to 40 drops three to four times at day, is used to eliminate flatulence and colic; a decoction is taken as tonic and galactogogue (Morton, 1981).

Anise is also used for stomach pains, for phlegm and coughs, and for fever by a small ethnic group of Afro-Caribbean origin in Guatemala (Giron et al., 1991).

In a study of 1235 cases of examples of home remedies (remedios *caseros*), Trotter (1981) drew up an analysis to determine the characteristics of the ethnopharmacological resources currently in use in Mexican American communities along the eastern end of the United States–Mexico border: Anise is one of the most commonly encountered remedies. The *yerba aniz* is used to cure *susto* (fright or soul loss), as a mild sedative to relax nervousness, as a laxative to overcome constipation and to treat stomach cramps, insomnia, and mild infections.

In markets of central and northern Mexico and southwestern United States, *hierba anis*, a complex of six species, is found. It is used to treat gastrointestinal ailments and as a condiment, and one of the species is *P. anisum*. The *hierba anis* is used in folk remedies in the form of tea to treat liver obstruction and tuberculosis, menstrual pains and regulation, colic and stomach ailments, colds, and cough (Linares and Bye, 1987). In these countries, however, the same or similar terms indicate different plants that have some common uses: anis, anis en grano, anis verde, and hierba dulce (*P. anisum*, Apiaceae); Aneth odourant, anis, aniseed, fennel, fenoil, and hinojo (*Foeniculum vulgare*, Apiaceae); and anis, anis-anis, anis cimarron, anis de chucho, anis del campo, anisillo, and anisillo de monte (*Tagetes filifolia*, Compositae) (Morton, 1981).

In Morocco, the fundamental natural drugs today in use in medicine represent the heredity of ancestral therapeutic traditions. The particular geographical location of this country, at the crossroads of Europe and Africa, has favored meetings between men and cultures: Phoenicians, Punics, Roman, Jews, Muslims, etc. The richness of the Moroccan pharmacopoeia rises from this cultural background. *P. anisum* is cultivated in Morocco, but the fruits are also imported; the main therapeutic indications of the drug are as an aphrodisiac, antidote poison, and aperitive for aerophagie, and to counter digestive difficulties (Bellakhdar et al., 1991); the fruits also are widely used—generally in infusions of 10 to 15 g per liter as a carminative, or as galactogogue (50 g/L) (Sijelmassi, 1993).

Only a small amount of anise is produced in different parts of India, and the aniseed and its oil for internal requirements are imported from Europe and Southeast Asia. However, the Indian terms *saunf* and *badyan* are used to indicate indifferently the fruits of *P. anisum*, *Foeniculum vulgare*, and *Illicium verum*; this lack of discrimination is probably because of the similar odor and flavor of the three drugs. Aniseed is used for flavoring food and beverages and is considered a stimulant, carminative, and diuretic. Anise oil is an aromatic carminative (Ilyas, 1980; Jain and Jain, 1992).

The particular location of Afghanistan has made it, since antiquity, one of the most important centers in Asia. Conquered by Alexander the Great in 330 B.C., Afghanistan was influenced by Hellenistic culture, followed by the Indian and subsequently the Islamic civilizations. This rich cultural and varied background explains the variety of medical traditions. *P. anisum* is cultivated, and its fruits, according to the information obtained from traditional healers of the different provinces of the country, are used as a spasmolytic and antiemetic (Younos et al., 1987).

The ancient therapeutic indications of anise fruits and of the essential oil obtained from them have remained unchanged with the passing of the centuries, even though they have been stripped of the fame that the people's fancy assigned to this drug. Even today, we can find in the traditional medicine (not only of the countries of the Mediterranean basin but also in the those where the plant was acclimatized and naturalized) that it is principally used for its carminative, stomachic, expectorant, antispasmodic, emmenagogue, and galactogogue activities; at high doses it is also used for its antispastic and antiseptic properties. In Arabic traditional medicine, anise is still used as an analgesic and antiinflammatory agent (Twaij et al., 1987) and in the treatment of epilepsy, tremors, and nervous disorders (Abdul-Ghani et al., 1987).

Apart from being used as a corrective for taste and smell in the pharmaceutical, food, and liquor industries, anise is also used topically for revulsive friction and against parasites.

From examination of the empirical use of anise in traditional medicine, to confirm or clarify different activities, a whole series of chemical, pharmacological, and toxicological studies received a stimulus that has given rise to a whole series of interesting results that justify using of the drug in the ways in which it has been — and is still — employed.

9.2 PHARMACOLOGICAL PROPERTIES

The first studies on the biological activity of anise were carried out in the second half of the last century. They essentially deal with the toxic effect of anise-containing drinks that were the object of numerous studies.

Magnan (1868, 1869a,b) was the first to investigate the biological activity and toxicity of anise oil, the exact composition of which was not known. In a series of studies, Magnan showed that up to a dose of 20 g in a dog weighing 16 kg, anise oil did not cause any epileptic crises; at a dose of 12 g, it did not kill cats. However, the cats experienced "cardiopalmus, tiredness, much thirst and an increase in diuresis" (Strumpf ref. by Cantani, 1887).

Subsequently, Cadeac and Meunier (1889) undertook a comparative study, both on man and on dogs, on the effects of fennel and anise oils, one or the other of which were used indifferently in the composition of absinthe. Cadeac and Meunier concluded that anise oil has a considerable irritant action on humans, determining a significant papular eruption that persists for 15 to 20 days. At therapeutic doses, anise oil causes slight, uncertain, transitory excitation, whereas at high levels it causes seizures that appear suddenly after a relative calm. Whatever the absorption route, anise oil induces sleep and muscular resolution in man and in animals. Cadeac and Meunier conclude that to breathe an anise plant causes sleep, which recalls one of the pieces of information reported by Pliny (Plinii Caii Secundi, 1831): hung on the pillow, it alleviates insomnia to those who smell it.

In addition, anise oil produces abundant salivation, has a slow but painless carminative effect, increases the appetite, facilitates digestion, and increases bile secretion; it has a diuretic activity; it decreases cardiac rate and increases inotropism and slightly reduces arterial blood pressure; and in small doses it modifies neither frequency nor width of respirations, whereas at toxic doses it leads to respiratory arrest.

In view of the wide use that was made of this drug in the preparation of liquors, the studies that followed mainly dealt with the toxicity of anise oil (Daremberg, 1895; Lalou, 1903; Varenne *et al.*, 1903; Varenne, 1904; Kobert, 1906; Lesieur, 1906). In the second half of this century, the studies had a new drive and different objectives because it was possible to separate and assay the components of essential oil, and in particular the two stereoisomers *cis* and *trans* of anethol (Naves et al., 1958; Naves, 1960). The studies thus had to deal with the pharmacokinetic and pharmacodynamic of *trans*-anethole; that is, the principal component of essential oil.

In a monographic memoir, Le Bourhis (1973a) published the results of a series of studies on *trans*-anethol that were conducted by his work group, concerning the metabolism, psychotropic activity and acute and chronic toxicity of the stereoisomer. In Hänsel (1994), we find a wide-ranging monograph on *Pimpinella* sp.

9.3 PHARMACOKINETICS

Meyer and Meyer (1959) reported that no percutaneous absorption of anise oil or anethol occurred within 2 hours through the intact, shaved abdominal skin of the mouse.

Fritsch et al. (1975) used a method of *in situ* perfusion to study the gastrointestinal absorption of four additives, including *trans*-anethol, in the rat. Because of its liposolubility, this compound is largely absorbed from the digestive tract by a mechanism of passive diffusion.

Vignoli et al. (1960b) determined the quantities of *trans*-anethol that is orally administered to a rabbit that can be eliminated by the kidneys, the lungs, and the intestine. The researchers found that the quantity of unchanged anethol eliminated in urine does not exceed 10% of the dose given and that about 10% of this quantity is eliminated in sulpho-conjugated form. In the feces, the unchanged anethol was absent and the exhaled quantity was extremely weak. The authors concluded that it was probable that most of *trans*-anethol administered was destroyed or used by the organism. To confirm this hypothesis, Le Bourhis (1968) examined the kinetics of the disappearance of *trans*-anethol from the organism and its distribution in the different organs and looked for its presence in the emunctories. The author used mice for his research work, and his search for *trans*-anethol in the emunctories was carried out on rabbits and rats. This study has shown that *trans*-anethol was not fixed in the organism (on the contrary, it disappeared quite rapidly, whatever its mode of administration [oral or intravenous route]); that ethanol did not alter the kinetics of *trans*-anethol as, after intravenous introduction, *trans*-anethol was not uniformly spread throughout the organism but was concentrated in the liver, lungs, and brain; that after intake by mouth, absorption was only slight and most of the substance remained in the stomach; and finally, that the emunctories contained only traces of *trans*-anethol. Thus, *trans*-anethol was rapidly metabolized.

Le Bourhis (1973a) then made an overall study, in mice and rats, of the absorption and elimination of *trans*-anethol administered intravenously or orally; with successive doses he followed its disappearance from blood and the elimination, through the different emunctories, of the unmodified drug. In contrast to the work of Vignoli et al. (1960b), the accumulation of *trans*-anethol in the organism was shown to be absolutely transitory, because it is rapidly metabolized and the degradation products are completely eliminated via the kidneys and lungs. A considerable quantity is transiently fixed in the brain, where, however, there are no anethol metabolites; these are instead to be found in considerable quantity in the liver, where the biotransformation of *trans*-anethol occurs.

To confirm the rapid metabolism of *trans*-anethol, and considering the studies of Bray et al. (1953), Le Bourhis (1970) had previously searched for possible catabolites of the substance in the urine of men, rabbits, and rats. In the urine of all three groups tested, Le Bourhis found that 35 to 50% of the dose ingested was present as anisic acid, a metabolite less toxic than *trans*-anethol, and 4 to 5% was present as para-hydroxybenzoic acid, both of which acids were conjugated in variable proportions with glucuronic acid, sulfuric acid, or glycocoll. The metabolites of the remaining 50% of the dose administered are unknown.

Le Bourhis and his coworkers performed a series of studies using ^{14}C-labeled *trans*-anethol on the side-chain or on the nucleus to verify whether the metabolites that had not been found could have been eliminated via the intestine or whether, instead, they had been transformed into simple substances that could not be identified with the means used for analysis: When the side-chain of the molecule was marked, 50% of the ^{14}C in relation to the radioactivity was eliminated via the kidneys, whereas when the molecule was marked in the nucleus, the percentage eliminated via that kidneys was 90%. The air expired by the animals that received *trans*-anethol labeled in the nucleus did not contain radioactive substances; in the air expired by animals that received the molecule marked in the side-chain, $^{14}CO_2$ was eliminated immediately after administration, and its quantity increased progressively up to the twelfth hour and represented on the whole 15% of the dose ingested.

In the feces, the radioactivity emitted did not exceed 1% of the *trans*-anethol administered, whether it was labeled on the nucleus or on the side-chain. The residual radioactivity in animals killed 48 hours after ingestion of the labeled product was very weak.

After ingestion of *trans*-anethol labeled on the nucleus, lower levels of the parent compound were found in rat blood brain and heart compared to the rest of organism, whereas the quantity present in the stomach was higher and remained so for the 24 hours following administration, decreasing rapidly afterward, whereas the ^{14}C fixed itself electively in the liver. When *trans*-anethol was marked on the side-chain, it was more uniformly distributed in the organism even though the radioactivity of the liver and stomach was always higher than that of other organs.

The pathways of anethol metabolism and the influence of dose on the metabolism route have been investigated in rats and mice by Sangster et al. (1984a,b). *Trans*-[*methoxy*^{14}C]-anethol as a single 50 mg/kg dose was given orally to female Wistar albino rats and intraperitoneally to male CD-1 mice. *Trans*-anethol is completely metabolized and eliminated from the body within 72 hours. Both rats and mice metabolize the propenyl side-chain via a similar route, and the variation between the two species is mainly quantitative.

The major routes of elimination were urine and expired air (as $^{14}CO_2$); excretion of ^{14}C in the feces and as volatile compounds in expired air comprised only <2% of the dose. Eleven labeled metabolites were identified in the urine, the majority as glucuronides. In addition, an unlabeled metabolite, 4-hydroxypropylbenzene, was identified in the urine as glucuronide. These metabolites arose from side-chain oxidation, side-chain cleavage, and various conjugations. 4-methoxybenzoic acid was a major metabolite in rat urine and was the most important metabolite in mouse urine. At low doses, more than 50% of the elimination was via expired air as $^{14}CO_2$, whereas at high doses, urine was the major route. A species-specific regioselectivity has been marked in the side-chain oxidation of anethol: In the rat, the preferred route is via epoxidation to the diols, resulting in the urinary elimination of two 1-(4-methoxyphenyl)propane-1,2-diol isomers, making up from 2% to 15% of the dose over the dose range studied. In the mouse ω-oxidation is favored, yielding cinnamyl compounds and 4-methoxybenzoic and 4-methoxyhippuric acids, with the elimination of the latter two compounds making up from 10% to 42% of the dose range studied. The use of very high doses of *trans*-anethol might cause a metabolic overload, thereby distorting the pharmacokinetics of the drug.

By virtue of their electrophilic nature, the yielded epoxides are an important determinant of anethol toxicity. The data obtained in this study cannot be extrapolated to man to assess the risk occurring from the use of anethol at low levels in human foodstuffs. The design of toxicity tests based on metabolic criteria may give more information (Sangster et al., 1984b).

In confirmation of this, Caldwell et al. (1983) have found that in rats, the fate of *trans*-anethol is influenced by dose size. At low doses, O-demethylation leads to $^{14}CO_2$ excretion; as the dose increases, this pathway becomes less important, putting an emphasis on oxidation of the side-chain, and consequently increasing the urinary excretion of *trans*-anethol as diols and p-methoxybenzoylglycine. Later, Caldwell and Sutton (1988) investigated, in human volunteers, the influence of dose size on the metabolic fate of *trans*-anethol. The dose administered approximated the amount of daily intake in the diet from food or beverage (1, 50, and 250 mg). Dose level had no systematic effect on either rate or route of excretion, and the pattern of urinary metabolite was unaffected.

Sangster et al. (1987) investigated the metabolism of [methoxy-^{14}C]-labeled *trans*-anethol and estragole, administered in human volunteers at the doses encountered in the diet (1 mg and 100 µg, respectively); both drugs were readily absorbed from the gastrointestinal tract. The major routes of elimination of ^{14}C were in the urine and in the expired air as $^{14}CO_2$; about 90% of the *trans*-anethol and 65 to 70% of the estragole dose were recovered within 48 hours. Nine ^{14}C metabolites were found in the urine after administration of *trans*-anethol and five after estragole, and all are products caused by oxidative O-demethylation and various oxidations of the side-chain. Of the *trans*-anethol dose, 56% was eliminated as glycine conjugate of 4-methoxyhippuric acid and 3% as two isomers of 1-(4-methoxyphenyl)propane-1,2-diol. The elimination of the 0.2 to 0.4% estragole dose as 1-hydroxyestragole is very important in safety assessment because this metabolite is the proximate carcinogen responsible for the hepatocarcinogenicity of estragole in mice (Drinkwater et al., 1976; Phillips et al., 1981; Miller et al., 1983).

However, it has to be considered that, in a normal diet, man is exposed to very small quantities of this metabolite, in comparison with the high doses of estragole administered to rodents, with the consequent extensive exposure to higher amounts of the carcinogenic metabolite (Sangster et al., 1987).

9.4 PHARMACODYNAMICS

As previously mentioned, most pharmacological activity studies deal with *trans*-anethol. Vignoli et al. (1960a) studied the toxicity and some of the activities of this compound, which up to then had rarely been investigated. The authors found that *trans*-anethol had a slight anthelmintic effect and an antiseptic activity comparable to those of gomenol and eucalyptol; an antispasmodic action on isolated intestine; and adrenolytic and noradrenolytic effects, moderate choleretic action accompanied by a contracting effect on the Oddi sphincter, and light diuretic action in chloralosed dogs.

Aniseed and anise oil have been used since ancient times to treat gastrointestinal disorders and respiratory ailments. Reiter and Brandt (1985) observed that a few studies on the effects of volatile oils and their constituents on isolated intestinal segments have been conducted but that investigation of the oils' action on bronchial or tracheal smooth muscle is scarce; therefore, Reiter and Brandt systematically studied the relaxant effects, on tracheal and on ileal smooth muscle, of volatile oils from 22 plants of eleven different families, including anise oil and some of its constituents.

All of the investigated oils had relaxant effects on tracheal smooth muscle, which effects developed shortly after the addition of the oils to a bath; at a concentration of 60 mg/L, anise oil and fennel oil produced, in some of the muscles, an increase in resting force in concentrations lower than the relaxing ones. Both oils produced, however, complete relaxation at 200 mg/L. Anethol, the main component of anise and fennel oils, could not be regarded as a relaxant; in effect, anethol with an EC_{50} of 10 μmol/L increased the resting force of tracheal muscle, with a maximal increase of 40% at 40 μmol/L. At a concentration of 130 μmol/L, the muscles developed undulations of their resting force at a somewhat reduced level.

Anise and fennel oils did not inhibit the phasic contractions of the electrically stimulated ileal myenteric plexus-longitudinal muscle preparation, but increased the contraction force with an EC_{50} of 6 to 7 mg/L; this positive inotropic effect led to contraction forces of up to 50% above the control. In addition, these oils, in concentrations of 20 to 40 mg/L, led to an increase in resting force of 20 to 40% of the uninfluenced contraction force. A similar effect of anise oil on dog's intestine had already been reported by Muirhread and Gerald (1916).

At a concentration of 130 μmol/L, anethol increased the peak force of the phasic contractions of ileal longitudinal muscle by 30%; at higher concentrations, it reduced peak force and induced, at 400 μmol/L, irregular spontaneous contractions in three out of eight experiments.

An increase in the motor activity of the small intestine of the mouse, caused by fennel oil, had also been observed by Imaseki and Kitabatake (1962). A similar observation of the effect of anethol on rabbit intestine was made much earlier by Stross (1922). Niiho et al. (1977) confirmed a dose-dependent stimulating effect on rabbit stomach motility, *in situ*, of combined stomachic vegetable drugs whose effective ingredient was fennel. The stomach movements, depressed by sodium pentobarbitone, were also stimulated by the combined stomachic. Employing the guinea pig ileum and using acetylcholine and histamine as spasmogens, Forster et al. (1980) demonstrated the antispasmodic activity of the alcoholic extract of fennel.

After all, according to Binz (1888), the term carminative might derive from *Carum*, the antique name for cumin (*Carum carvi*), an Apiaceae having this activity. According to others (ref. by Binz, 1988), it might come from the Latin *carminare* (to card wool), because the gastroenteric tube, when torpid and loaded with gas, stimulated by the essential oils present in these vegetable drugs returns to normal conditions after having expelled the gas.

Therefore, *trans*-anethol is probably responsible for the positive inotropic effect of anise and fennel oils; as these oils are used as carminatives, it seems likely that the symptomatic relief of intestinal disorders can be achieved not only by oils, usually referred to as antispasmodics, which diminish motor activity in the gastrointestinal tract, but also by those oils that stimulate motor activity (Reiter and Brandt, 1985).

Albunquerque et al. (1995) studied, on muscle preparations of the toad and rat, the action of *Croton zehntneri* oil, which is composed mainly of estragole (59%) and anethol (27%) and has a strong and pleasant odor of anise and clove. The extracts of bark and leaves of *Croton zehntneri* are used as sweeteners in food and in drinks, and in folk medicine they are employed as sedatives and for the relief of gastrointestinal disorders.

Albunquerque and colleagues also studied the actions of pure anethol and estragole. In the toad sciatic nerve-gastrocnemius preparation, estragole and anethol reduced the gastrocnemius twitch amplitude elicited by sciatic nerve stimulation.

In the rat phrenic nerve-diaphragm preparation, the response to direct (applied pulses) and indirect (via nerve) stimulation were judged by the amplitude of the muscle contraction. Anethol and estragole markedly inhibited the response to indirect stimulation, but the estragole was more potent. In the case of direct stimulation, anethol did not induce significant alteration of the twitch, and estragole always reduced it.

Anethol and estragole reduced the amplitude of contractions induced by acetylcholine on the toad rectus abdominis muscle, and anethol caused a greater reduction than estragole. Studying the effects of anethol and estragole on the concentration-response curve for the amplitude of acetylcholine-induced contractions, Albunquerque and colleagues observed that larger concentrations of acetylcholine were necessary to achieve a given intensity of effect and that the maximal tension development was depressed.

In addition, caffeine contractions of toad rectus abdominis were enhanced by anethol and estragole, and this enhancement was blocked by procaine. Lowering the temperature to 4°C in the presence of anethol and estragole provoked a rapid cooling contraction that could be blocked by procaine. Each of the substances alone caused contractions, and occurrence of this response was increased by elevation of calcium and other divalent cations that were also effective.

The authors concluded that anethol and estragole may have two sites of action on muscle fibers: one on the postjunctional membrane, acting by blocking neuromuscular transmission, and the other on the sarcoplasmic reticulum, acting by increasing myoplasmic calcium.

Together with his coworkers, Le Bourhis (1973a) examined the activity of the two isomers, *cis-* and *trans-*, of anethol in various experimental models. First, the researchers studied modifications in cerebral electrical activity in the rabbit following intraperitoneal administration of the two isomers. At higher doses of the *trans-* isomer, only slow and wide waves appeared, characteristic of a state of drowsiness; sleeping time was doubled with respect to controls at doses ranging from 300 to 700 mg/kg body weight. The *cis*-anethol did not modify sleeping time.

In the mouse and rat *trans*-anethol, at doses of 25 or 50 mg/kg intraperitoneally, potentiates the hypnotic action of methylphenobarbitone or of chloral hydrate, so that lethargy occurs at doses that alone would have had no effect (Le Bourhis, 1973a).

Seto and Keup (1969), investigating the effect of alkylmethoxybenzene and alkylmethylendioxybenzene present in essential oils, found that, in general, the compounds with the allyl side-chain were more active in potentiating pentobarbitone sleeping time than their propenyl isomers, but that anethol and estragole had a lower activity. The observation that many essential oils were effective in potentiating the pentobarbitone sleeping time, but ineffective compared with ethanol, indicates that the effect on the pentobarbitone sleeping time is the result of interference with the metabolism of pentobarbitone.

Boissier et al. (1969) had previously evidenced the hypnotic and anticonvulsant properties of *trans*-anethol, and in view of the fact that this substance is frequently consumed in the presence of alcohol as an anisated drink, the authors studied the effects resulting from this association on acute toxicity and on hypnotic and antidepressant activities. They concluded that, although each one of these substances had very similar effects on the central nervous system, acting as hypnotics having anticonvulsant properties, the effects of the two associated substances were additive but were not potentiated or degraded significantly.

The anticonvulsant activity of *trans*-anethol with respect to fluorotil, pentylenetetrazole, and electroshock was made manifest at doses equal or over 50 mg/kg; the depressive effect occurred at doses of at least 100 mg/kg, lethargy happened at 250 mg/kg, and death came about at 400 mg/kg (Le Bourhis, 1973a).

Many studies have been carried out to evaluate biological activities of crude drugs that have played an important role in traditional Arabic medicine, and Abul-Ghani et al. (1987) investigated the effects of plants used in the Arabic villages and towns of Palestine to treat epilepsy, tremors, and nervous disorders. Among the plants studied, the aqueous extracts of leaves and stems of *P. anisum*, prepared imitating the methods used by the local community, were not able to completely prevent the onset of convulsions when administered intraperitoneally in female mice 45 minutes before injection of a convulsant dose of picrotoxin, but the extracts significantly delayed the onset of picrotoxin-induced seizures and decreased the mortality rate. The γ-aminobutyric acid content, detected in brain cortical tissue at the onset of picrotoxin-induced convulsion in mice pretreated with anise extract, increased slightly. Therefore, Abul-Ghani et al. concluded that, as was the result for the other plants tested, the anticonvulsivant effect of anise plant was weak and did not justify its use in folk medicine; the *Rosmarinus officinalis* extract was the only plant that showed an effect that justified continuation of the studies.

The anticonvulsant effect of the essential oil of anise, used in the Iranian traditional medicine, is confirmed by Pourgholami et al. (1999): They observed that this essential oil suppressed tonic convulsions induced by pentylenetetrazole or by maximal electroshock in male mice, and it also elevated the threshold of pentylenetetrazole-induced clonic convulsions in mice.

Twaij et al. (1987) carried out a screening of five medicinal plants, used in folk medicine in Iraq in the form of beverages, for analgesic activity. The aqueous extract of the *P. anisum* plant showed a significant analgesic activity against benzoquinone-induced writhing and in thermal tests.

The use for centuries of anise fruits and anise oil in traditional medicines of different countries as an expectorant, in cough mixtures, and in lozenges, prompted a methodological approach to the study of this therapeutic use of anise.

In 1953, Van Dongen and Leusink described a method to study the role of ciliary movements in the air passage of cats and to study the effect of the expectorant drugs on inhibition of ciliary movement resulting from morphine, codeine, and a mixture of the six main opium alkaloids (Opial). Among the drugs assayed were anise oil and anisated ammonia spirit (anise oil 4, ammonia liquida 20, spiritus 76), administered by gavage in an emulsion with gummi arabicum. Two anise oil drops neutralized the inhibition of the ciliary movements caused by the minimal active dose of the opium preparations, and a hypersecretion could be observed. If the anisated ammonia spirit was administered, at doses of 0.25 to 0.5 g/kg, hypersecretion in the air passages was present, but the acceleration of the ciliary movements was scarcely found in these doses, which contained only 4% of anise oil, an amount apparently too small to cause an acceleration of ciliary movement. Higher doses caused vomiting.

Boyd and coworkers (1944, 1946, 1954, 1968, 1970, 1971) have reviewed the effects of various volatile oils on the output and composition of respiratory fluid and the mechanism of their action. Anise oil in doses below 10 mg/kg body weight produced a nonstatistically significant increase in the output of respiratory tract fluid in rabbits anaesthetized with urethane.

In albino rats, anaesthetized with urethane, 0.0015 mL/kg body weight of anise oil increased the output of respiratory tract fluid by a mean of 28.4% (Boyd and MacLachlan, 1944), but anise oil did not affect the chloride concentration or specific gravity of the respiratory tract fluid (Boyd and Pearson, 1946).

Paregoric (Canforated Opium Tincture), in oral doses of from 0.5 to 10 mL/kg body weight, had an expectorant action when assayed in cats, guinea pigs, rabbits, hens, and especially in albino rats anaesthetized with urethane (Boyd and MacLachlan 1944); this effect was more pronounced when the samples of paregoric had aged to a brown color. Some ingredients of paregoric, at an oral dose per kilogram, indicated in parentheses, had an expectorant effect in albino rats: tincture of opium (0.025 mL), camphor (1.5 mg), 60% ethanol (0.5 mL), anise oil (0.0015 mL), and possibly benzoic acid (2.5 mg). However, the expectorant effect of opium tincture was not caused by its morphine content (Boyd, 1954).

The steam inhalation of anise oil on respiratory tract fluid in rabbits did not appreciably affect the volume output until doses of 729 mg/kg were reached. The highest doses produced 20% death and local irritation of the respiratory tract lining, which appeared as congestion after 6 hours and as leucocytic infiltration and destruction of the ciliated mucosa after 24 hours. Specific gravity was generally decreased, and total solids tended to increase, particularly after inhalation of the higher doses; the insoluble mucus did not change significantly (Boyd and Sheppard, 1968). The authors commented that, as anise oil affected respiratory tract fluid only when given in toxic and lethal doses, addition of this oil to steam inhalation probably added little except aroma.

Certain volatile agents, when inhaled in very small amounts, may have a direct bronchomucotropic action on the respiratory tract; when these agents are given orally to laboratory animals, the dose required to produce the same effect is on the order of a thousand times that absorbed by inhalation (Boyd and Pearson, 1946).

In an experiment on urethanized rabbits, an ethanolic solution of anisaldehyde was added to the water vaporizer in a dose corresponding to 1 mL/kg body weight. The endotracheal inhalation increased the volume output and mucus content of the respiratory tract fluid, probably because of the direct stimulation of cells lining the respiratory tract (Boyd and Sheppard, 1970). The anise oil is characterized by a content of anisaldehyde of 0.1 to 3.5% (*Pharmacopée Européenne*, 1997), and that could contribute to the expectorant effect of anise oil.

Boyd and Sheppard (1968, 1970, 1971) were the first to investigate and report a seasonal variation in the mucotropic action of inhalants. The inhalation of anethol in urethanized rabbit did not affect volume but produced a dose-dependent decrease in the specific gravity of respiratory tract fluid in months other than those of autumn; during the autumn, the volume output of respiratory tract fluid is augmented. Anethol, therefore, had a limited bronchomucotropic effect.

Among the plants used for human food, the estrogenic principles of anise were not identified as isoflavone derivatives, but as a compound related to stilbene, the anethol (Liener, 1980). Anise has been used since prehistoric times for its effects on the reproductive system (Albert-Puleo, 1980; Cheeke, 1989). As already mentioned, in Greco-Roman traditional medicine, as in other cultures, anise has been reputed to draw down milk, to child birth make easier, to expel the menstrue, and to increase libido.

The estrogenic activity of products of the vegetable kingdom has long been investigated. Zondek and Bergmann (1938) extended the experiment to a series of natural essential oils, particularly to those obtained from the Apiaceae. The study was carried out on adult castrated mice, infantile noncastrated rats, and adult castrated rats, with fennel oil, anise oil, dill oil, and other essential oils. Fennel and anise oils demonstrated definite estrogenic activities, which are to be ascribed to anethol, but the authors concluded that as fennel oil is more active than anethol, its estrogenic activity cannot be ascribed exclusively to that substance, but rather to another, still-unknown constituent of the natural oil.

Sharaf and Goma (1965) examined the effects of anise oil in ovariectomized mice and rats, using the vaginal smear technique, and in immature rabbits or ovariectomized rats, using the uterine weight method. An estrogenic reaction was obtained in ovariectomized rats, evidenced by cornification of the vaginal cells. The uterine development in the rats treated by anise oil was greater than in untreated controls; moreover, other experiments showed that anise oil blocked the effect of testosterone on the seminal vesicles and that of progesterone on the vagina.

A hypothesis formulated by Albert-Puleo (1980) is that the estrogenic activity of anise might be caused by two dimers, dianethol and photoanethol, resembling the estrogenic agent stilbene and related compounds. This theory was not confirmed, however, because the presence of dianethol, after long conservation of anise oil, has not been proved.

Miething et al. (1990) took into consideration both De Varda's (1891) report of a fluorescent compound — a polymer named *fotoanetolo* — obtained from anethol under exposure to sunlight, which Höring Grälert (1909) identified as dimethoxystilbene and that Steinegger and Hänsel (1972) described as a dimer of anethol, dianethol, formed by autooxidation, whose structure was given as 4-4-dimethoxydiethyl-stilbene, and Hennig-Niklasch's (1959) inability to confirm the formation of any photoanethol. Miething et al. thus made an investigation to clarify whether photoanethol was formed in anise oil under exposure to light and what its structure might be. The study results were that 4-4-dimethoxystilbene is formed in anise oil stored in a white glass bottle for 6 months under normal light conditions. The latter substance can not be found under normal gas chromatographic condition, as it undergoes chemical degradation. An enriched fraction obtained from the stored oil by preparative layer chromatography contained a compound identified as 4-4-dimethoxystilbene, using high-performance liquid chromatography with photodiode-array detection.

This identification is very important because the dimethyl ether of diethylstilbestrol showed a potent oestrogenic activity when administered orally, at levels of 20 or more mg per pound of feed, to domestic fowl and turkey, whereas the dimethylether of dihydroxystilbene at these levels was estrogenically inactive (Japp, 1945).

Further study on the chemical structure both of the natural compound and of the products formed in the stored oil with reference to estrogenic activity would be of interest to clarify beyond doubt this age-old therapeutic use of the drugs.

Anise oil, anethol, and several related compounds inhibit the deposition of glycogen by adrenal cortical extract in the liver of fasted, adrenalectomized rats; the effect is reversible by increased levels of adrenal extract. Anethol also blocks the effect of adrenal cortical hormones in reducing insulin sensitivity of adrenalectomized rats and increasing adrenal hypertrophy of unilaterally adrenalectomized rats. These compounds might compete for an active carbohydrate intermediate formed during glycogenogenesis. Coté et al. (1951) concluded that further investigations are needed on the possibility of a disturbance of hormone metabolism resulting from liver damage.

The antimicrobial activity of essential oils has long been known (Cavel, 1918; Morel and Rochaix, 1928, 1929; Morel et al., 1933; Vignoli et al., 1960a; Vignoli and Morel, 1967), and anise is widely used as an antimicrobial in folk medicine in different countries.

Because of this antimicrobial action, the essential oils are used also as food preservers, and the aromatic waters are frequently used in the pharmaceutical industries as flavors, perfumes, and preservatives. However, anise water does not have a very pronounced inhibitory effect on the growth and survival of *Pseudomonas aeruginosa* (Ibrahim and Ogunmodele, 1991).

In a screening of higher plants seeds and fruits, the acetone extract of anise fruits showed *in vitro* antimicrobial activity, and *Candida albicans* and *Streptomyces venezuelae* were most susceptible (Maruzzella and Freundlich, 1959); anise oil has been reported to possess antimicrobial properties, but terpeneless oil has a less powerful antibacterial activity than the corresponding volatile oil (Maruzzella and Lichtenstein, 1956).

In a screening for antibacterial activity of Argentine folk medicinal plant water extracts against *Salmonella typhi,* Perez and Anesini (1994) found that anise fruit had no activity. Fyfe et al. (1998) described the synergistic inhibition of combination of anise oil with methyl-paraben against *Salmonella enteriditis*, whereas *Listeria monocytogenes* was less sensitive to inhibition by this combination.

Anise oil is examined for antimicrobial activity against a wide range of food-borne Gram-positive and Gram-negative bacteria, yeast, and molds; it showed antimicrobial activity against most of the organisms tested, with the exception of *Flavimonas oryzihabitans* and *Pseudomonas* species (Lachowitz et al., 1998).

To explain the mechanism of action of essential oil on bacterial cell metabolism, Knobloch et al. (1986) examined, using a membrane preparation from purple bacterium *Rhodopseudomonas sphaeroides* (Rhodospirillaceae), the influence of some terpenoids on the primary energy metabolism, which is of fundamental importance to each living cell. Anethol seems to inhibit ATP generation more than the respiratory electron flow, which may cause a more pronounced uncoupling effect. The authors assumed that the essential oil components might interfere with the phospholipid bilayer of biological membranes.

Kubo and Taghigughi (1988) Kubo and Himejima (1991, 1992), Kubo et al. (1993), and Himejima and Kubo (1992) investigated the synergic effect of anethol on antifungal activity, against *Candida albicans* and *Saccharomyces cerevisiae,* of a sesquiterpene dialdehyde, polygodial, isolated from the sprouts of *Polygonum hydropiper* and from various other plant sources (McCallion et al., 1982).

It is known that polygodial increases the permeability of the antibiotics, such as actinomycin D, through the plasma membrane, as its primary mode of action seems to be damage of the plasma membrane structure, but unexpectedly, it did not enhance the antifungal activity of anethol. In contrast, a remarkable increase occurred in the antifungal activity of polygodial when it was combined with a sublethal amount of anethol. This synergic effect seems very important, as the control of systemic infections caused by fungi, especially in patients with impaired host defense mechanisms, is becoming increasingly serious.

The antifungal activity of nagilactones, three natural nor-diterpene dilactones, isolated from various parts of 15 species of the gymnosperm genus *Podocarpus*, also showed only moderate to weak activity against *Candida albicans*, *Saccharomyces cerevisiae*, and *Pityrosporum ovale*, but the activity of the nagilactones was significantly increased by 1/2 MIC of anethol: Anethol enhanced the activity of nagilactone E as much as 128-fold (Kubo et al., 1993).

Reichling et al. (1991) investigated the biological activity of some rare phenylpropanoids of the genus *Pimpinella*. Their results show that compounds with epoxy groups have a negative effect on the growth of plants in herbicide tests *in vitro*; these compounds also have insecticidal and acaricidal activities. However, they are ineffective against the phytopathogenic fungi tested.

The antifungal and antiaflatoxigenic activity of anethol was studied against two strains of *Aspergillus parasiticus*; up to a concentration of 400 mg/mL, when complete inhibition was observed, anethol delayed growth and reduced mycelial weight but showed a stimulative effect on the toxin production of both strains (Karapinar, 1990).

Anise oil has been, and is still, used in folk medicine against parasites (Wichtl, 1989). In 1887, a treatise on *Farmacologia clinica per medici, farmacisti e studenti* (Cantani) reported that anise oil, diluted with alcohol or in the form of ointment applied locally, surely killed lice and cockroaches.

Using pure anise oil was not recommended, especially on delicate parts of the body, because, as a result of its irritant activity, the oil could provoke acute edema and erysipelas with serous exudate. In addition, the remedy was considered much safer than mercurial ointment, which, smeared on children's heads, could cause salivation, tooth decay, general cachexia, and so forth. Küchenmeister counseled using pure anise oil against scabies, and for the same purpose, Abl recommended a mixture of anise and rosemary (references made in Cantani, 1887).

Marcus and Lichtenstein (1979, 1982) studied the toxicity of components of anise oil against houseflies. They found that *trans*-anethol was the major insecticidal agent present in anise oil, with a LD_{50} of 75 µg/fly when topically applied to houseflies. *Trans*-anethol and anisaldehyde increased the toxicity of other insecticides, such as parathion or paraoxon, when applied simultaneously, apparently because of an increased penetration of the insecticide into the insect body and because of retardation of the pesticide's degradation to nontoxic, water-soluble metabolites. *Trans*-anethol, injected intraperitoneally into rats with sublethal doses of [^{14}C] fonofos or [^{14}C] parathion, had no significant effect on the metabolism and excretion of the insecticides, but after administration of *trans*-anethol in the rat diet (1%), the degradation of [^{14}C] parathion by liver microsome was increased. Marcus and Lichtenstein also studied the interaction of anise oil and *trans*-anethol with pentobarbitone; when they were injected simultaneously, Marcus and Lichtenstein found a significantly increased pentobarbitone sleeping time in mice. In general, these results indicate that anise oil and *trans*-anethol interact with drug-metabolizing enzymes, although the interaction did not seem to be of sufficient magnitude or duration to be of practical concern. These natural compounds can compete with pentobarbitone detoxication as alternate substrates for the mixed-function microsomal enzymes, or by blocking the degradation of the pentobarbital by some other mechanism.

This would explain why the test materials had an effect only when administered simultaneously with the pentobarbitone.

Six phenylpropenyl esters were identified and isolated in extracts of fruits from *Pimpinella* species, among which were *P. anisum*, *P. major*, and *P. saxifraga*. Those esters with epoxypropenyl groups, tested for antigermination activity, were active against seeds from several different species (Kleiman et al., 1988).

The liquid extract, tincture, or oil of anise fruits is used as flavoring in pharmaceutical formulations containing bitter drugs with an unpleasant odor, or drugs having a burning or piquant taste; anisated ammonia spirit is added to liquids with alkaline taste (Capra, 1958).

The characteristic flavor of some toothpastes, mouthwashes, and cleaning gums is obtained from a phenolic ether volatile oil from anise (Lewis, 1977). Anise oil may be responsible for lesions at the corners of the mouth that are often confused with cheilitis caused by vitamin A deficiency (Weiner and Bernstein, 1989). Lovema (1983) reported of a case of the sensitivity of mucous membrane and the skin to anise oil.

Trans-anethol, a phenylpropanoid, occurs at a high percentage in anise oil (84 to 93%; *Pharmacopée Européenne*, 1997), and also in oils of other *Apiaceae*; that is, in fennel oil. The *trans*-anethol is the main isomer present in anise oil, and it determines the odor and sweet taste of the oil; it is widely used as a flavoring agent in beverages, candy, and so forth. In addition, in an attempt to find noncaloric sweetening compounds from plants, Hussain et al. (1988, 1990) found during follow-up laboratory studies that *trans*-anethol is nearly 13 times sweeter than sucrose.

Estragole (1-methoxy-4-allylbenzene), another phenyl propanoid, which occurs in a low percentage (0.5 to 6%; *Pharmacopée Européenne*, 1997) in anise oil, is a sweet flavoring substance and is also synthesized for use as a food additive. Because it has been determined that estragole is a hepatocarcinogen in the mouse, Hussain et al. (1988) decided not to use a human taste panel to evaluate the sweetness intensity of this substance relative to sucrose, and taking into consideration the close structural relationship of anethol with estragole, the authors suggested that the assumption of safety of anethol should be reevaluated. In addition, it may be noted that the use of anethol in the pharmaceutical or food industry is considerable and that in Mediterranean countries, both alcoholic and nonalcoholic beverages (pastis, pernod, ouzo, anisette, mistrà, sambuca, boonekamp, goldwasser, anice lattante, etc.) containing aniseed extract or anise oil are widely used.

At any event, the amount of flavors or sweetening agents consumed in medicinals is generally little, so their use as formulation agents is not likely to be responsible for significant adverse toxic effects. In addition, anethol is quite toxic in animals but is considered safe in humans in doses of 2.5 mg/kg (Weiner and Bernstein, 1989).

Williams and Barry (1989) investigated, using excised human skin, the penetration-enhancing activities of anise oil toward the permeation of 5-fluorouracil, as a model of polar penetrant, but anise oil had little activity, causing only a threefold increase.

The effect of a beverage containing anise extract on iron absorption was tested in tied-off intestinal segments of rats (el-Shobaki et al., 1990). The results show that anise enhances the absorption of iron; therefore, el-Shobaki et al. recommended the administration of this beverage to children and also to adults as a preventive agent against iron-deficiency anemia and also recommended using it for the preparation of bioavailable medicinal iron.

9.5 TOXICOLOGY

The toxic effects of anise were known in the past, and more recent initial studies were focused on its acute and chronic toxicity, principally of the essential oil, even though it was widely used for the preparation of alcoholic beverages. It was observed that excessive use of these alcohols provoked drunkenness, tremors, and epileptiform convulsions and caused muscular resolution, analgesia, and sleep (Cadeac and Meunier, 1889). Varenne et al. (1903) found no toxic effect after daily administration of anise fruits for 3 years; they instead observed that an animal treated with anise fruits rapidly recovered from gastroenteritis. The term *anisism* indicated chronic intoxication, which has a similar symptomatology to alcoholism (Garnier, 1961).

Many toxic effects are reported in literature. Anise oil can cause dermatitis (erythema, scaling, and vesiculation). Mitchell and Rook (1979; ref. by Duke, 1986) reported that aniseed caused cheilitis and stomatitis and that anethol used in toothpaste has produced contact sensitivity. In addition, in the book *Narcotic Plants*, Emboden (1972) reported that anise oil contains compounds that can be aminated *in vivo*, resulting in a series of three dangerous hallucinogenic amphetamines.

Anise fruits added to bakery products may produce acute allergic reactions and may often elude detection because of the small quantity that is ingested but that can yet have marked clinical effects (Liener, 1980). In 102 patients with a initial diagnosis of idiopathic anaphylaxis, Stricker et al. (1986) performed skin-prick tests with 79 selected food antigens; in seven subjects, 10 different antigens, including anise fruits, provoked anaphylaxis. All reactions to anise antigen occurred within 4 hours of ingestion, but after the antigen was eliminated from the diet, no patient had another episode of anaphylaxis. It has also been observed that cross-reactivity among the Apiaceae is the cause of the many positive results obtained with anise, carrot, parsley, fennel, and caraway (Wüthrich and Dietschi, 1985).

Contact dermatitis in food handlers from anethol is reported by Garcia-Bravo et al. (1997), and contact allergy to anethol in toothpaste associated with loss of taste is reported by Franks (1998). Jensen-Jarolim et al. (1997) identified in the anti-Bet v1 and antiprofilin allergens, in addition to higher-molecular weight allergenic molecules, the responsibility for type 1 allergy to anise, fennel, coriander.

Taking into consideration the fact that that the *Umbelliferae* (*Apiaceae*) is the family richest in species containing furocoumarins, Ceska et al. (1987) analyzed a few umbellifer fruits using HPLC analysis and an ultrasensitive photobiological assay (Ashwood-Smith et al., 1983; Ceska et al., 1986) based on the inhibition of growth of two strains of *Escherichia coli*; one having normal capacity for repair of DNA damage, and the other being a DNA-repair-deficient mutant having little capacity for repair. Aniseed had no furocoumarins detectable by HPLC analysis, but using the bioassay, it was possible to detect traces of them — approximately <0.005 µg/g of 5-methoxypsoralen and 8-methoxypsoralen.

Caujolle and Meynier (1958) carried out comparative studies on the toxicity of *trans*-anethol and *cis*-anethol, the two isomers that had been separated and characterized by Naves et al. (1958; Naves, 1960), who had also supplied Caujolle and Meynier with the samples to be assayed. Caujolle and Meynier found that *cis*-anethol was 10 to 20 times more toxic than *trans*-anethol. The European Pharmacopoeia (1997) prescribes that the *cis*-anethol content in anise oil must be less than 0.5%.

Taylor et al. (1964) investigated the toxicity of a number of allyl, propenyl, propyl, and related compounds in the rat, using as criteria the acute oral toxicity and the hepatotoxicity in a short-term, high-dose study. All the compounds tested, except piperonal, were depressants. Deaths were delayed with allyl alcohol, the shorter chain allyl esters, propyl butyrate, and most of the aromatic compounds. Anethol was less toxic than the allyl analogue, estragole. Anethol death time was from 4 hours to 4 days, and LD_{50} was 2090 mg/kg; the toxic signs were depression with low doses, coma with higher portions. Estragole death times were from 4 hours to 8 days, and its LD_{50} was 1820 mg/kg; the toxic signs were a marked depression, coma in some rats for 24 hours, rough fur, wet posterior, and porphyrin-like deposit around the eyes. Anisaldehyde, contained in small amounts (0.1 to 3.5%; *Pharmacopée Européenne*, 1997) in the anise oil, was the most toxic substance tested and had death times of less than 24 hours and a LD_{50} of 1510 mg/kg.

The nature of other substituents on the benzene ring influenced the hepatotoxicity of the aromatic allyl, propenyl, and propyl compounds, and hepatotoxicity appeared to be independent of the nature of the three-carbon side-chain. Anethol caused slight hydropic changes in the hepatic cells (male only) when fed at a level of 1% in the diet to rats for 15 weeks, but no hepatic damage was seen on a dose of 0.25% added to the diet for 1 year.

Vignoli et al. (1960a; Vignoli and Morel, 1967) determined that the LD_{50} of *trans*-anethol, administered intraperitoneally in the mice, was 1.63 g/kg; they observed that doses lower than 0.50 g/kg did not affect the behavior of the animals. Starting from a dose of 0.70 g/kg, the animals became drowsy, and with an increase in dose, the drowsiness became irregular and the reflexes slower. From 1 g/kg, Straub's sign became manifest, indicating a catatonic effect. On an increase in the dose, sleep was irregular and not proportional to the dose, and animals died within 24 hours.

In the course of a study on the toxicity of food additives, Le Bourhis and his group (1973a) examined *trans*-anethol. First of all they found that LD_{50}, determined 24 hours after oral administration of this compound, was 3200 ± 300 mg/kg in the rat and 5000 ± 300 mg/kg in the mouse; intraperitoneally, it was 900 ± 45mg/kg and 650 ± 36 mg/kg, respectively. *cis*-Anethol is more toxic than the *trans*-isomer; in fact, the LD_{50} given orally in the mouse was 240 ± 9 mg/kg; intraperitoneally, it was 93 ± 4 mg/kg in the rat and 135 ± 11 mg/kg in the mouse.

In studies on chronic toxicity, Le Bourhis found that rats that ingested food daily containing 0.5% anethol had no toxic effects, whereas among those that ingested food containing 1 or 2% anethol, young rats had a retarded growth, and adult rats suffered a reduction in the adipose layer. This result was linked to a bad absorption of the food and had to be put in relation with the effect of anethol on the glucidic metabolism (Coté et al., 1951). The authors supposed that the different diet, especially as regards the protein content, may have prevented hepatic damage.

In any case, the damage caused by *trans*-anethol was not long-lasting, because as soon as they stopped consuming it, animals put on weight again and were able to reproduce normally. At any rate, the animals could ingest, with their food, doses of anethol that would have been mortal if administered by gavage; this is because anethol is rapidly metabolized in the organism. Le Bourhis (1973b), thus, suggested that the acceptable daily dose of *trans*-anethol should be increased to 25 mg/kg.

Miller et al. (1983) assayed some naturally occurring and synthetic alkenylbenzene derivatives, structurally related to safrole and estragole, for their hepatocarcinogenicity in mice. Estragole and its proximate carcinogenic metabolite, 1-hydroxyestragole — previously shown to induce hepatic tumors when administered to male CD-1 mice only during the preweaning period — also induced hepatic tumor on administration for 12 months in the diet of female CD-1 mice. Anethol was inactive in this assay and was also inactive when administered intraperitoneally during the preweaning period at total doses of up to 9.45 μmol/mouse to male CD-1 or B6C3F mice.

Leclerc et al. (1987) examined the relationship between the type of alcohol consumed (wine, beer, cider, aniseed spirit or *pastis*, fortified wines, whisky, and other spirits) and the precise location of cancer in 2443 French men with cancer in the upper respiratory and digestive tract. Their results indicate that different alcoholic beverages might not produce the same effect; for the *larynx supraglottis* location (47% for epilarynx, 26% for glottis), more drinkers of anise spirits than expected were found. It is to be considered, however, that heavy drinkers almost always drink several beverages, including wine.

In a chronic feeding study carried out by Truhaut et al. (1989) with *trans*-anethol administered in the diet at concentrations of 0, 0.25, 0.5, and 1% for 117 to 121 weeks, to male and female Sprague-Dawley rats, no apparent treatment-related reaction was noted; the only effect was a transient retardation of body-weight gain. Histological examination revealed certain nonneoplastic and neoplastic lesions common in older rats; the incidence of some hepatic lesions was significantly higher in some treated groups than in controls. Truhaut et al. observed altered cell foci (females group 1%), nodular hyperplasia (males group 0.5% and males and females groups 1%), and benign and malignant tumors (females group 1%). The authors stressed that the low incidence of hepatocarcinomas is restricted to a single species and sex and to the highest dose tested (concentration 1%). They added that the changes observed are not thought to be of genetic origin and, consequently, *trans*-anethol does not constitute a significant carcinogenic risk to man.

A study was also carried out by al-Harbi et al. (1995) on the influence of anethol treatment on Ehrlich ascites tumor in the paws of Swiss albino mice. The results showed that anethol reduced the tumor weight and volume and increased the survival time of the Ehrlich ascites tumor-bearing mice; it also caused a significant cytotoxic effect in Ehrlich ascites tumor cells in the paws, reduced the levels of nucleic acid and malondialdheyde, and increased glutathione concentration. The histopathological changes after treatment with anethol were comparable to the standard cytotoxic drug cyclophosphamide. Anethol can be considered an anticarcinogenic, mitodepressive, and non-clastogenic drug.

These results are interesting, and further studies are needed to explore the mode of action and safety of anethol for its possible uses in cancer therapy. Perhaps it might be opportune to reconsider how, in past centuries, anise was used as a remedy for various forms of tumor (Hartwell, 1982).

Anethol has been shown to block inflammation and carcinogenesis, but it is not known how these effects are mediated. The results of a study by Chainy et al. (2000) demonstrating that anethol blocks both early and late cellular responses transduced by tumor necrosis factor may explain its role in the suppression of inflammation and carcinogenesis.

In 1977 Gheershbein found that a subcutaneous injection of anise oil and its components, anethol and estragole, stimulated liver regeneration in partially hepatectomized rats. He observed an increase in size of the hepatic cells, but because of the short period of investigation, morphological findings in the liver were not remarkable even though the doses administered were massive.

In a study (Swanson et al., 1979) couintering mutagenic activity for *Salmonella typhymurium* Ta 1535, TA100, TA 98, estragole and anethol appeared to have very weak activity; the mutagenicity of the 1-hydroxy metabolites was higher, but it was still low as compared with the parent compounds.

In a primary screening test *in vitro* of food additives used in Japan, anise (fruits, oil) has proved negative in reverse mutation assays in *Salmonella typhimurium* (Ames test), DNA-repair with *Bacillus subtilis*, and chromosomal aberration tests with mammalian cells in culture (Ishidate et al., 1984).

Anise and cumin seed pyrolyzed without oil were tested for their genotoxic effect on somatic and germ cells of Swiss mouse; they showed a moderate frequency of chromosomal aberrations (aberrant methaphase), development of micronuclei in polychromatic erythrocytes, and sperm head abnormalities (Balachandran et al., 1991).

The administration of anethol with coffee enhanced the antigenotoxic effect of coffee against cyclophosphamide and other genotoxins in the mouse bone narrow micronucleus test (Abraham, 1996).

The Committee of FAO-OMS, experts on food additives, have repeatedly examined the issue of the evaluation of flavor innocuity, about which numerous factors must be considered. *Trans*-anethol was appraised for the first time at the eleventh meeting of the committee, which had attributed to the substance an acceptable daily dose (DJA), with reservations, of 0 to 1.25 mg/kg body weight. This dose was successively reexamined at the twenty-third meeting, and while waiting for the results of a long-term study, the DJA was raised to 0 to 2.5 mg/kg body weight. A new reexamination during the 31st and 33rd meetings caused the committee to delay the temporary DJA until 1988, but reducing it to 0 to 1.2 mg/kg body weight at the 33rd meeting. At the 37th meeting, the committee examined the results of 3 years of histological investigations on rat liver in the course of a long-term study: at a dose of 10 mg/kg in food, the researchers observed an increase in the incidence of adenomas and hepatocellular carcinomas in the female, but this did not happen at at lower doses; in males, a slight increase in the incidence of hepatocellular adenomas, but not carcinomas, was noticed. In addition, for all doses assayed, both in male and female rats, an increase in proliferative nonneoplastic lesions of the liver was found. The results of the two studies on genotoxicity were negative.

Not having sufficient data to evaluate the significance of the malignant tumors of the liver in female rats, the committee concluded that further study on the metabolism was necessary, and in particular on the pharmacokinetics in mice, rats, and human. Moreover, given the positive results obtained using *in vitro* assays of the mutation of bacteric genes, the committee deemed it desirable to carry out both studies on chromosomic aberrations and *in vitro* assays on genetic mutations in mammalian cells.

Subsequently, the necessity for an epidemiological study on the effects of the consumption of food rich in *trans*-anethol on reproduction (teratogenicity) will be examined. In the meantime, the DJA has been further reduced to 0 to 0.6 mg/kg body weight.

As during the thirty-ninth meeting, held in 1992, the data available were not considered sufficient to attribute a definite DJA to *trans*-anethol, the committee extended the temporary DJA (0 to 0.6 mg/kg body weight) up to 1997. As for estragole, found in small quantities in anise oil, because it has been demonstrated that both the compound and its metabolites are mutagenic in bacterial systems and cause the appearance of hepatoma in a sensible strain of mice, the committee did not set a DJA, and the substance is not included among acceptable additives anymore.

Newberne et al. (1999) published a series of safety evaluations performed by the Expert Panel of the Flavour and Extract Manufacturers' Association; this panel initiated, in 1993, a program to reevaluate the safety of more than 1700 GRAS flavoring substances under conditions of intended use. The production of a hepatotoxic metabolite, anethol epoxide, is used to observe pathological changes in different species and different sexes of laboratory rodents in chronic and subchronic studies. At low doses, *trans*-anethol is efficiently detoxicated in rodents and humans by O-demethylation and omega-oxidation, respectively, and epoxidation is a minor pathway. At high doses, there is an increased epoxidation associated with a more pronounced hepatotoxicity, but not genotoxicity, in females compared to that in male rats. The hepatocarcinogenic effects in the female rats are secondary to hepato-toxicity caused by continuous exposure to high hepatocellular concentrations of anethol epoxide.

Because *trans*-anethol undergoes efficient metabolic detoxication in humans at the low levels of exposure derived from alimentary use, the neoplastic effects in rats associated with dose-dependent hepatotoxicity are not indicative of any significant risk to human health from the use of *trans*-anethol as a flavoring substance.

9.6 THERAPEUTIC INDICATION OF *P. ANISUM*

Anise fruits and anise oil are mainly used as a mild mucolitic expectorant and carminative. They are indicated specifically in flatulent colic, in bronchial catarrh, and in spasmodic tracheitis with a persistent cough. Anise extract or tincture, anise oil, and *trans*-anethol are also widely used as flavoring in many pharmaceutical formulations or in preparations for oral hygiene.

We find anise fruits or oil in numerous galenical preparations present up to the present day in Pharmacopoeias of different countries, as camphorated opium tincture (Paregoric), anisated ammonia spirit, *Species pectorales*, and *Species purgativae* (i.e., the compound senna powder or the Senna infusion with manna as adjuvant to prevent tormina from cathartics). Anise fruits or oils are also present in many multiingredient pharmaceutical speciality carminatives, digestives, laxatives, diuretics, and depuratives and in cough mixtures and lozenges.

The dose for powdered anise fruits is 0.25 g in cachets or capsules, one after meals; the dose for the 2% infusion is 200 mL after meals. A tablespoon of the infusion is given for dyspepsia many times a day; as a mucolytic, a cup should be taken in the morning or in the evening; and to the unweaned, a dessert spoon of infusion can be put in the bottle. In flatulent colic, anise is also associated with peppermint.

The dose for concentrated anise water dose 0.3 to 1 mL; it is used as a flavored vehicle. The dose for anise spirit is 0.3 to 1.2 mL, as a flavoring agent sometimes used as a stomachic and carminative. Anise tincture, always taken after meals, is used at an adult dose of 20 drops; it is often employed in pediatrics at doses of 5 to 10 drops. Anise essential oil or tincture have to be administered on a lump of sugar or be diluted to avoid mucosa irritation.

Anise oil as carminative may be given in a dose of 0.1 mL. It is a common ingredient of cough preparation; in bronchial catarrh, an association of anise oil (4 g) and sulfur (1 g) (balm of anisated sulfur) is used at a dose from 6 to 10 drops in a potion. Anisated ammonia spirit, an ammoniacal, alcoholic solution containing 3% of anethol by volume, is used orally at a dose of 10 to 20 drops and 1 to 5 g daily—it acts as a reflex stimulant if inhaled. Canforated opium tincture, containing 0.3% anise oil, is used in the treatment of cough or as an intestinal sedative (costive) at doses from 2 to 10 mL; the average dose is 4 mL.

For topical use, liniments or ointments containing 5 to 10% anise oil are employed for external use as a revulsive and against parasites; anise oil in combination with sassafras oil (1% of each) is used in an ointment base for scabies.

In folk medicine, anise fruit is used as emmenagogue, galactogogue, and aphrodisiac.

9.7 OTHER *PIMPINELLA* SPECIES

With respect to *P. anisum,* there is much less information and research work on the other species belonging to the genus *Pimpinella* (Apiaceae). Among these, *P. saxifraga L.* and *P. major* (L.) Hudson (*P. magna* L.) still enjoy a certain reputation in traditional medicine. For other species of *Pimpinella,* use in traditional medicine is limited and only a few — insufficient — scientific studies are known.

It is known that *P. alpina* it is used in Java as an aphrodisiac, for its great strengthening qualities, and also as a diuretic (Lewis, 1977). The decoction of the aerial part of *P. epibracteata* Bak. is used empirically in Madagascar as a febrifuge to treat this symptom of malaria (Rasoanaivo et al., 1992). An essential oil, with a high content of azulene, was obtained from *P. nigra* Willd. root (Gawlowska, 1964, 1967).

The essential oil of the fruits of *P. serbica* Benth. et Hook was investigated and was found to contain monoterpenes, sesquiterpenes (β-caryophyllene and its isomer α-lupulene), and two phenylpropan derivatives (dillapiole and nothoapiole); these two latter compounds and β-caryophyllene are the main components of the essential oil (Ivanic et al., 1983).

P. diversifolia DC. is a plant reputed in the Indian folk medicine of its places of origin (Afghanistan, Himalayas, India, Japan, and Pakistan) to possess antidiabetic activity (Rahman and Zaman, 1989); the fruits' (seeds) ethanolic extract, tested for fungitoxicity against *Aspergillus niger* and *Absidia ramosa,* showed the greatest activity, completely inhibiting the mycelial growth, but further investigation needs to prove its value in therapy (Pandey et al., 1983). Bottini et al. (1986) identified the major components of the root essential oil and reported that the extract of the whole plant possesses spermicidal activity in rat semen (Setty et al., 1977, ref. by Bottini et al., 1986).

P. major and *P. saxifraga* are two related species, each very variable in height, leaf-dissection, and flower color. This variation justifies their taxonomic treatment as one or two species, each with several or without subspecies; *P. nigra* Willd. is considered to be a subspecies of *P. saxifraga* (*P. saxifraga* ssp. *nigra* Willd.). Further investigation is thus necessary on these various taxa, as at present, information is conflicting and inadequate.

P. major L. (Hudson) is distributed mostly in Europe, except in the extreme north and south (Tutin, 1968). The earliest information on this plant goes back the sixteenth century, a period in which so many and such powerful virtues were attributed to it, that a whole volume could have been filled about it (Fournier, 1947). It was not only considered to be a remedy against all ills, including the plague and cholera, but it was also reputed to have magical properties.

P. saxifraga L. is distributed in most of Europe, except the extreme south and most of the islands, as far as Asia. In the Mediterranean region it is used as a tonic to improve well-being and mental harmony (Lewis, 1977).

Stahl and Herting (1976) have found that the roots of *P. major, P. peregrina,* and *P. saxifraga* do not contain any of the furocumarins; that the main components of the roots and fruits of *P. major* and *P. saxifraga* are isoeugenol epoxy tiglic esters, isoeugenol 2-methyl butanoic ester, and 4-(1,2epoxypropyl)-phenyl tiglic ester; and that the fruits of *P. major* contain also β-bisabolene and larger amounts of bisabolangelone.

Phenylpropanoids of *Pimpinella* species root were reexamined by Martin et al. (1985), and a structure of a fatty acid ester of a phenylpropanoid was elucidated.

Kubeczka et al. (1986) investigated the chemical volatile constituents of the main species of *Pimpinella* that might be useful for a taxonomic differentiation of the different species belonging to this genus. They found that *P. major* essential oil is characterized by the phenylpropanoids trans-epoxy-pseudoisoeugenyl tiglate (approximately 56% of the oil) and besides by pregeijerene (approximately 10%), a neosesquiterpene; the analogue phenylpropanoid-2-methyl butyrate has been detected in traces. The main constituent of *P. saxifraga* root oil is the epoxy-pseudoisoeugenyl 2-methylbutyrate (approximately 46%), whereas the analogue tiglate can be found in traces; the root oil also contains progeijerene (approximately 9%). Oil of *P. nigra* contains more than 70% terpene hydrocarbons and only small amounts of phenylpropanoids; 1,4-dimethylazulene (approximately 16%) is responsible for the dark blue color of the oil. These results have been confirmed by Bohn et al. (1989) for *P. major.*

The phototoxic and photomutagenic potency of *P. major* root, containing bifunctional furocumarins as isopimpinellin, has been investigated, using an arginine-requiring mutant of *Chlamydomonas reinhardii*. The extracts of this root, containing a bifunctional furocumarin (isopimpinellin), exhibited a weak phototoxic activity and a lower mutagenic potency than the *Angelica* root and *Imperatoria* rhizome (Schimmer et al., 1980). The aqueous extract of *P. major* roots was investigated in antibacterial screening, and it exhibited a weak activity.

Isopimpinellin is toxic to *Schistosoma*-carrying snails and also to fish, and it tales tubercolostatic actions against *Micobacterium tuberculosis* (Harborne and Baxter, 1996).

In *The Encyclopedia of Pure Materia Medica*, Allen (1877) supplied a complete and accurate record of the pharmacological and toxicological symptoms recorded in controlled experiments made on healthy individuals, observed in the sick after administration of the drug, or observed after poisonous doses. Among the vegetable drugs, we found much information about *P. saxifraga*, mainly originating from Schelling (ref. by Allen, 1877), who recorded the symptoms observed on chewing the fresh root or taking two drops of root tincture, and from Berridge (1873, ref. by Allen, 1877), who had taken 10 drops of the first homeopathic dilution of tincture. Some symptoms are an hypersecretion of saliva during and after chewing; heaviness of head and inclination to sleep; pain in the occiput while reading; on walking, oppression of breath; weary pain in limbs; while writing, stinting in last joint of right ring finger; and in addition to heat and rush of blood to head, vertigo, sweat, and acid eructations. Many other general symptoms, both localized and not definitely localized, are described.

In his *Dictionary of Pratical Materia Medica*, which is a schematized list of the symptoms of remedies recorded in homeopatic literature, in the description of *P. saxifraga* root, Clarke (1900) referred to similar information to that already reported by Allen.

The pimpernel root name includes the drug originating from both *P. major* and *P. saxifraga*, which have the same therapeutic indications: the infusion or decoction at 3 to 4% at a dose of 3 cups a day, before or 3 hours after meals, used as a cough sedative, mild expectorant, and broncho-secretolytic, and used in affections of the upper respiratory tract (hoarseness, pharyngitis, tracheites, and angina); as infusions or tinctures the roots are used as a gargle in phlogosis of the oro-pharyngeal cavity. The alcoholic extracts can be found in some products for oral hygiene. In traditional medicine they are also used as a stomachic, as tonic-astringents in pyrosis and in stomach cramps, and also as a diuretic in calculosis and in gravel (Wichtl, 1989).

Chun-shu Yang, Professor in the Beijng University of Chinese medicine and Pharmacology, gave me the following information about some therapeutic uses of *Pimpinella* species in various districts of China:

P. arguta Diels: Distributed in Shaanxi, Gansu, Henan, Hubei, Guizhou. The root and entire herb are used as drug for invigorating blood circulation and eliminating blood stasis.

P. brachycarpa (Kom.) Nakai: Grows in Jilin, Liaoning, Hebei, Guizhou. The entire herb is used to relieve influenzal stomach ache.

P. candolleana Wight et Arn.: Harvested in Guangxi, Sichuan, Guizhou, Yunnan. The root and entire herb are used as an antirheumatic and as a drug for invigorating blood circulation, as well as for the treatment of stomachache, rheumatic or rheumatoid arthritis, trauma, and scrofula.

P. caudata (Franch.) Wolff: Cultivated in Hubei, Sichuan, Yunnan. The root is used to treat lumbago caused by rheumatism.

P. coriacea (Franch) de Boiss.: Distributed in Guangxi, Sichuan, Guizhou, Yunnan. The root and entire herb are used as an antirheumatic.

P. diversifolia DC.: Distributed in China, middle-south China, and Shaanxi, Gansu. The root and entire herb are used to treat cold, dysentery, jaundice, and injury (by external use).

P. rhomboides Diels: Harvested in Shaanxi, Gansu, Henan, Hubei, Sichuan. The root is used to promote blood circulation to induce the substance of swelling.

P. thellungiana Wolf: cultivated in Shaanxi, Gansu, Henan, Hubei, Sichuan. The root and entire herb are used as drug for dispelling internal cold and for the treatment of papitation, shortness of breath, and cough.

References

Abdul-Ghani, A.-S., El-Lati, S.G., Sacaan, A.I., Suleiman, M.S., and Amin, R.M., Anticonvulsivant effects of some Arab medicinal plants, *Int. J. Crude Drug Res.*, 25, 39–43, 1987.

Abraham, S.K., Anti-genotoxic effects in mice after the interaction between coffee and dietary constituents, *Food Chem. Toxicol.*, 34, 15–20, 1996.

Albert-Puleo, M., Fennel and anise as estrogenic agents, *J. Ethnopharmacol.*, 2, 337–344, 1980.

Albunquerque, A.A.C., Sorenson, A.L., and Leal-Cardoso, J.H., Effects of essential oil of *Croton zehntneri*, and of anethole and estragole on skeletal muscles, *J. Ethopharmacol.*, 49, 41–49, 1995.

al-Harbi, M.M., Qureshi, S., Raza, M., Ahmed, M.M., Giangreco, A.B., and Shah, A.H., Influence of anethole treatment on the tumour induced by Ehrlich ascites carcinoma cells in paw of Swiss albino mice, *Eur. J. Cancer. Prev.*, 4, 307–318, 1995.

Alibert, J.L., *Nuovi Elementi di Terapeutica e di Materia Medica*, Vol. 3, 4th ed., Marotta, Naples, 1823.

Allen, T.F., *The Encyclopedia of Pure Materia Medica*, Vol. 7, Gregg Press, Vaduz Liechtenstein, 1877.

Alojsio, P., *Farmacopea Messinese*, Fiumara e Nobolo, Messina, 1815.

Antidotario Romano Latino, e Volgare, I. Ceccarelli, Trans., Valuasense, Venice, 1678.

Ashwood-Smith, M.J., Poulton, G.A., Ceska, O., Liu, M., and Furniss, E., An ultrasensitive bioassay for the detection of furocoumarins and other photosensitizing molecules, *Phytochem. Photobiol.*, 38, 113–118, 1983.

Balachandran, B., Narender Sivaswamy, S., and Sivaramakrishnan, V.M., Genotoxic effects of some foods and food components in Swiss mice, *Indian J. Med. Res.*, B94, 378–383, 1991.

Bellakhdar, J., Claisse, R., Fleurentin, J., and Younos, C., Repertory of standard herbal drugs in the Moroccan pharmacopoea, *J. Ethnopharmacol.*, 35, 123–143, 1991.

Binz, C., *Lezioni di Farmacologia Sperimentale*, Jovene, Naples, 1888.

Boerhaave, H., *Libellus de Materia Medica et Remediorum Formulas quae Serviunt Aphorismis de Cognoscendis et Curandi Morbis*, 6th ed., Floteront, Nicaeae, 1772.

Bohn, I., Kubeczka, K.-H., and Schultze, W., The essential root oil of *Pimpinella major*, *Planta Med.*, 55, 489–490, 1989.

Boissier, J.-R., Simon, P., and Le Bourhis, B., Action psychotrope du trans-anethole et de l'éthanol administrés conjointement à la souris, *Ann. Nutr. Alim.*, 23, 215–222, 1969.

Bottini, A.T., Dev, V., Garfagnoli, D.J., Mathela, C.S., Melkani, A.B., Miller, A.A., and Sturm, N.S., Oxiranylphenyl esters from *Pimpinella diversifolia*, *Phytochemistry*, 25, 207–211, 1986.

Boyd, E.M., Expectorants and respiratory tract fluid, *Pharmacol. Rev.*, 6, 521–542, 1954.

Boyd, E.M. and Hicks, R.N., Oil of anise, anethole and respiratory tract fluid, unpublished data.

Boyd, E.M. and MacLachlan, M.L., The expectorant action of paregoric, *Can. Med. Assoc. J.*, 50, 338–344, 1944.

Boyd, E.M. and Pearson, G.L., On the expectorant action of volatile oils, *Am. J. Med. Sci.*, 211, 602–610, 1946.

Boyd, E.M. and Sheppard, E.P., The effect of steam inhalation of volatile oils on the output and composition of respiratory tract fluid, *J. Pharmacol. Exp. Ther.*, 163, 250–256, 1968.

Boyd, E.M. and Sheppard, E.P., Inhaled anisaldehyde and respiratory tract fluid, *Pharmacology*, 3, 345–352, 1970.

Boyd, E.M. and Sheppard, E.P., An autumn-enhanced mucotropic action of inhaled terpenes and related volatile agents, *Pharmacology*, 6, 65–80, 1971.

Boyd, E.M., Perry, W.F., and Stevens, M.E.T., The effect of damage to the tracheal mucosa upon the drainage of respiratory tract fluid, *Am. J. Physiol.*, 140, 467–473, 1944.

Bray, H.G., James S.P., Thorpe W.V., and Wasdell, M.R., The Methabolism of ether in the rabbit. I. Anisole and diphenyl ether. *Biochem. J.*, 54, 547–551, 1953.

Brugnatelli, *Farmacopea Generale*, Naples, 1808.

Cadeac, M.M. and Meunier, A., Action comparative de l'essence de fenouil et de l'essence d'anis, *Lyon Med.*, 62, 336–338, 1889.

Caldwell, J. and Sutton, J.D., Influence of dose size on the disposition of *trans*-[*methoxy*–^{14}C] anethole in humans volunteers, *Food Chem. Toxiol.*, 26, 87–91, 1988.

Caldwell, J., Farmer, P.B., Sangster S.A., and Smith, R.L., Dose-dependent disposition of *trans*-anethole in the rat, *Br. J. Pharmacol.*, 79, 212P, 1983.

Cantani, A., *Manuale di Farmacologia Clinica (Materia Medica e Terapeutica)*, 2nd ed., Vallardi, Milan, 1887.

Capra, C., I correttivi dei medicamenti, *Il Farmaco, Ed. Prat.*, 13, 499–524, 1958.

Cartheuser, J.F., *Pharmacologia Theoretico-Pratica-Rationi et Experientiae Superstructa*, Deregni, Venetiis, 1756.

Catanuto, N., *Catanense Dispensatorium sive Antidotarium*, Catanae, 1666, p. 64, 281–282.

Caujolle, F. and Meynier, D., Toxicité de l'estragole ed des anétholes (*cis* et *trans*), *C.R. Acad. Sci.*, 246, 1465–1468, 1958.

Cavel, L., Sur la value antiseptique de quelques huiles essentielles, *C.R. Acad. Sci.*, 166, 827–829, 1918.

Ceska, O., Chaudhary, S.K., Warrington, P.J., and Ashwood–Smith, M.J., Furocoumarins in the cultivated carrot, *Daucus carota, Phytochemistry*, 25, 81–83, 1986.

Ceska, O., Chaudhary, S.K., Warrington, P.J., and Ashwood-Smith, M.J., Photoactive furocoumarins in fruits of some Umbellifers, *Phytochemistry*, 26, 165–169, 1987.

Chainy, G.B., Manna, S.K., Chaturvedi, M.M., and Aggarwal, B.B., Anethole blocks both early and late cellular responses transduced by tumor necrosis factor: effect on NF-kappaB, AP-1, JNK, MAPKK and apoptosis, *Oncogene*, 19, 2943–2950, 2000.

Cheeke, P.R., *Toxicant of Plant Origin. Vol. IV Phenolics*, CRC Press, Boca Raton, FL, 1989.

Chevallier, A. and Richard, A., *Dizionario delle Droghe Semplici e Composte*, Vol. 1, Tasso, Venice, 1830, p. 200–201, 255.

Clarke, J.H., *A Dictionary of Practical materia Medica,* Vol. 3, Health Science Press, Rustington, Sussex, United Kingdom, 1900.

Codex Medicamentarius seu Pharmacopoea Parisiensis, Cavelier, Paris, 1748, p. 130.

Comité mixte FAO/OMS d'experts des additifs alimentaires, 11th rapport. *Normes d'identité et de pureté et évaluation toxicologique des additifs alimentaires: divers aromatisants et édulcorants non nutritifs.* Série Rapport techniques 383, OMS, Geneva, 1968.

Comité mixte FAO/OMS d'experts des additifs alimentaires, 23rd rapport. *Evaluation de certains additifs alimentaires.* Série Rapport techniques 648, OMS, Geneva, 1980.

Comité mixte FAO/OMS d'experts des additifs alimentaires, 31st rapport. *Evaluation de certains additifs alimentaires et contaminants.* Série Rapport techniques 759, OMS, Geneva, 1987.

Comité mixte FAO/OMS d'experts des additifs alimentaires, 33rd rapport. *Evaluation de certains additifs alimentaires et contaminants.* Série Rapport techniques 776, OMS, Geneva, 1989.

Comité mixte FAO/OMS d'experts des additifs alimentaires, 37th rapport. *Evaluation de certains additifs alimentaires et contaminants.* Série Rapport techniques 806, OMS, Geneva, 1991.

Comité mixte FAO/OMS d'experts des additifs alimentaires, 39th rapport. *Evaluation de certains additifs alimentaires et de certains produits toxiques naturels.* Série Rapport techniques 828, OMS, Geneva, 1992.

Cortesi, I.B., *Pharmacopoeia seu Antidotarium Messanense, in quo tum sunplicia tum composita Medicamenta, usu recepta accurate examinatur*, Brea, Messanae, 1629.

Coté, L., Hughes, J., Oleson, J.J., Williams J.H. (1951) Antagonism of anethole and adrenal extract on glycogen formation in adrenalectomized rats. *Arch. Biochem.*, 34, 352–359.

Daremberg, G., La toxicité des boissons alcooliques mesurée à l'aide d'injections intraveineuses chez le lapin, *Arch. Med. Exp.*, 7, 719–752, 1895.

De Rochefort, D., *Cours Élémentaire de Matière Médicale Suivi d'un Prècis de l'Art de Formuler*, Vol. 2, Méquignon, Paris, 1789.

De Varda, G., Azione della luce sull'anetolo, *Gazz. Chim. Ital.*, 21, 183–188, 1891.

Dorvault, F.-L.-M., *L'Officine ou Répertoire Général de Pharmacie Pratique*, 10th ed., Asselin, Paris, 1880.

Drinkwater, N.R., Miller, E.C., Miller, J.A., Pitot, H.C., Hepatocarcinogenicity of estragole (1-allyl-4-methoxybenzene) and 1-hydroxyestragole in the mouse and mutagenicity of 1-acetoxyestragole in bacteria, *J. Natl. Cancer Inst.*, 57, 1323–1331, 1976.

Duke, J.A., *Handbook of Medicinal Herbs*, 3rd ed., CRC Press, Boca Raton, FL, 1986.

Durante, C., *Herbario nuovo*, Bonfadino, Rome, 1685, p. 31–32.

Edwards, H.M. and Vavasseur, P., *Manuale di Materia Medica ossia Breve Descrizione de' Medicamenti*, Vol. 1, Minerva, Naples, 1829.

el-Shobaki, F.A., Saleh, Z.A., and Saleh, N., The effect of some beverage extracts on intestinal iron absorption, *Ernahrungswiss*, 29, 264–269, 1990.

Emboden, W., *Narcotic Plants*, Studio Vista, London, 1972.

Flückiger, F.-A. and Hanbury, D., *Histoire des Drogues d'Origine Végétale*, Vol. 1, Doin, Paris, 1878.

Forster, H.B., Niklas, H., and Lutz, S., Antispasmodic effects of some medicinal plants, *Planta Med.*, 40, 309–319, 1980.

Fournier, P., *Le Livre des Plants Médicinales et Vénéneuses de France*, Lechevalier, Paris, 1947.

Franks, A., Contact allergy to anethole in toothpaste associated with loss of taste, *Contact Dermatitis*, 38, 354–355, 1998.

Fritsch, P., De Saint Blanquat, G., and Derache, R., Absorption gastro-intestinale, chez le rat, de l'anisole, du *trans*-anethole, du butylhydroxyanisole et du safrole, *Food Cosmet. Toxicol.*, 13, 359–363, 1975.

Fyfe, L., Armstrong, F., and Stewart, J., Inhibition of *Listeria-monocytogenes* and *Salmonella-enteriditis* by combinations of plant oils and derivatives of benzoic-acid. The development of synergistic antimicrobial combinations, *Int. J. Antimicrobial Agents*, 9, 195–199, 1998.

Galenus, C., *Quinta Classis Eam Medicinae Partem Quae Ad Pharmaciam Spectat Exponens, Simplicium Medicamentorum*, Iuntas, Venetiis, 1550.

Garcia-Bravo, B., Perez Bernal, A., Garcia-Hernandez, M.J., and Camacho, F., Occupational contact dermatitis from anethole in food handlers, *Contact Dermatitis*, 37, 38, 1997.

Garnier, G., Bézanger-Beauquesne, L., and Debraux, G., *Ressources Médicinales de la Flore Française*, Vol. 2, Vigot Frères, Paris, 1961.

Gawlowska, M., *Pimpinella nigra* Willd. in Poland, *Dissertationes Pharm.*, 16, 541–545, 1964.

Gawlowska, M., *Pimpinella nigra* Willd. in Poland. Part II. Azulene in the essential oil obtained from Radix *Pimpinellae nigrae*, *Dissertationes Pharm.*, 19, 181–186, 1967.

Gervasius Redivivus seu Nicolai Gervasii Dispensatorium Panormitanum, opera et studio Augustini Gervasi, Panormi, 1772.

Gheershbein, L.L., Regeneration of rat liver in the presence of essential oils and their components, *Food Cosmet. Toxicol.*, 15, 173–181, 1977.

Giacosa, P., *Trattato di Materia Medica, Farmacologia e Tossicologia*, 2nd ed., Bocca, Turin, 1901.

Giron, L.M., Freire, V., Alonzo, A., and Caceres, A., Ethnobotanical survey of the medicinal flora used by the Caribs of Guatemala, *J. Ethnopharmacol.*, 34, 173–187, 1991.

Gunther, R.T., *The Greek Herbal of Dioscorides*, Hafner, London, 1968.

Hänsel, R., Keller, K., Rimpler, H., and Schneider, G., Eds., *Hagers Handbuch der Pharmazeutischen Praxis*, Vol. 6, Springer, Berlin, 1994.

Harborne, J.B. and Baxter H., *Dictionary of Plants Toxins*, Wiley and Sons, Chichester, United Kingdom, 1996.

Hartwell, J.L., *Plants Used Against Cancer. A survey*, Quarterman, Lawrence, MA, 1982.

Hebert, *History of the Twelve Great Livery Companies of London*, 1834.

Hennig-Niklasch, F., *Untersuchungen zur autoxidation von aetherischen olen, insbesondere von anisöl und dessen hauptbestandteil. Anethol.* Diss., Frankfurt, 1959.

Himejima, M. and Kubo, I., Antimicrobial agents from *Licaria puchuri-Major* and their synergistic effect with polygodial, *J. Nat. Prod.*, 55, 620–625, 1992.

Hörings, P. and Grälert, K.P., Über das sogenannte photoanethol. Ein beitrag zur chemischen lichtwirkung, *Ber. Dtsch. Chem. Ges.*, 42, 1204–1207, 1909.

Hussain, R.A., Kinghorn, A.D., and Soejarto, D.D., Sweetening agents of plant origin: literature search for candidate sweet plants, *Econ. Bot.*, 42, 267–283, 1988.

Hussain, R.A., Poveda L.J., Pezzuto, J.M., Soejarto, D.D., and Kinghorn, A.D., Sweetening agents of plant origin: phenylpropanoid constituents of seven sveet-tasting plants, *Econ. Bot.*, 44. 174–182, 1990.

Ibrahim, Y.K.E. and Ogunmodele, M.S. (1991) Growth and survival of *Pseudomonas aeruginosa* in some aromatic water. *Pharm. Acta Helv.*, 66, 286–288.

Ilyas, M. (1980) Spices in India. III, *Econ. Bot.*, 34, 269–277.

Imaseki, I. and Kitabatake, Y. (1962),Yakugaku Zasshi, 82, 1326 (ref. by Reiter, M. and Brand W. (1985).

Ineichen, G. (1962) *El libro Agregà de Serapiom, volgarizzamento di Frater Jacobus Philippus de Padua*, Istituto per la collaborazione culturale, Venezia–Roma.

Ishidate, M.Jr., Sofuni T., Yoshikawa, K., Hayashi, M., Nohmi, T., Sawada, M., Matsuoka, A. (1984) Primary mutagenicity screening of food additives currently used in Japan. *Food Chem. Toxic.*, 22, 623–636.

Isolement et caractérisation du *cis*-anéthole, *Bull. Soc. Chim. France.*, 4, 566–569.

Ivanic, R., Savin, K., and Robinson F.V., Essential oil from *Pimpinella serbica* fruits, *Planta Med.*, 48, 60–61, 1983.

Jain, S.C. and Jain, R., Plants acting on gastrointestinal tract, *Proceedings of the 1st Symposium on Natural Drugs and the Digestive Tract*, Naples, July 1992.

Japp, R.G., Activity of synthetic estrogens on oral administration in the domestic fowl and turkey, *Endocrinology*, 37, 369–376, 1945.

Jensen-Jarolim, E., Leitner, A., Hirschwehr, R., Kraft, D., Wuthrich, B., Scheiner, O., Graf, J., and Ebner, C., Characterization of allergens in Apiaceae spices: anise, fennel, coriander and cumin, *Clin. Exp. Allergy*, 27, 1299–1306, 1997.

Karapinar, M., Inibitory effect of anethole and eugenol on the growth and toxin production of *Aspergillus parasiticus*, *Int. J. Food Microbiol.*, 10, 193–199, 1990.

Kleiman, R., Plattner, R.D., and Weisleder D., Antigermination activity of phenylpropenoids from the genus *Pimpinella*, *J. Nat. Prod.*, 51, 249–256, 1988.

Knobloch, K., Weigand, H., Weis, N., Schwarm, H.-M., and Vigenschow, H., Action of terpenoids on energy metabolism, in *Progress in Essential Oil Research*, Brunke, E.J., Ed., de Gruyter, Berlin, 1986.

Kobert, R., *Lehrbuch der Intoxikationen*, Enke, Stuttgart, 1906.

Kubeczka, K.-H., Bohn, I., and Formacek, V., New constituents from the essential oils of *Pimpinella* species, in *Progress in Essential Oil Research*, Brunke, E.J., Ed., de Gruyter, Berlin, 1986.

Kubo, I. and Himejima, M., Anethole, a synergist of polygodial against filamentous microorganism, *J. Agric. Food Chem.*, 39, 2290–2292, 1991.

Kubo, I. and Himejima, M., Potentiation of antifungal activity of sesquiterpene dialdehydes against *Candida albicans* and two other fungi, *Experientia*, 48, 1162–1164, 1992.

Kubo, I. and Taghigughi, M., Polygodial, an antifungal potentiator, *J. Nat. Prod.*, 51, 22–29, 1988.

Kubo, I., Muroi, H., and Himejima, M., Combination effects of antifungal nagilactones against *Candida albicans* and two other fungi with phenylpropanoids, *J. Nat. Prod.*, **56**, 220–226, 1993.

La Sacra Bibbia, annotated by G. Ricciotti, Salani, Florence, 1940.

Lachowicz, K.J., Jones, G.P., Briggs, D.R., Bienvenu, F.E., Wan, J., Wilcock, A., and Coventry, M.J., The synergistic preservative effects of the essential oils of sweet basil (Ocimum basilicum L.) against acid-tolerant food microflora, *Lett. Appl. Microbiol.*, 26, 209–214, 1998.

Lalou, S.D., *Contribution à l'étude de l'essence d'absinthe et de quelques autres essences (hypose, tanaise, sauge, fenouil, coriandre, badiane)*, Naud, Paris, 1903.

Le Bourhis, B., Recherches préliminaires sur le méthabolisme du trans-anethole, *Ann. Biol. Clin.*, 26, 711–715, 1968.

Le Bourhis, B., Identification de quelques métabolites du trans-anéthole chez l'homme, le lapin et le rat, *Ann. Pharm. Franc.*, 28, 355–361, 1970.

Le Bourhis, B., *Propriétés Biologiques de l'Anéthole*, Maloine, Paris, 1973a.

Le Bourhis, B., Propriétés biologiques du trans-anéthole. Essai de détermination de la dose journalière acceptable, *Parf. Cosm. Sav. France*, 3, 450–456, 1973b.

Leclerc, A., Brugère, J., Luce, D., Point, D., and Guenel, P., Type of alcoholic beverage and cancer of the upper respiratory and digestive tract, *Eur. J. Cancer Clin. Oncol.*, 23, 529–534, 1987.

Lemery, N., *Dizionario ovvero Trattato Universale delle Droghe Semplici*, Bertella e Perlini, Venice, 1742.

Lesieur, C., Nouvelles recherches sur la toxicité expérimentale des essences usuelles, *Arch. Med. Exp.*, 18, 803–817, 1906.

Levey, M., *The Medical Formulary or Aqrabadhin of Al-Kindi Translated with a Study of its Materia Medica*, University Wisconsin, Madison, 1966.

Lewis, W.H., *Medicinal Botany*, Wiley-Interscience, NewYork, 1977.

Liener, I.E., *Toxic Constituents of Plant Foodstuffs*, 2nd ed., Academic, London, 1980.

Linares, E., and Bye, R.A. Jr, A study of four medicinal plant complexes of Mexico and adjacent United States, *J. Ethnopharmacol.*, 19, 153–183, 1987.

Lovema, A.B., Report of a case of sensitivity of mucous membranes and the skin to oil of anise, *Arch. Dermatol.*, 37, 70, 1983.

Magnan, V., Epilepsie alcoolique: action spéciale de l'absinthe. Epilepsie absinthique, *C.R. Soc. Biol.*, 1868, 156–161, 1868.

Magnan, V., Alcoolisme aigu: épilepsie causée par l'absinthe, alcoolisme chronique, accident épileptiforme, symptomatique des lésions organiques, *C.R. Acad. Sci.*, 68, 825–828, 1869a.

Magnan, V., Conférences cliniques sur les maladies mentales et nerveuses, alcoolisme, alcool et absinthe. Epilepsie absinthique, *Gazette des hôpitaux*, 79, 310, 1869b.

Marcus, C. and Lichtenstein, E.P., Biological active components of anise: toxicity and interactions with insecticides in insects, *J. Agric. Food Chem.*, 27, 1217–1223, 1979.

Marcus, C. and Lichtenstein, E.P., Interactions of naturally occurring food plant components with insecticides and pentobarbital in rats and mice, *J. Agric. Food Chem.*, 30, 563–568, 1982.

Martin, R., Reichling, J., and Becker, H., Reinvestigation of the phenylpropanoids from the roots of *Pimpinella* species, *Planta Med.*, 51, 198–202, 1985.

Maruzzella, J.C. and Freundlich, M., Antimicrobial substances from seeds, *J. Am. Pharm. Assoc.*, 48, 356–358, 1959.

Maruzzella, J.C. and Lichtenstein, M.B., The *in vitro* antibacterial activity of oils, *J. Am. Pharm. Assoc.*, 45, 378–381, 1956.

Mattioli, M.P.A., *I Discorsi di M. Pietro Andrea Mattioli Medico Sanese, ne i Sei Libri della Materia Medicinale di Pedacio Dioscoride Anazarbeo*, Valgrisi e Costantini, Vinegia, 1557.

McCallion, R.F., Cole, A.L.J., Walker, J.R.L., Blunt, J.W., and Munro, M.H.G., Antibiotic substances from New Zealand plants. II. Polygodial, an anti-*Candida* agent from *Pseudowintera colorata*, *Planta Med.*, 44, 134–138, 1982.

Meyer, F. and Meyer, E., Percutaneous absorption of essential oil and their constituents, *Arzneimittel-Forsch.*, 9, 516–519, 1959.

Miething, H., Seger, V., and Hänsel, R., Determination of photoanethole from a stored essential oil of anise fruits as 4,4-dimethoxystilbene by high performance liquid chromatography-ultraviolet coupling, *Phytother. Res.*, 4, 121–123, 1990.

Miller, E.C., Swanson, A.B., Phillips, D.H., Fletcher, T.L., Liem, A., and Miller, J.A., Structure-activity studies of the carcinogenicities in the mouse and rat of some naturally occurring and synthetic alkenylbenzene derivatives related to safrole and estragole, *Cancer Res.*, 43, 1124–1134, 1983.

Mitchell, J.C. and Rook, A., *Botanical Dermatology*, Greenglass, Vancouver, 1979.

Moldenke, H.N. and Moldenke, A.L., *Plants of the Bible*, Ronald, New York, 1952.

Morel, A. and Rochaix, A., Recherches sur le pouvoir infertilisant de quelques essences végétales, *Bull. Sci. Pharmacol.*, 35, 631–634, 1928.

Morel, A. and Rochaix, A., Sur la nature des propriétés des essences végétales et de leurs constituants définis, *C.R. Soc. Biol.*, 101, 1152–1153, 1929.

Morel, A., Rochaix, A., and Perrot, L., Sur les pouvoirs antiseptique et infertilisant de quelques phénols et éthers-oxydes de phénols à fonction allylique, propénylique ou aldéhydique, *C.R. Soc. Biol.*, 114, 1118–1120, 1933.

Morton, J.F., *Atlas of Medicinal Plants of Middle America — Bahamas to Yucatan*, Thomas, Springfield, Illinois, 1981.

Muirhead, A.L. and Gerald, H.F., *J. Pharmacol. Exper. Ther.*, 8, 253, 1916.

Naves, Y.-R., Études sur les matières végétales volatiles. CLXV Synthèse du *cis*-anéthole à partir du *trans*-anéthole, *Helv. Chim. Acta*, 43, 230–232, 1960.

Naves, Y.-R., Ardizio P., and Favre, C., Étude sur les matières végétales volatiles CL (I), 1958.

Newberne, P., Smith, R.L., Doull, J., Goodman, J.I., Munro, I.C., Portoghese, P.S., Wagner, B.M., Weil, C.S., Woods, L.A., Adams, T.B., Lucas, C.D., and Ford, R.A., The FEMA GRAS assessment of trans-anethole used as a flavouring substance, *Food Chem. Toxicol.*, 37, 789–811, 1999.

Niiho, Y., Takayanagi, I., and Takagi, K., Effects of a combined stomachic and its ingredients on rabbit stomach motility in situ, *Japan J. Phamacol.*, 27, 177–179, 1977.

Odier, L., *Lezioni di Medicina Pratica*, 2nd ed., Vol. 2, Silvestri, Milan, 1813.

Pandey, D.K., Tripathi, N.N., Tripathi, R.D., and Dixit, S.N., Antifungal activity of some seed extracts with special reference to that of *Pimpinella diversifolia* DC, *Int. J. Crude Drug Res.*, 21, 177–182, 1983.

Perez, C. and Anesini, C., *In vitro* antibacterial activity of Argentine folk medicinal plants against *Salmonella typhi*, *J. Ethnopharmacol.*, 44, 41–46, 1994.

Pharmacopée Européenne, Conseil de l'Europe, Strasbourg, 1997.

Pharmacopoeia Augustana Renovata revisa et appendice Aliquot Medicamentorum Selectiorum Aucta, Lotteri, Augustae Vindelicorum, 1734.

Phillips, D.H., Miller, J.A., Miller E.C., and Adams, B., Structures of the DNA adducts formed in mouse liver after administration of the proximate hepatocarcinogen 1'-hydroxy-estragole, *Cancer Res.*, 41, 176–186, 1981.

Plinii Caii Secundi, *Historia Naturalis liber XX/LXXII-LXXIII, ex recensione I. Harduini et recentiorum adnotationibus,* tomus VI, Pomba, Turin, 1831.

Pourgholami, M.H., Majzoob, S., Javadi, M., Kamalinejad, M., Fanaee, G.H., and Sayyah, M., The fruit essential oil of *Pimpinella anisum* exerts anticonvulsant effects in mice, *J. Ethnopharmacol.*, 66, 211–215, 1999.

Rahman, A.U. and Zaman, K., Medicinal plants with hypoglycemic activity, *J. Ethnopharmacol.*, 1989, 1–55, 1989.

Rasoanaivo, P., Petitjean, A., Ratsimamanga-Urverg, S., and Rakoto-Ratsimaga, A., Medicinal plants used to treat malaria in Madagascar, *J. Ethnopharmacol.,* 37, 117–127, 1992.

Regimen Sanitatis Salerni or *Regola Sanitaria Salernitana*, Canesi, Rome, 1963.

Reichling, J., Merkel B., and Hofmeister, P., Studies of the biological activities of rare phenylpropanoids of the genus *Pimpinella*, *J. Nat. Prod.*, 54, 1416–1418, 1991.

Reiter, M. and Brandt, W., Relaxant effects on tracheal and ileal smooth muscles of the guinea pig, *Arzneim.-Forsch.*, 35, 408–414, 1985.

Reus Dispensatiorum Universale, Argentorari, 1686.

Ricettario Fiorentino (El) del Arte, et Universita de Medici, et Spetiali della Città di Firenze, Torrentino, Florence, 1550.

Sangster, S.A., Caldwell, J., Hutt, A.J., Anthony, A., and Smith R.L., The metabolic disposition of [*methoxy-* ^{14}C] labelled *trans*-anethole, estragole and p-propylanisole in human volunteers, *Xenobiotica*, 17, 1223–1232, 1987.

Sangster, S.A., Caldwell, J., and Smith, R.L., Metabolism of anethole. I. Pathways of methabolism in the rat and mouse, *Food Chem. Toxicol.*, 22, 695–706, 1984a.

Sangster, S.A., Caldwell, J., and Smith, R.L., Metabolism of anethole. II. Influence dose size on the route of metabolism of trans-anethole in the rat and mouse, *Food Chem. Toxicol.*, 22, 707–713, 1984b.

Schimmer, O., Beck, R., and Dietz, U., Phototoxizität und photomutagenität von furocumarinen and furocumarindrogen bei *Chlamydomonas reinhardii*, *Planta Med.*, 40, 68–76, 1853.

Semmola, G., *Trattato di Farmacologia e Terapeutica generale*, Vitale, Napoli, 1853.

Seto, T.A. and Keup, W., Effects of alkylmethoxybenzene and alkylmethylene-dioxybenzene essential oils on pentobarbital and ethanol sleeping time, *Arch. Int. Pharmacodyn.*, 180, 232–240, 1969.

Setty, D.S., Kamboj, V.P., and Khanna, N.M., *Indian J. Exp. Biol.,* 15, 231, 1977.

Sharaf, A. and Goma, N., Phytoestrogens and their antagonism to progesterone and testosterone, *J. Endocrinol.*, 31, 289–290, 1965.

Sijelmassi, A.I., *Les Plants Médicinales du Maroc*, 3rd ed., Le Fennec, Casablanca, 1993.

Stahl, E. and Herting, D., Die verteilung von inhaltsstoffen in drei *Pimpinella*-arten, *Phytochemistry*, 15, 999–1001, 1976.

Steinegger E. and Hänsel, R., *Lehrbuch der Pharmakognosie*, 3rd ed., Springer, Berlin, 1972.

Stricker, W.E., Anorve-Lopez, E., and Reed, C.E., Food skin testing in patients with idiopathic anaphylaxis, *J. Allergy Clin. Immunol.*, 77, 516–519, 1986.

Stross, W., Beiträge zur pharmakologie des kampfers, *Arch. f. exp. Path. u. Pharm.*, 95, 304–336, 1922.

Swanson, A.B., Chambliss, D.D., Blomquist, J.C., Miller E.C., and Miller, J.A., the mutagenicities of safrole, estragole, eugenol, *trans*-anethole, and some of their known or possible metabolites for *Salmonella typhimurium* mutants, *Mutation Res.*, 60, 143–153, 1979.

Taylor, J.M., Jenner, P.M., and Jones, W.I., A comparison of the toxicity of some alkyl, propenyl and propyl compounds in the rat, *Toxicol. App. Pharmacol.*, 6, 378–387, 1964.

Theatrum Sanitatis, Codice 4182 della R. Biblioteca Casanatense, Libreria dello Stato, Rome, 1940.

Thomson, *Chronicles of London Bridge*, 1827.

Trotter, R.T. II, Folk remedies as indicators of common illnesses: examples from the United States-Mexico border, *J. Ethnopharmacol.*, 4, 207–221, 1981.

Truhaut, R., Le Bourhis, B., Attia, M., Glomot, R., Newman, J., and Caldwell, J., Chronic toxicity/carcinogenicity study of *trans*-anethole in rats, *Food Chem. Toxicol.*, 27, 1989.

Tschirch, A., *Handbuch der Pharmacognosie*, Band I/2, Tauchnitz, Leipzig, 1910.

Tutin, T.G., Heywood, V.H., Burges, N.A., Moore, D.M., Valentine, D.H., Walters, S.M., and Webb, D.A., Eds., *Flora Europaea*, Vol. 2, Cambridge University Press, Cambridge, 1968.

Twaij, H.A.A., Elisha, E.E., Khalid, R.M., and Paul, N.J., Analgesic studies on some Iraqi medicinal plant, *Int. J. Crude Drug. Res.*, 25, 251–254, 1987.

Van Dongen, K. and Leusink, H., The action of opium-alkaloids and expectorants on the ciliary movements in the air passages, *Arch. Int. Pharmacodyn.*, 93, 261–276, 1953.

Varenne, E., *Etude sur l'anéthol et l'estragol*, Eyméoud, Paris, 1904.

Varenne, E., Roussel, J., and Godefroy, L., Action de l'anéthol sur l'organism, *C.R. Acad. Sci.*, 137, 1294–1296, 1903.

Vignoli, L. and Morel, M.-C., *L'Anis*, Studer, Geneva, 1967.

Vignoli, L., Mercier, J., Dessaigne, S., Morel, M.-C., Simon, O., Contribution à l'étude pharmacodynamique du *trans*-anethol, *Pharmacia Mediterranea*, 3, 587–604, 1960a.

Vignoli, L., Morel, M.-C., and Simon, O., Contribution à l'étude du dosage et de l'elimination du *trans*-anethol, *Pharmacia Mediterranea*, 3, 332–348, 1960b.

Wüthrich, B. and Dietschi, R., Das *sellerie-karotten-beifuss-gewurz-syndroms*: hauttest- und RAST-ergebnisse, *Schweiz. Med. Wschr.*, 115, 258–364, 1985.

Weiner, M. and Bernstein, I.L., *Adverse Reaction to Drug Formulation Agents*, Dekker, New York, 1989.

Wichtl, M., *Piante Officinali per Infusi e Tisane*, Della Loggia, OEMF, Milano, 1989.

Williams, A.C. and Barry, B.W., Essential oils as novel human skin penetration enhancers, *Int. J. Pharmac.*, 57, R7–R9, 1989.

Younos, C., Fleurentin, J., Notter, D., Mazars, G., Mortier, F., and Pelt, J.-M., Repertory of drugs and medicinal plants used in traditional medicine of Afghanistan, *J. Ethnopharmacol.*, 20, 245–290, 1987.

Zondek, B. and Bergmann, E., Phenol methyl ethers as oestrogenic agents, *Biochem. J.*, 32, 641–645, 1938.

10 Monographs and Formulations of the Genera *Pimpinella*, *Illicium*, and *Foeniculum*

María Dolores Contreras and Juan Carlos Orte

CONTENTS

0-415-32246-4/04/$0.00+$1.50

10.1 INTRODUCTION

Anise and fennel have historically been used as effective remedies against certain illnesses in various civilizations. Anise appears to have been used in Chinese medicine since early times (Mességué and Bontemps, 1986) and in Indian medicine since the fifth century B.C. Fennel was considered the ideal tonic in Greek medicine.

Treatment methods involving these substances have been handed down from generation to generation. These plant-based medicines are currently very popular and easily obtained. In some countries, they can even be bought in street markets,

Anise and fennel are of interest to the pharmaceutical industry not only for their medicinal properties but also for their use in other fields, such as the food industry. As a consequence, numerous pharmacopoeias and formularies include monographs on both substances, their direct derivatives (essential oils), or their main component, anethole.

10.2 MONOGRAPHS

The pharmacopoeias only accept the medicinal uses of the following species of anise and fennel: *Pimpinella anisum* (Umbelliferae) and *Illicium verum* (Magnoliaceae), *Foeniculum vulgare* Miller sp. *vulgare* var. dolce (Miller) Thellung, and *F. vulgare* Miller sp. *vulgare* var. *vulgare*. They include monographs on aniseed, star anise, anise oil, fennel fruit, and fennel oil.

To be used for treatment, both the plant-based medicine and the essential oil must satisfy the conditions included in the monographs of the pharmacopoeias. These specify the part of the plant used and the minimum content of essential oils, the macroscopic and microscopic characteristics. The pharmacopeias also include methods for identification of the principal component (anethole) or other components (fenchone and estragole) and for evaluation and other requisites such as the content of water or total ash. Monographs on essential oils include identification and evaluation tests for their components, and the limits for each.

Information on aniseed, star anise, anise oil, fennel fruit, and fennel oil found in the pharmacopoeia is detailed below.

10.2.1 ANISEED

Depending on the country, aniseed is known as anise, anice, anis verde, anis vert, anise fruit, anisi fruit, anisi fructus, fructus anisi vulgaris (Martindale, 1999), matalaúva, and matalahúga.

It is found in the pharmacopoeias of Germany, Argentina, Australia, Belgium, Great Britain, Czechoslovakia, Egypt, France, Hungary, Italy, Portugal, Rumania, Russia, Spain, Switzerland (Martindale, 1996) and also in the *European Pharmacopoeia* (2002).

British Pharmacopoeia (2000), *European Pharmacopoeia* (2002), and *Real Farmacopea Española* (2002) specify that aniseed consists of the whole dry cremocarp of *P. anisum* L. It contains not less than 20 mL/kg essential oil and has an odor reminiscent of anethole.

The fruit is a cremocarp and is generally entire; a small fragment of the thin, rigid, slightly curved pedicel is frequently attached. The cremocarp is ovoid or pyriform and slightly compressed laterally. It is yellowish-green or greenish-grey, 3 mm to 5 mm long, and up to 3 mm wide, surmounted by a stylopod with two short, reflexed stylar points. It must not contain starch.

Essential oils can be extracted from the powdered fruit and the anethole identified using thin-layer chromatography. They can also be evaluated following the general procedures found in the pharmacopoeias. The oils undergo alteration through oxidation, and the pharmacopoeias recommend that it is stored in a well-closed container, protected from light.

Aniseed's action and use are carminative.

10.2.2 Powdered Aniseed

Powdered aniseed (*British Pharmacopoeia,* 1988) is greenish-yellow or greenish-brown and must fulfill all the requirements of aniseed. Its action and use are carminative.

10.2.3 Star Anise

Depending on the country, star anise is known as badian, anis étoile, anisum badium, anisum stellatum, badiana, badiane de chine, star anise fruit, strenanis, and anís estrellado (Martindale, 1999). It is included in the pharmacopoeias of Argentina, Australia, Brazil, China, France, Great Britain, Portugal, and Spain (Martindale, 1996), and in the *European Pharmacopoeia* (2002).

British Pharmacopoeia (2000), *European Pharmacopoeia* (2002), and *Real Farmacopea Española* (2002) specify that star anise consists of the dried composite fruit of *I. verum* Hooker fil. It contains not less than 70 mL/kg essential oil, with reference to the anhydrous drug.

Star anise is a composite fruit, consisting of follicles. Each follicle is about 12 to 20 mm long and 6 to 11 mm high. The seeds are up to 8 mm long. The fruit carpels are brown and have an odor of anethol.

The pharmacopoeias include methods to determine the presence of *I. anisatum* L., a toxic substance that is often mixed fraudulently with star anise. It can be identified by microscopy: the fruit should not contain follicles with an upwardly turned beak and straight fruit stalks that are not curved at the distal end (these are the fruits of *I. anisatum* L., [toxic] syn *I. religiosum*). It can also be identified by thin-layer chromatography: chromatograms are developed with the "problem" solution and two reference solutions of anethole and myristicine. The chromatogram obtained with the test solution shows no zone corresponding to myristicine.

Store star anise in a well-closed container protected from light.

10.2.4 Anise Oil

Common names for anise oil (Martindale, 1999) include anisi aetheroleum, anise oil, aniseed oil, esencia de anís, essence d'anis, and oleum anisi.

Monographs are included in the pharmacopoeias of Germany, Argentina, Australia, Belgium, Great Britain, Czechoslovakia, Egypt, France, Hungary, Yugoslavia, Mexico, Poland, Portugal, Rumania, Russia, Spain, and Switzerland (Martindale, 1996), and also in *European Pharmacopoeia* (2002) and *National Formulary 18* (1994).

British Pharmacopoeia (2000), *European Pharmacopoeia* (2002), and *Real Farmacopea Española* (2002) specify that anise oil is obtained by steam distillation from the dry ripe fruits of *Pimpinella anisum* L. or *I. verum* Hook. fil. It is a clear, colorless or pale-yellow liquid that solidifies on cooling, is practically insoluble in water, and is miscible with alcohol, ether, light petroleum, and methylene chloride. Its relative density is 0.978 to 0.994, its refractive index is 1.552 to 1.561 and its freezing point is 15° to 19°C.

Identification of anise oil can be carried out via thin-layer chromatography, using three different reference solutions (anethole, anisaldehyde and linalol), or via gas chromatography. The chromatogram obtained with the test solution shows six peaks similar in retention time to the six peaks in the chromatogram obtained with the reference solution. The reference solution consists of a mixture of fixed quantities of linalol, estragole, α-terpineol, *cis*-anethole, anethole, and anisaldehyde. The profiles obtained for the oil of *I. verum* and the oil of *P. anisum* are slightly different. The percentages of each of these six components in the chromatogram should fall within the limits indicated in Table 10.1.

TABLE 10.1
Percentages of Components in the Gas Chromatography of Anise Oil

Substance	Percentage
Linalol	0.1–1.5
Estragole	0.5–6.0
α-Terpineol	0.1–1.5
cis-Anethole	<0.5
trans-Anethole	84–93
Anisaldehyde	0.1–3.5

Sources: British Pharmacopoeia, Her Majesty's Stationery Office, London, 2000; *European Pharmacopoeia*, 4th ed., Directorate for the Quality of Medicines of the Council of Europe, Strasbourg, France, 2002; and *Real Farmacopea Española*, 2nd ed., Ministerio de Sanidad y Consumo, Madrid, 2002.

Store in a well-filled, airtight container, protected from light and heat. *British Pharmacopoeia* (1988) gives 25°C as the maximum temperature of storage. If the oil has crystallized before use, it should be melted and mixed.

According to the pharmacopoeias, it should be specified whether the anise oil is obtained from *P. anisum* or from *I. verum.*

The action and use of anise oil are for flavor.

10.2.5 Fennel Fruit

Fennel fruit is known as fennel, fenchel, fennel fruits, fennel seed, fenouil, fenouil doux, foeniculum (Martindale, 1999), fruto de hinojo, and funcho. It is found in the pharmacopoeias of Germany, Australia, Belgium, China, Czechoslovakia, Egypt, France, Hungary, Poland, Portugal, Rumania, Russia, and Switzerland (Martindale, 1989), in addition to those of Great Britain (*British Pharmacopoeia,* 2000), Spain (*Real Farmacopea Española,* 2002), and *European Pharmacopoeia* (2002). The latter includes specific monographs for sweet fennel and bitter fennel, which are summarized below.

10.2.6 Sweet Fennel

Sweet fennel consists of the dry cremocarps and mericarps of *F. vulgare* Miller sp. *vulgare* var. *dulce* (Miller) Thellung. It contains not less than 20 mL/kg essential oil, calculated with reference to the anhydrous drug. The oil contains not less than 80.0% anethole.

Sweet fennel is pale green or pale yellowish-brown. The fruit of sweet fennel is a cremocarp of an almost cylindrical shape with a rounded base and a narrowed summit crowned with a large stylopod. It is generally 3 to 12 mm long and 3 to 4 mm wide. The mericarps, usually free, are glabrous. Each bears five prominent, slightly carenated ridges. When cut transversely, four vittae on the dorsal surface and two on the commissural surface may be seen with a lens.

Anethole can be identified by thin-layer or gas chromatography. In both cases, prior extraction of the essential oil from the recently powdered fruit is carried out. The essential oil contains not more than 10.0% estragole and not more than 7.5% fenchone. Both substances are evaluated by gas chromatography.

Store sweet fennel in a well-closed container, protected from light and moisture.

Its action and use are carminative.

TABLE 10.2
Physicochemical Characteristics and Toxicity of the Anethole Isomers

	trans-Isomer	*cis*-Isomer
Solid crystalline	20 to 21°C	
mp	21.4°C	
Liquid	>23°C	
Density (4, 20)	0.9883	0.9878
Bp 2,3	81 to 81.5C°	79 to 79.5°
nd20	1.56145	1.55455
UV max (ethanol)	259 nm (ε 22.30)	253.5 nm (ε 18500)
LD_{50} i.p. in rats	900 mg/kg (Boissier)	93 mg/kg (Boissier)

10.2.7 Bitter Fennel

Bitter fennel consists of the dry cremocarps and mericarps of *F. vulgare* Miller sp. *vulgare* var. *vulgare.* It contains not less than 40 mL/kg essential oil, calculated with reference to the anhydrous drug. The oil contains not less than 60.0% anethole and not less than 15.0% fenchone. Bitter fennel is greenish-brown, brown, or green. The essential oil obtained must not contain more than 5.0% estragole. Store in a well-closed container, protected from light and moisture.

The action and use for bitter fennel are carminative.

10.2.8 Powdered Fennel

Powdered fennel contains (*British Pharmaceutical Codex* 1973) not less than 1% v/w volatile oil. Store in a cool dry place; the powdered drug should be stored in airtight containers. Protect from light.

10.2.9 Fennel Oil

Other names for fennel oil are aetheroleum foeniculi, essence of fennel, essencia de fauncho, and oleum foeniculi. It is found in the pharmacopoeias of Germany, Argentina, Australia, Czechoslovakia, Egypt, Hungary, Japan, Nordica, Poland, Romania, Switzerland, and Yugoslavia (Martindale 1996), and also in *National Formulary 18* (1994).

National Formulary 18 (1994) defines fennel oil as a colorless or pale-yellow oil, with the characteristic aromatic odor and taste of fennel, obtained by distillation from fennel. Soluble with alcohol 1:1 (90%). If solid matter separates, it should be melted and mixed before use. Store in airtight containers. Fennel oil is used as an aromatic carminative.

10.2.10 Anethole

Anethole, also known as anethol, is p-propenylanisole, (E)-1-methoxy-4-(prop-1-enyl) benzene (Martindale, 1996).

Anethole (O'Neil et al., 2001) is the main constituent of anise, star anise, and fennel oil and is responsible for the characteristic odor and flavor of anise. The presence of an asymmetric C in its chemical structure gives rise to two isomers (*cis* and *trans*) with similar physicochemical and ultraviolet absorption characteristics (Table 10.2). The isomers' commercial use is conditioned by their different toxicity: The *trans* isomer is used in medicine and the *cis* in cosmetics.

Only the Pharmacopoeia of Brazil and the *National Formulary 18* (1994) include a monograph for this substance. The latter specifies that anethole is obtained from the essential oil of anise from the species *Pimpinella anisum* L. or from *I. verum* Hook. fil., other sources, or by chemical synthesis.

It is a yellow liquid with a smooth texture and an anethole aroma, containing less than 23% anethole. Anethole from anise oil, star anise, or fennel contains only traces of the *cis* isomer. All the pharmacopoeias limit its concentration (see Table 10.1).

Anethole should be stored in a cool place and protected from the light. The label should indicate whether it has been obtained from natural sources or prepared synthetically. It has the same uses as anise oil.

10.3 PHARMACEUTICAL USES OF ANISE AND FENNEL

10.3.1 Pharmaceutical Uses of Anise, Anise Oil, and Anethole

Anise, anise oil, and anethole are used for two different medicinal purposes, pharmacological actions, and organoleptic correction of drugs.

10.3.2 Pharmacological Actions

Anise, anise oil, and anethole possess carminative, digestive, expectorant, and spasmolytic qualities, particularly at a respiratory and digestive level. They have been shown to have weak diuretic and diaphoretic effects. Furthermore, anise oil possesses antibacterial (Patel ,1968; Fyfe et al., 1998; De et al., 2002) and antifungal activity (Soliman and Badeaa, 2002).

10.3.3 Therapeutic Indications

The main properties of anise and related substances (Castaño et al. 1998) are derived from its carminative and expectorant activity. As an expectorant, it is frequently associated with other medicines (del Pozo 1977). These substances are also indicated for cases of lack of appetite, gastrointestinal spasms, colic in nursing mothers, halitosis, catarrh, bronchitis, and asthma, and to be used against constipation. As a folk remedy, they are used to prevent flatulence and against a dry cough.

10.3.4 Side Effects and Toxicity

Taken orally at high doses, anise oil exhibits (Castaño et al., 1998) narcotic effects, delirium, anaesthesia, and convulsions. In high doses and with prolonged treatment, it acts as a narcotic, slowing circulation and producing muscular paresia, cerebral congestion, and other organic alterations. Cases of intoxication caused by overdose have been described in children (Montoya, 1990). To avoid this, it is recommended that children not be given a dose of more than five drops a day (Castaño et al., 1998). Anise oil may also cause contact dermatitis, probably because of its anethole content.

The most common intoxication occurs as a result of substituting star anise for another species, *I. religiosum*, with neurotoxic sesquiterpene lactones (anisatin) and toxic alkaloids such as shikimin and shikimitoxin (Castaño et al., 1998). In Spain in 2002, several young children were hospitalized as a result of ingesting infusions of anise (Garzo et al., 2002; Gil et al., 2002). This led to the removal by health authorities of over-the-counter products containing anise. At present, only products fulfilling the specifications of the *Real Farmacopea Española* (2002) with regard to anise can be sold in Spain. Recently, the European Community Commission established (Official Diary L 033 from 2/2/2002, Directive 2002/75/CE) a strict guideline for the importation of star anise to avoid adulteration with *I. religiosum.*

10.3.5 Oral Formulations

Anise is normally taken orally in the form of infusions or as capsules, syrups, and elixirs for more precise dosage. Some formulations are included in pharmacopoeias and formularies. The preparations described are classified according to whether they are obtained from the fruit, the oil, or anethole.

10.3.5.1 Dosage Forms Obtained from the Fruit

When the fruit is the only component, it is taken orally as a powder or dry extract in rigid gelatin capsules, with an expectorant activity (Castaño et al., 1998).

Star anise capsules:
- Badian powder, 300 mg for one capsule
- Dry extract of Badian, 60 mg for one capsule

The dry extract of Badian is not included in any pharmacopoeia. However, the *Real Farmacopea Española* (2002), *British Pharmacopoeia* (2000), and *United States Pharmacopoeia 23* include monographs on extracts, preparation, and requisites for its use in therapeutics.

The oral liquid preparations are well known, easy to administer, and especially useful for children in suitable doses. Some of these, included in pharmacopoeias, are:

Infusion (Castaño et al., 1998):
- Prepared with 1% badian
- Therapeutic properties: Carminative, eupeptic, spasmolytic, and antidiarrheal
- Dose: Usually 2 to 6 g/day

Tisane:
- According to Griffith (1954a), the tisane is made with a cup of water and one third a spoonful of anise
- Therapeutic indications: Carminative and for stomach problems
- Dose: For children, up to one quarter of a teaspoonful

The *British Pharmacopoeia* (2000), *European Pharmacopoeia* (2002), and *Real Farmacopea Española* (2002) specify that plant-based tisanes must be prepared with drugs that comply with the specifications of the individual monograph, and the recommendations for the microbiological quality of the preparation must be considered. The tisanes are prepared immediately before use.

Liquid extract of anise (1:1):
- Dose: Usually 10 to 20 drops, two or three times a day

Liquid extract of anise is not included in any pharmacopoeia. However, preparation and requisites for therapeutic use of liquid extracts in general can be found in *British Pharmacopoeia* (2000), *European Pharmacopoeia* (2002), *Real Farmacopea Española* (2002), and *United States Pharmacopoeia 23*.

Anise liquid (Castaño et al., 1998):
- Badian powder, 40 g
- Cinnamon powder, 1 g
- Sugar, 500 g
- Distilled water, 1 L

Star anise tincture (*Formulario Español de Farmacia Militar,* 1975; Castaño et al., 1998):
- Coarsely powdered fruit of *I. verum* Hook. *f.*, 20%
- Alcohol, 80%, sufficient to produce 100%
- Properties: Carminative and eupeptic
- Dose: 5 to 20 g/day

The *Formulario Español de Farmacia Militar* (1975) defines tincture of star anise as a colored liquid obtained by maceration.

10.3.5.2 Dosage Forms Obtained from Anise Oil

When used as a carminative (Castaño et al., 1998), anise oil is taken orally in doses of one to five drops, one to three times per day, on a sugar lump or in capsules of 25 to 50 mg, one to three times per day. Anise oil is also common in liquid drugs to be taken orally. The most common use is as anisated ammonia spirit, used as an expectorant. It is currently available commercially in some countries.

Anisated ammonia spirit (*Farmacopea Oficial Española,* 1954):
- Anise oil, 3 parts
- Ammonia, 17 parts
- Alcohol, 80 parts
- Protect from light
- Properties: Expectorant

Furthermore, anisated ammonia spirit is usually added to potions containing antitussives or a variety of water-soluble expectorants in a proportion of 1 to 1.5% (del Pozo, 1977).

Potion:
- Anisated ammonia spirits, 1.5 g
- Sodium benzoate, 5 g
- Polygale infusion, 120 g
- Syrups, 30 g

Anise oil forms part of some formulations, together with spasmolytic substances, to take advantage of its gentle carminative activity.

Carminative for children (syrup):
- Anise oil, 0.5 g
- Luminal, 0.2 g
- Vichy salt, 0.5 g
- Belladonna tincture, 10 drops
- Simple syrup, 30 g
- Distilled water, 120 g
- Filter

Because of the use of this formulation for children, and so as not to reinforce the depressant properties of the luminal, the anise is commonly dissolved without ethanol or mixtures of ethanol and glycerine. Thus, the anise oil is dissolved in a recommended suitable oral tensioactive (Tween 80, maximum 0.2% of total weight). In *Farmacopea Oficial Española* (1954), it also includes Belladonna tincture: 30 mg of alkaloids of Belladonna in 100 mL. The current editions of some pharmacopoeias (*British Pharmacopoeia,* 2000; *European Pharmacopoeia,* 2002; *Real Farmacopea Española,* 2002) only include general monographs on tinctures. Vichy salt is the traditional name for sodium bicarbonate, which, in this formula, is used to dissolve the luminal.

10.3.5.3 Dosage Forms Obtained from Anethole

The use of anethole in formulations is less common than that of the fruit or oil. Only the NF XV (del Pozo, 1977) includes an elixir of anise obtained from anethole.

Elixir of anise:
- Anethole, 3.5 cc
- Fennel oil, 0.5 cc
- Bitter almond spirit, 12 cc
- Ethanol, 50 cc
- Syrups, 625 cc
- Distilled water, sufficient to produce 1000 cc
- Filter
- Uses: Carminative and for its pleasant flavor
- Dose: For nursing mothers, 1 cc.

10.3.6 Other Therapeutic Indications

10.3.6.1 Diuretic

Some diuretic species, including anise, are common folk remedies in the form of extemporaneous tisanes (del Pozo, 1978b), such as:

Tisane:
- Squill bulb, 3 g
- Birch leaf, 10 g
- Indian kidney tea leaf (Indian tea), 20 g
- Anise fruit (powdered), 10 g
- Equisetum arvense, 20 g
- Juniper berries (bruised), 27 g
- Lovage root, 10 g

10.3.6.2 Laxative

Mixtures of anise and other natural laxatives are used in powdered form, often in rigid gelatin capsules.

Anise powder with rhubarb and magnesia (Griffith, 1954b):
- Powdered rhubarb, 35 g
- Heavy magnesium oxide, 65 g
- Anethole, 8 cc
- Alcohol, 10 cc
- The alcoholic anethole solution is added to the rest of the ingredients
- Dose: 0.3 g for nursing mothers

Another formulation (per capsule):
- Senna powder, 12 mg
- Anise fruits powder, 12 mg
- Lactose, 40 mg
- Colloidal silica, 36 mg

10.3.6.3 Demulcent

Anise has weak demulcent activity and is often recommended for a dry cough. It is administered in the form of tisanes and infusions, as tablets to dissolve slowly in the mouth, by inhalation systems, and as an aerosol (Castaño et al., 1998).

Wet inhalation:
Anise oil, 5 to 10 drops
Hot water, 500 mL
Aerosol:
Anise oil, 1.2 g
Suitable excipients, sufficient to produce 50 mL

To relieve coughs in children, the World Health Organization (WHO, 2001) recommends simple linctus (*British Pharmacopoeia,* 2000) and pediatric simple linctus (*British Pharmacopoeia,* 2000) (both are detailed in flavoring vehicles). The ingredients include a small quantity of anise oil. Both preparations are harmless and cheap.

10.3.7 Anise and Its Derivatives as Organoleptic Correctors of Medicines

Anise and anise oil are generally not used for their therapeutic properties in pharmacy, but for preparation of vehicles with pleasant organoleptic characteristics for later addition to a medicine that is suitable in terms of solubility and stability—these vehicles are known as flavoring vehicles—and for the correction of unpleasant smell or taste of medicine.

10.3.8 Flavoring Vehicles

The traditional aromas used in pharmacy are natural oils. The pharmacopoeias include monographs on vehicles containing anise oil at subtherapeutic concentrations, as aroma or taste are considered more important than pharmacological activity. The most important preparations in this group are the waters, syrups, and elixirs described below.

10.3.8.1 Waters

These are very simple vehicles, as they are saturated solutions of aromatic substances in water. Because of their low solubility in water, their aromatic concentration is also low (Ansel and Popovich, 1990). These vehicles have the advantage of not containing sucrose or alcohol, for which reason they are recommended for some treatments (Reilly, 1998).

Anise water (Dodge, 1954):
Anise oil
Distilled water
Use: A vehicle with a pleasant taste, especially for children.

Dodge (1954) defines anise water as a clear saturated solution of anise oil in water.

10.3.8.2 Syrups

Syrups are extremely sweet and contain little or no alcohol. They are solutions of volatile aromatizing oil in syrup or glycerin in a proportion of 1:500. They are recommended as vehicles for water-soluble medicines, giving rise to uniform, stable preparations (Reilly 1998).

Syrup (Reilly 1998):
Anise oil, 2 mL
Alcohol, 6 ml
Glycerine or syrups, sufficient to produce 1000 mL

The anise oil content of this syrup is very low. The low proportion of ethanol is necessary to dissolve the oil.

10.3.8.3 Elixirs

Elixirs are solutions that contain 25% alcohol. The high alcohol content is a problem standing in the way of their widespread use. The alcohol may sometimes interact with the pharmacologic activity of the active ingredient. Elixirs are hydroalcoholic vehicles suitable for medicines soluble in water or in diluted alcohol. Some of these vehicles are described below.

Aromatic elixir (or simple elixir) (Reilly, 1998):
- Orange oil, 2.4 mL
- Lemon oil, 0.6 mL
- Coriander oil, 0.24 mL
- Anise oil, 0.06 mL
- Talc, 30 g
- Syrups, 375 mL
- Alcohol, purified water, sufficient quantity to make 1000 mL

Aromatic elixir is one of the most widely used vehicles for the preparation of medicines. The *National Formulary 18* (1994) includes a simpler formulation for an aromatic elixir:

Aromatic elixir:
- Anise oil, q.s.
- Syrup, 375 mL
- Talc, 30 g
- Alcohol, purified water, sufficient quantity to make 1000 mL
- Alcohol content between 21 and 23%

Talc is present in both elixirs and other preparations made for purely technological purposes. Anise and the other oils of this vehicle are soluble in 96% alcohol, are insoluble in water, and are of varying solubility in hydroalcoholic mixtures. Purified talc is frequently included as a component of the formulation to divide the total volume of the oils into small fractions and to increase the solution rate in the hydroalcoholic mixture. The talc is removed from the final formulation by filtration.

In some flavoring vehicles the alcohol content is greater than 23%. These vehicles can then be diluted, as in the case of compound orange spirit and concentrated anise water. The latter is diluted for the preparation of Simple Linctus and Pediatric Simple Linctus. All are included in pharmacopoeias such as *British Pharmacopoeia* (2000).

Compound orange spirits:
- Terpeneless orange oil, 2.5 mL
- Terpeneless lemon oil, 1.3 mL
- Anise oil, 4.25 mL
- Coriander oil, 6.26 mL
- Ethanol (90%), sufficient to produce 1000 mL
- Ethanol content: 86 to 90% v/v

Orange and lemon oils present problems in their conservation, as they develop an unpleasant taste of trementine. The oils' stability is increased by eliminating the majority of the terpenes.

These terpeneless oils have a stronger aroma, are more water-soluble, and have greater stability (Irache et al., 1997).

Concentrated anise water:
- Anise oil, 20 mL
- Ethanol (90%), 700 mL
- Water, q.s. 1000 mL
- Add 50 g of previously sterilized purified talc, and leave for a few hours, shaking occasionally, and filter
- Extemporaneous preparation
  - Weight per mL: 0.898 to 0.908
  - Ethanol content: 60 to 64% v/v

Simple linctus:
- Citric acid monohydrate 2.5%, 1.25 mL
- Concentrated anise water, 0.5 mL
- Amaranth solution, 0.3 mL
- Chloroform spirit, 0.3 mL
- Syrup to 5 mL
- Diluent syrup

Simple linctus is an oral solution containing 2.2% w/w citric acid mono in a suitable vehicle with anise flavor. Pediatric simple linctus (*British Pharmacopoeia,* 2000) is more dilute.

Pediatric simple linctus:
- Simple linctus (as above), 1.25 mL
- Syrup to 5 mL
- Diluent syrup

10.3.9 Correction of Smell or Taste of Medicines

Anise oil and anethole are used to correct the taste or smell of many drugs because of their favorable organoleptic properties (sweet and volatile).

Because of their antagonist properties (sweet taste), anise oil and anethole are useful for correcting bitterness (morphine) or saltiness (salts) in some drugs (Adjei et al., 1992). They thus form part of numerous extemporaneous preparations and oral liquid specialties such as syrups and suspensions, often at concentrations above 3000 ppm. Anise oil and anethole are also used in solid preparations of solution (effervescent granules and tablets) before ingestion. In general, the aromatizing agent does not exceed 0.1% of the total weight of the formula.

As a result of its frequent use, the taste and aroma of anise are often associated with analgesics, antiacids, and laxatives. Some examples of the use of anise to correct taste and aroma in extemporaneous liquid preparations found in the pharmacopoeias are described below:

Aromatic cascara fluid extract (del Pozo, 1978a; *United States Pharmacopoeia 23*:
- Cascara, in very coarse powder, 1000 g
- Magnesium oxide, 120 g
- Pure licorice extract, 40 g
- Saccarine, 2 g
- Anise oil, 0.65 mL
- Coriander oil, 0.15 mL

Methyl salycilate, 0.10 mL
Ethanol (95%), 200 mL
Purified water, sufficient to make1000 mL
Treatment: laxopurgative

Syrup composed of senna (*Farmacopea Oficial Española,* 1954):
Senna, 100 g
Distilled water, 475 g
White sugar, 475 g
Manna, 150 g
Alcohol (95%), 50 g
Anise oil, 5 drops
Amine valerianate, 0.5 g
Lactic acid, 5 g
Treatment: laxative

Camphorated opium tincture (BF 2000) (elixir paregoric):
Opium tincture, 30 mL
Benzoic acid, 5 g
Racemic camphor, 3 g
Anise oil, 3 mL
Ethanol (60%), sufficient to make 1000 mL
Extemporaneous preparation
Ethanol content: 56 to 60% v/v
Anhydrous morphine content: 0.045 to 0.055% w/v

Paregoric elixir is currently used to treat addiction to opiates in newborn infants (Blondel et al., 1993; Theis et al., 1997; Wagner et al., 1998) at varying dosages.

Concentrated camphorated opium tincture (*British Pharmacopoeia,* 2000):
Opium tincture, 400 mL
Benzoic acid, 40 g
Racemic camphor, 24 g
Anise oil, 24 mL
Ethanol (96%), 400 mL
Water, sufficient to produce 1000 mL
Extemporaneous preparation
Ethanol content: 54 to 59% v/v
Anhydrous morphine content: 0.36 to 0.44% w/v

Effervescent granules (*Farmacopea Oficial Española* 1954):
Sodium sulphate, 8%
Potassium sulphate, 2%
Effervescent excipients, 50 g
Anise oil, q.s.

Anethole is used as an aromatizing agent in diphenhydramine hydrochloride elixir (Reilly, 1998).

The use of anise as a flavoring agent in chewable tablets (Mendes et al., 1989) deserves a special mention. This includes initial effect, flavor, posterior effects, and association of olfactory sensation.

Anise is recommended for flavoring tablets that have an initial sour or bitter taste, in concentrations of 0.1 to 3 w/w, depending on the use of the tablet. For this purpose, anise is used in the form of

- Ethanolic solutions, sometimes with the addition of small amounts of propylene glycol
- Aroma in powder (very frequent), diluted with sugars, starches, or gums, or encapsulated in cyclo-dextrins, which increases its stability (Irache et al., 1997); essential oil content is usually 20%
- Adsorbed powders, usually on silica gel; essential oil content can be up to 70% (Mendes et al., 1989)

In these cases, the flavoring agent is released in the mouth by dissolution, dispersion of the emulsifying agent, or desorption of the oils in the saliva. Because of their instability, powdered aromas should not be used after 6 to 12 months storage.

Anise is also used as a flavoring agent in microcapsules, which have the advantage of being very stable; loss of volatile material with time is minimal, and incompatibilities with other components of the medicine are avoided; microcapsules release their aroma in different ways: rupture of the covering by internal or external means, heat, dissolution, and in some cases, by biodegradation (Bakan, 1994)

The volatile and unstable nature of anise oil and its high content in chewable tablets conditions (Mendes and Anaebonam, 1990).

- The technology of the drug: oils cannot be incorporated before granulation and must be well mixed with the other components (uniform distribution).
- The requirements of stability apply to aspects that are not generally considered for other types of tablets: stability of the aroma, although changes in the aroma may not affect the activity of the drug, they may have negative effects on the consumer (Mendes et al., 1989).
- The temperature at which the drug is conserved: oils should not be exposed to high temperatures.

Mendes et al. (1989) propose the following formulation for a chewable tablet:

Chewable tablet:
- Mannitol USP, 720 mg
- Sodium saccharin, 6 mg
- Acetaminophen NF, 120 mg
- Binder solution, 21.6 mg
- Peppermint oil, 0.5 mg
- Syloid 244, 0.5 mg
- Banana, 2 mg
- Sodium chloride, 6 mg
- Magnesium stearate, 27.4 mg
- Quantity per tablet

10.3.10 Other Pharmaceutical Uses of Anise and Its Derivatives

Anise has shown activity in fields different than those mentioned previously and is included in a wide range of formulations, most of which are patented. It has been shown to have significant antibacterial (De et al., 2002) and antifungal activity (Soliman and Badeaa, 2002) and is a component of nasal sprays in hyperosmotic solutions of sodium chloride (Zellner, 1999). Its effect against candidiasis when taken orally is also being investigated (Naito et al., 2001), and it forms part of several combinations of preservatives suitable for the elaboration of drugs (Fyfe, 2000).

Anise facilitates the transdermal penetration of other medicines (Williams and Barry, 1989; Kararlia et al., 1995), and is thus included in topical plasters with gentamicin, suitable for acne and retinoid creams (Pocalyko et al., 2001).

Anise oil and anethole show insecticidal and insect-repellant activity (Kim et al., 2003) and act synergically with some insecticides (Clark and Shivik, 2002). Anethole, the principal component of anise oil, has shown activity against leishmaniases and other parasites (Mikus et al., 2000). Anise oil also has stronger antioxidant properties (Singh et al., 1998) than those of more usual products such as butylhydroxy anisole (BTA) and butylhydroxytoluene (BTH) and has thus been proposed for commercial use in the production of oily drugs.

10.3.11 Anise and Its Derivatives in Cosmetic Products

Anise oil and anethole are traditionally used in the manufacture of toothpastes (Gonzalez, 1983) and of elixirs for mouthwashes. The aroma and taste left in the mouth after use of toothpaste is extremely important commercially. The most commonly used flavoring is mint or aromatic combinations with other aromatic oils, including anise. Menthol gives a sensation of freshness, and the anise leaves a pleasant aftertaste. It is important that the flavor components remain unaltered during the manufacture of the toothpaste and its conservation.

Elixir for mouthwash:
- Nonionic tensioactive, 13%
- Demineralized water, 3%
- Glycerine, 18%
- Mint oil, 1%
- Anise oil (aromatizer), 1%
- Saccharin at 1%, 10%
- Ethanol (96%) sufficient to produce 100 g

Toothpaste:
- Sodium lauryl sulphate, 1%
- Anise oil (aromatizer), 1%
- Sorbitol 70%, 5%
- Glycerine, 15%
- Preservative, 0.05%
- Colouring, 0.001%
- Carbopol 940, 1.25%
- Triethanolamine, 1%
- Water, 100 g

The use of anise in cosmetics is currently being widened, with very different uses being found in addition to those mentioned above. For example, Eliaz and Gonen (1999) propose the topical use of patented mixtures of anise oil with other plant extracts to promote hair growth, minimize hair loss, and intensify or restore color. Tsuchikura et al. (2001) include anethole in deodorants because of its antiseptic properties.

In cosmetic aromatherapy, anise is recommended, among others, for decreasing inflammation and eczema of the skin and to combat dry or greasy skin (Fotinos, 2000). When used for this purpose, it forms a part of several highly volatile vehicles immersed in a polymeric matrix.

10.3.12 Pharmaceutical Uses of Fennel and Fennel Oil

The fruit and oil of fennel have high anethole content and are used in pharmacy for the same purposes as anise: for their pharmacological activity and for the correction of organoleptic properties of drugs.

10.3.12.1 Pharmacological Activity

It possesses carminative, digestive, expectorant, and spasmolytic activity, particularly at respiratory and digestive level. It also shows estrogenic activity. It is also used externally, in which situation it acts as an anti-inflammatory and possesses keratoplastic activity.

10.3.12.2 Therapeutic Indications

Similar to anise, the main properties of fennel and fennel oil are derived from their carminative and expectorant activities. Fennel is also useful in amenorrea and dismenorrea (Castaño et al., 1998). It should not be employed in cases of hyperestrogenism, pregnancy, or lactation, as it generally shows the same side effects as anise.

10.3.12.3 Oral Formulations

Preparations analogous to those of anise and anise oil should be taken orally: carminative, expectorant, and flavoring agent of medicines. Dosage is higher than that of anise when taken as a carminative or expectorant. The formulations are summarized below:

- Infusions:
  - Fennel, 10 to 30 g
  - Water, 1 L
  - Dose: 1 cup after meals
- Infusions for nursing mothers:
  - Fennel, 5 to 10 fruits
  - Water, 50 cc
- Liquid extract (1:1):
  - Dose: 20 to 30 drops, three times per day.
- Tincture (1:10):
  - Dose: 50 drops, one to three times per day.
- Capsules (*Formulario Español de Farmacia Militar,* 1975):
  - Powdered fennel, 300 to 500 mg
  - For one capsule
  - Dose: 1 to 4 g per day.
- Fennel oil:
  - Dose: 0.1 mL as carminative (Reilly, 1998)
  - Dose: One to three drops, two to three times per day in infusion, on a sugar lump, or in capsules (25 to 50 mg/capsule, two or three per day)
- Dry extract (5:1):
  - Dose: 0.3 to 2 g per day

Few formulations for fennel are included in the pharmacopoeias, probably because of its similarity to anise, both in terms of physicochemical characteristics and medicinal properties. It is sometimes found together with anise, and only the *British Pharmaceutical Codex* (1973) includes licorice powder, a compound that does not include anise.

Licorice powder, compound (*British Pharmaceutical Codex,* 1973):
- Licorice, peeled, in powder, 160 g
- Sublimed sulfur, 80 g
- Senna lead, in powder, 160 g
- Fennel, in powder, 80 g
- Sucrose, in powder, 540 g
- Properties: same as anise
- Dose: 5–10 g

Fennel oil is used in the pharmaceutical industry for the same organoleptic purposes as anise. In cosmetics, it is used as a hydrating agent for dry and flaky skin, a decolorant, a dermopurifier, and a stimulant (Parra and Pons, 1995).

References

Adjei, A.L., Doyle, R., and Reiland, T., Flavors and flavor modifiers, in *Encyclopedia of Pharmaceutical Technology*, Vol. 6, Swarbrick, J. and Boylan, J.C. Eds., Marcel Dekker, New York, pp. 101–140, 1992.

Ansel, H.C. and Popovich, N.G., Miscellaneous preparations, in *Pharmaceutical Dosage Forms and Drug Delivery Systems,* 5th ed., Lea & Febiger, Philadelphia, 1990.

Bakan, J.A., Microencapsulation, in *Encyclopedia of Pharmaceutical Technology*, Vol. 9, Swarbrick, J. and Boylan, J.C., Eds., Marcel Dekker, New York, 1994.

Blondel, S., Lefebvre, P., Tondeur, M., and Blum, D., Neonatal abstinence syndrome: current and future aspects, *Rev. Méd. Bruxelles*, 73–77, 1993.

British Pharmacopoeia, Her Majesty's Stationery Office, London, 1988.

British Pharmacopoeia, Her Majesty's Stationery Office, London, 2000.

British Pharmaceutical Codex, The Pharmaceutical Press, London, 1973.

Castaño, M.T., Ruiz, L., and Vidal, J.L., in *Monografías farmacéuticas*, Colegio Oficial de Farmacéuticos de la Provincia de Alicante, Gráficas Olmedilla, Alicante, 1998.

Clark, L. and Shivik, J., Aerosolized essential oils and individual natural product compounds as brown treesnake repellents, *Pest Managed Sci.*, 58, 775–783, 2002.

De, M., De, A.K., Sen, P., and Banerjee, A.B., Antimicrobial properties of star anise (*Illicium verum Hook f*), *Phytother. Res.*, 16, 94–95, 2002.

del Pozo, A., Expectorantes Fenólicos, in *Farmacia Galénica Especial*, Tomo II, Romargraf, Barcelona, 1977.

del Pozo, A., Laxopurgantes y anesosmóticos, in *Farmacia Galénica Especial,* Tomo II, Romargraf, Barcelona, 1978a.

del Pozo, A., Preparados de acción diurética a base de drogas vegetales, in *Farmacia Galénica Especial,* Tomo II, Romargraf, Barcelona, 1978b.

Dodge, A.A., Aceites volátiles, in *Farmacia Práctica de Remington*, Chapter 70, Fulleton, E., and Martín, E.W. Ed., Unión Tipográfica Hispano Americana, Mexico, 1954.

Eliaz, I.G. and Gonen, S., Method and product for promoting hair growth and treating skin conditions. Patent PCT Int. Appl. WO 9742964 A1, 1999.

European Pharmacopoeia, 4th ed., Directorate for the Quality of Medicines of the Council of Europe, Strasbourg, France, 2002.

Fabregas, J.L. and Moragues, J., Better tasting solid oral pharmaceutical formulations for aluminum and magnesium basic carbonate, Spain, Patent ES 2112747 A1, 1999.

Farmacopea Oficial Española, 9th ed., Real Academia Nacional de Medicina, Madrid, 1954.

Formulario Español de Farmacia Militar, 7th ed., Ministerio del Ejercito, Madrid, 1975.

Fotinos, S., Transdermal devices comprising essential oils for aromatherapy. Patent PCT Int. Appl. WO 2000016752 A2, 2000.

Fyfe, L., Synergistic bactericides comprising preservatives and plant oils. Patent PCT Int. Appl. WO 9909836 A1, 2000.

Fyfe, L., Armstrong, F., and Stewart, J., Inhibition of *Listeria monocytogenes* and *Salmonella enteriditis* by combinations of plant oils and derivatives of benzoic acid: the development of synergistic antimicrobial combinations, *Int. J. Antimicrobial Agents*, 9, 195–199, 1998.

Garzo, C., Gomez, P., Barrasa, A., Martinez, R., Ramirez, R., and Ramon, F., Cases of neurological symptoms associated with star anis consumption used as a carminative, *An. Españoles Pediatria*, 57, 290–294, 2002.

Gil, M., Perez, J.L., and Ibarra, I., Convulsive status secondary to star anis poisoning in a neonate, *An. Españoles Pediatria*, 57, 366–368, 2002.

Gonzalez, J., Higiene de la cavidad bucal, in *Cosmetología Teorico-Práctica*, Consejo General de Colegios Oficiales Farmacéuticos, Ed., Heliotipa Artística, Madrid, 1983, p. 181–203.

Griffith, I., Preparadas líquidos elaborados por maceración o lixiviación, in *Farmacia Práctica de Remington*, Chapter 39, Fulleton, E. and Martín, E.W., Eds., Unión Tipográfica Hispano Americana, Mexico, 1954a, p. 296–309.

Griffith, I., Preparados sólidos que se confeccionan sin maceración ni lixiviación, in *Farmacia Práctica de Remington*, Chapter 41, Fulleton, E. and Martín, E.W., Eds.,Unión Tipográfica Hispano Americana, México, 1954b, p. 314–319.

Irache, J.M., Ygartua, P., and Renedo, M.J., Correctivos y colorantes, in *Tecnología Farmacéutica. Vol. II: Formas Farmacéuticas*, Vila, J.L., Ed., Editorial Síntesis, Madrid, 1997, p. 347–378.

Kararlia, T.T., Kirchoffa, C.F., and Penzotti S.C., Enhancement of transdermal transport of azidothymidine (AZT) with novel terpene and terpene-like enhancers: in vivo-in vitro correlations, *J. Controlled Release*, 34, 43–51, 1995.

Kim, S.I., Roh, J.Y., Kim, D.H., Lee, H.S., and Ahn, Y.J., Insecticidal activities of aromatic plant extracts and essential oils against *Sitophilus oryzae* and *Callosobruchus chinensis*, *J. Stored Products Res.*, 39, 293–303, 2003.

Martindale. The Extra Pharmacopoea, 31st ed., Reynolds, J.R.F., Parfitt, K., Parson, A.V., and Sweetman, S.C., Eds., Royal Pharmaceutical Society, London, 1996, p. 1667–1768.

Martindale. The Complete Drug Reference, 32nd ed., Parfitt, K., Ed., Pharmaceutical Press, London, 1999, p. 1541–1646.

Mendes, R.W., Anaebonam, A.O., and Daruwala, J.B., Chewable tablets, in *Pharmaceutical Dosage Forms. Tablets. Vol. 1*, 2nd ed., Lieberman, H.A., Lachman, L., and Schwartz, J.B., Eds., Marcel Dekker, New York, 1989, p. 367–417.

Mendes, R.W. and Anaebonam, A.O., Chewable tablet, in *Encyclopedia of Pharmaceutical Technology*, Vol. 2, Swarbrick, J., and Boylan, J.C., Eds., Marcel Dekker, New York, 1990, p. 397–417.

Mességué, M. and Bontemps, M., *Guerir naturellement*, Editions Robert Laffont, and Editions Génerique, Paris, 1986.

Mikus, J., Harkenthal, M., Steverding, D., and Reichling, J., In vitro effect of essential oils and isolated mono- and sesquiterpenes on *Leishmania major* and *Trypanosoma brucei*, *Planta Med.*, 66, 366–368, 2000.

Montoya, M.A., Poisoning by star anise (*Illicium verum*) tea, *Gaceta Med. Mexico*, 126, 341–342, 1990.

Naito, Y., Wu, C.C., Seal, M.G., Gelosa, F., Yoshioka, M., Safran, P., and Marotta, F., Protective effect of a polygodial/anethole-containing natural product against candida albicans gastrointestinal colonization and dissemination, *Int. Med. J.*, 8, 3–9, 2001.

National Formulary 18, The United States Pharmacopoeia, The National Formulary, United States Pharmacopoeial Convention, Rockville, MD, 1994.

O'Neil, M.J., Smith, A., Heckelman, P.E., Obenchain, J.R., Gallipeau, J.A.R., D'Arecca, M.A., and Budavari, S., Eds., Index Merck, in *The Index Merck*, 13th ed., Merck & Co., Whithouse Station, NJ, 2001, p. 647.

Parra, J.L. and Pons, L., Ingredientes cosméticos, in *Ciencia Cosmética. Bases fisiológicas y criterios prácticos*, Consejo General de Colegios Oficiales Farmacéuticos, Ed., Heliotipa Artística, Madrid, 1995, p. 227–426.

Patel, R.N., Antimicrobial activy of the fruits of *Illicium verum. Hook F, Indian J. Pharm.*, 30, 43–44, 1968.

Pocalyko, D.J., Pillai, S., Mahajan, M.N., and Granger, S.P., Cosmetic compositions containing anise extract and retinoids. Patent PCT Int. Appl. WO 2001035921 A1, 2001.

Real Farmacopea Española, 2nd ed., Ministerio de Sanidad y Consumo, Madrid, 2002.

Reilly, W.J., Pharmaceutical necessities. Coloring, flavoring, and diluting agents, in *Remington Farmacia*, Chapter 80, Gennaro A.R., Ed., Editorial Medica Panamerica, Madrid, 1998, p. 2115–2136.

Singh, G., Kapoor, I.P.S., and Pandey, S.K., Studies on essential oils. Part thirteen: Natural antioxidant for sunflower oil, *J. Sci. Industrial Res.*, 57, 139–142, 1998.

Soliman, K.M. and Badeaa, R.I., Effect of oil extracted from some medicinal plants on different mycotoxigenic fungi, *Food Chem. Toxicol.*, 40, 1669–1675, 2002.

Theis, J.G., Selby, P., Ikizler, Y., and Koren, G., Current management of the neonatal abstinence syndrome: a critical analysis of the evidence, *Biol. Neonate*, 71, 345–356, 1997.

Tsuchikura, K., Koga, M., and Mori, S., Deodorants for masking ammonia-related unpleasant odor in hair wave-setting preparations, Japan, Patent JP 2000344629 A2, 2001.

United States Pharmacopoeia 23, The National Formulary, United States Pharmacopoeial Convention, Rockville, MD, 1994.

Wagner, C.L., Katikaneni, L.D., Cox, T.H., and Ryan, R.M., The impact of prenatal exposure on the neonate, *Obstet. Gynecol. Clin. N. Am.*, 25, 169–194, 1998.

Williams, A.C. and Barry, B.W., Essential oils as novel human skin penetration enhancers, *Int. J. Phar.*, 57, R7–R9, 1989.

WHO, Cough and cold remedies for the treatment of acute respiratory infections in young children. Department of Child and Adolescent Health and Development, Document WHO/FCH/CAH/01.02, 2001, World Health Organization, Geneva.

Zellner, G., Nasal spray liquid, Patent DE 19731472 A1, 1999.

11 Economic Importance and Market Trends of the Genera *Pimpinella, Illicium,* and *Foeniculum*

Antonio Rapisarda

CONTENTS

11.1 PLANTS AND THEIR PRODUCTS

11.1.1 ANISE

Pimpinella anisum L. (sin. *Anisum vulgare* Gaertn.; *A. officinarum* Moench. Family: Apiaceae). International names: anise, anise seed, aniseed, sweet cumin.

The cremocarps of *P. anisum* L. are known as anise or aniseed. There are about 150 species belonging to this genus in Eurasia and in Africa. *P. anisum* is native to the eastern coast area of the Mediterranean, to Asia Minor, and to Egypt. Spontaneously growing anise may only be found on a few isles of the Aegean Sea. Anise was used in ancient times by the Egyptians. It is mentioned by Plinio, Pitagora, and Dioscoride as a seasoning and medicinal plant named anneson (in Greek, ανισον = unequal, irregular). Anise is still a popular spice today. Its fruit (*Anisi vulgaris fructus*) is officially regarded as a drug by the Pharmacopoeias of several states of the world (Hornok, 1992).

The fruit (misnamed a seed) is ovoid or pyriform, laterally compressed, 3 to 5 mm long, 2 to 3 mm broad, and greyish-green to greyish-brown. The individual mericarp is broadly ovoid and five-ridged, with short hair and numerous vittae. Spanish and Italian anise is large and light colored, whereas German and Russian anise is smaller, ovoid, and dark (Ilyas, 1980). The mass of 1000 fruits is 1 to 4 g, and the fruit is similar to the poisonous hemlock fruit. The characteristic odor and the sweet, spicy taste of the fruits originate from the contained essential oil that is accumulated in 30 small oil passages and 2 to 4 larger oil passages (Ilyas, 1980). The fruits contain 8 to 23% of a fatty oil and 2 to 7% of an essential oil that is a colorless or pale-yellow liquid obtained by

0-415-32246-4/04/$0.00+$1.50

steam distillation of the crushed fruit. Because of its characteristic and agreeable odor and because of the taste of the spice, it is used for flavoring in lieu of the fruits. The chief constituent (80 to 90%) of anise oil is anethole, which is responsible for the characteristic odor. Other constituents found in trace amounts are methyl chavicol, p-methoxyphenyl acetone, vanillin, anisaldehyde, anisalcohol, hydroquinone, cresol, eugenol, myristic, and anisic acid (Ilyas, 1980).

Because of its wide field of applicability, anise is cultivated in many areas. The most important countries cultivating anise are Spain, Bulgaria, Romania, Italy, the Russian Federation, Turkey, India, Mexico, and North Africa (Hornok, 1992). The cultivation of this plant needs attention. Unlike the cumin plant, this species did not suffer from any pest or disease.

11.1.2 Badian

Illicium verum Hook. f. (sin. *I. stellatum* Makino = Family: Illiciaceae). International names: badian, star anise, Chinese anise, Chinese star anise, Siberian cardamom.

Illicium verum Hook. f. has been variously put in the Magnoliaceae, Schizandraceae, Winteraceae, and Illiciaceae by different workers. It is an evergreen tree up to 10 m tall that is indigenous to southern China and Indochina, and according to Zeven and de Wet (1982), it is not known in the wild. Its name derives from the Latin *Illicium,* from *illicere,* to attract, for its tempting perfume; *verum* = true, genuine, so as to distinguish it from false fruits. It was known and used as an appetizer and digestive and for flavoring food in China about in thousand years BC, under the Sung dynasty, and it is mentioned in the ancient codex Pen-Ts'-ao. In 1588, still unknown to Greek and Latin populations, it was introduced in Europe by T. Cavendish, an English seaman, who imported it into England. Its presence in Europe was largely extended in the eighteenth century.

The star-shaped fruit of *I. verum*, consisting of the free carpels, constitutes the star anise of commerce. Because star anise reached Europe from eastern Asia through Siberia, it is also known there as Siberian cardamom. The aggregate of fruits (follicles) is termed a follicetum. Each follicle is boat-shaped, 12 to 17 mm long, and opens from above along the marginal suture. The pericarp is reddish brown, woody, and wrinkled. The carpel contains a compressed, ovoid, hard, smooth, shiny, brown seed (Ilyas, 1980).

The fruit of *I. verum* has an agreeable, aromatic, sweet taste and a pleasant odor resembling anise. It is harvested when ripe and is distilled immediately for the essential oil, or it is dried. An essential oil is present in the seed and the pericarp, which comprise approximately 30 to 35 g/kg of the fruits; 80 to 90% of this oil is anethole, whereas the rest consists of chavicol methyl ester, p-methoxyphenylacetone, and safrole. This oil, called star anise oil, has the characteristic odor and taste of true anise oil and is used in lieu of anise oil and the spice, as it displays the same properties (Ilyas, 1980).

As a substitute for anise, star anise is grown in Indochina, where it is native, and produces a crop known as Anisi stellati fructus (Hornok, 1992). It is propagated by seeds. The tree comes to bear fruits in its tenth year and continues to fruit for 80 to 100 years (Ilyas, 1980).

Small (1996) points out that the common names star anise, Chinese star anise, and Japanese star anise are used in contradictory ways among major reference works to refer to the comestible *I. verum* and the poisonous *I. anisatum* L. *Illicium anisatum* (sin. *I. religiosum* Sieb & Zucc.) is a small shrub or tree that grows to 8 m in height. It is also indigenous to southeastern Asia, and it is cultivated. Its seeds are very toxic and have been used as a fish poison. Claus and Tyler (1965) attribute this poisonous property to the seed's skimming content. Anisatin and related convulsant sesquiterpene lactones have been identified in *I. anisatum.* Nevertheless, the seeds of this species have been used medicinally in China; for example, to treat toothache and dermatitis. In Japan, cut branches are commonly used as Buddhist grave decoration on temple grounds. In North America, this toxic species can be grown outdoors as an ornamental (Bailey and Bailey, 1976).

Because both species have reputations as medicinal plants, the seeds of either plant could be encountered as imported items, or potentially in herbal practice, and an inadvertent substitution could occur. "Star anise" (so named for the eight-pointed fruit) is commonly used to refer to two economic species of *Illicium,* one with poisonous seeds and the other with seeds that are safely consumed. Although only one of the two species is considered toxic when taken orally, both are known to cause dermatitis in susceptible individuals (Mitchell and Rock, 1979). The two species are quite similar, danger arising because the fruits resemble each other closely (Claus and Tyler, 1965). The two species have often been confused for each other (Smith, 1947), and there have been several degrees of confusion between the two species in the literature; therefore, to avoid accidental poisoning, the unambiguous scientific nomenclature is preferable.

11.1.3 Fennel

Foeniculum vulgare Mill. (sin. *F. officinale* All. = Family: Apiaceae). International name: fennel.

The fruits of *F. vulgare* Mill. (Apiaceae) constitute commercial fennel. The plants are native to southern Europe and western Asia. Since Roman times, fennel fruits have been used as a flavoring for drinks. One example, sack, is mentioned in William Shakespeare's *The Tempest.*

There is a horde of wild and domesticated varieties and races that differ in size, odor, and taste of the fruits, but they are hardly distinguishable from one another. Different workers have viewed the different types of fennel as races or varieties or subspecies of *F. vulgare*. There are two major varieties of fennel in cultivation; namely, var. *vulgare* (Mill.) Thellung, and var. *dulce* (Mill.) Thellung. The former is cultivated all over the world and yields the sweet fennel oil. Cremocarps are green to yellow in the trade. The fruit are more aromatic and attractive, and hence they are preferred in the trade. The fruit is 7 to 8 mm long and 1.5 to 2.0 mm broad, and it is laterally compressed. The stylopodium persist on the fruit. The constituent mericarps have ridges and furrows running along their length on the outer wall. The pericarp has oil canals that contain an essential oil. The endosperm is rich in a fatty oil. The dried fruits (misnamed seeds) are aromatic and taste sweet, and they impart a cooling aftereffect in the mouth (Ilyas, 1980).

The fennel plant is pleasantly aromatic, so the leaves are added to various culinary preparations. Crushed fruits when distilled give 0.7 to 1.2% of a colorless or pale-yellow essential oil known as oil of fennel. The yield of Indian fennel oil is low compared with the 4 to 6% essential oil yield of European fennel. Fennel oil has the characteristic odor and flavor unique to fennel. In trade, two types of oil are known: sweet fennel oil from fruits of var. *dulce*, and bitter fennel oil from var. *vulgare*. The flavor and aroma of sweet fennel oil is superior to that of bitter fennel oil. The main constituent of the oil is anethole, while anisic acid, anisaldehyde, fenchone, methyl chavicol, pinene, camphene, phellkandrene, limonene, and foeniculin are present in small but variable amounts. The oil of sweet fennel is devoid of fenchone and may have up to 90% of anethole. It is responsible for the sweet taste, delicate flavor, and pleasant odor. The plant parts, on distillation, also yield an essential oil that is rich in phellandrene. The fennel fruits are classified according to their place of origin. Indian fennel oil is sweet and contains about 70% anethole and 6% fenchone (Ilyas, 1980).

On sale, the products of these plants often are not genuine, because they are obtained from or mixed with material from other plants. The fruits of *P. anisum* are adulterated with exhausted fruits, dusts, and other similar-looking fruits, whereas ground anise is adulterated with ground fennel, which resembles it in aroma and flavor and is considerably cheaper. In addition, anise essential oil is sometimes adulterated with the cheap star anise oil obtained from the fruits of *I. verum*, which are seldom adulterated with those of other species of *Illicium* (*I. religiosum* = Japanese star anise), are hardly aromatic, and have a disagreeable taste as a result of poisonous shikimic acid. In India, oil of fennel is also probably sold as anise oil; the other adulterants of this oil are turpentine oil and cedar wood oil (Ilyas, 1980).

In the U.S. market, as required by law, spices and herbs must be genuine products indicated by their common names on the labels. The identities of herbs and spices are established by their botanical names. If obtained from or mixed with material from other plants, the herbs and spices are both adulterated and misbranded. As a guide to the identities of food spice products, the Food and Drug Administration uses the following definition: "spices are aromatic vegetable substances used for the seasoning of food." They are true to name, and from them no portion of any volatile oil or other flavoring principle has been removed. In the absence of definitions and standards for the identity of spices, aniseed, star aniseed, and fennel are listed as Generally Regarded As Safe, and the following descriptions, established by the Food and Drug Administration, bring up to date the list of spices, following consultation with the American Spice Trade Association, and provide guidance concerning acceptable names for use in labeling spices and the food in which they are used.

- Anise (Aniseed): The dried, ripe fruit of *P. anisum* L., an annual herb of the parsley family. The dried seeds are greenish-grey in color, are crescent shaped, and possess a strong licorice-like odor. The principal active ingredient of the volatile oil is anethole. The quality characteristics are measured by the volatile oil and total acid insoluble ash.
- Star aniseed: The dried, ripe fruit of *Illicium verum* Hook f. The principal active ingredient of the volatile oil is anethole. The quality is measured by the amounts of volatile oil and total acid-insoluble ash.
- Fennel seed: The clean, dried, ripe fruit of *Foeniculum vulgare* Mill. The green- or yellowish-tan-colored seeds have a pleasant aromatic odor and possess a sweet anise-like taste. The quality is measured by the amounts of volatile oil, total and acid-insoluble ash, and moisture.

One must also remember that those spices or their oils that are in the United States Pharmacopoeia or the National Formulary are subject to the standards set forth in these compendia when used for drug purposes.

In the 15 European Union member states, anise, used extensively as a spice, is listed by the Council of Europe as a natural source of food flavoring (category N2). This category allows small quantities of aniseed to be added to foodstuffs, with a possible limitation of an active principle (as yet unspecified) in the final product (Council of Europe, 1981).

11.2 ECONOMIC IMPORTANCE

The fruits of *P. anisum, I. verum,* and *F. vulgare* are an important industrial source of essential oil, which is principally used today for the production of anethole and for the formulation of anise-flavored compounded oils widely employed for many purposes. The essential oils obtained by the above-mentioned three species are reciprocally employed. The economic importance of anise, badian, and fennel is the result of their use in the alimentary and pharmaceutical fields.

11.2.1 Food Flavoring Uses

The fruits of *P. anisum* have appetizer and digestive effects, and as a spice they are mainly used by bakeries, in confectionery, and by the liquor industry (Hornok, 1992). Anise is used for flavoring bread, cakes, curries, pastries, and candy in the producing countries. The fruits and their essential oil, extensively used for flavoring foods, beverages, and pharmaceuticals, mainly in temperate climates, are not popular as flavorings in Latin America (Morton, 1981).

In many Mediterranean countries there is at least one drink in which this flavor prevails, even if the taste generally does derive not only from the anise but also from fennel and badian. In France, particularly in southern regions, there are Pernod and other similar drinks, generically named pastis (Richard, Berger, etc.), and sweet liqueurs such as anisette. In Spain and North Africa there is a series of similar drinks. In Italy the alcoholic beverages are mistrà, anesone, anice forte, and sambuca (Molinari). Ouzo, raki, arrak, and so forth are found throughout eastern Mediterranean regions.

The unripe fruit of *I. verum* is chewed after meals as a digestive and breath sweetener, and dried fruits are employed for pickling and as a condiment for flavoring in curries, pickles, cookies, cakes, tea, coffee, and sweetmeats, particularly in Asia. The star anise fruits are also an important source of essential oil, principally used today for the production of anethole and for the formulation of anise-flavored compounded oils employed in liqueurs (bitter, brandy, etc.). The French liqueur mistrà and perhaps anisette and the Greek alcoholic beverage ouzo contain flavors obtained from *I. verum* or from a mixture of star anise and anise. The maximum level of essential oil is about 0.006% (570 ppm) in alcoholic beverages and 0.07% (681 ppm) in cookies, cakes, and so forth (Leung and Foster, 1999).

Fennel fruits are used in culinary preparations, confectionary, and liquors. Fennel fruit is used whole or ground as a spice in cooking to season bread, rolls, pastries, pickles, fish dishes, sauces, and so forth, and therefore it is much in demand in the small and great foodstuffs manufacturing industries. The flavor of fennel is like that of anise or licorice, and it is employed as a flavoring for condiments, soaps, creams, perfumes, and liqueurs.

There are several varieties of fennel; both common and sweet fennel are grown for their fruits' essential oil. Sweet fennel is also grown for the thickened bulb-like base of the leaf stems (often called anise), a 3- or 4-inch-wide structure that grows just above the ground. For foliage and fruits, the best fennel is found in the Mediterranean region. Although fennel grows wild in California, the climate is probably a little too moist there.

11.2.2 Medicinal Uses

Anise is said to possess expectorant, antispasmodic, carminative, and parasiticide properties. In traditional medicine, anise has been used for bronchial catarrh, pertussis, spasmodic cough, and flatulent colic, topically for pediculosis and scabies, and specifically for bronchitis, for tracheitis with persistent cough, and as an aromatic adjuvant to prevent colic from cathartics (*British Herbal Pharmacopoeia*, 1983). Anise has also been used as an estrogenic agent and has been reputed to increase milk secretion, promote menstruation, facilitate birth, alleviate symptoms of the male climacteric, and increase libido.

The chemistry of anise is well studied, and documented pharmacological activities support some of the herbal uses. Today, anise is used extensively as a spice and is widely used in conventional pharmaceuticals for its carminative, appetizer, cholagogue, digestive, spasmolytic, diuretic, stomachic, diaphoretic, expectorant, and flavoring properties.

Anise contains anethole and estragole, which are structurally related to safrole, a known hepatotoxin and carcinogen. Although both anethole and estragole have been shown to cause hepatotoxicity in rodents, aniseed is not thought to represent a risk to human health when it is consumed in the amounts normally encountered in foods (Newall et al., 1996). Anethole, anisaldehyde, and myristicin have exhibited mild insecticidal properties in animal studies (Leung, 1980). The bacteriostatic effect of anise essential oil (Aetheroleum anisi vulgaris) is widely known. Its pleasant odor and taste are also exploited by the medicine industry (Hornok, 1992). In addition, the fruits of *I. verum* and *F. vulgare* are used in conventional pharmaceuticals for their carminative, stomachic, and diuretic properties.

Anise, badian, and fennel essential oils are employed for to mask the bad or strong smell of medicinals and cosmetic products and as flavoring components in toothpastes, perfumes, soaps, creams, and lotions at a maximum level of 0.25% for anise essential oils and 0.4% for star anise essential oil (in the perfumes) (Leung and Foster, 1999).

In herbal medicine, the fruits of *P. anisum* L. are used as a digestive and after childbirth. Infusion of anise is used in particular by old people and children to increase appetite and as a digestive. Mixed with cumin and fennel it is given to women after childbirth to increase the flow of milk and to ease postpartum pains (Ghazanfar, 1994).

Bitter fennel (*F. vulgare* ssp *vulgare* var. *vulgare*) fruits are highly appreciated as a natural remedy and baby tea for support and sedative for the digestive organs. Bitter fennel is used for both pharmaceutical and food tea. Furthermore, it is used for flavoring of baker's goods, meat and vegetable meals and pickled cucumbers. The German pharmacopoeia (DAB 10) requires the following quality: 4% essential oil of the fruits, with at least 60% anethole, 15% fenchone, and no more than 5% estragole in the essential oil. The 1000-seed weight should be nearly 4 g for tea bags and 7 to 9 g for pharmaceutical use.

Sweet fennel (*F. vulgare* ssp *vulgare* var. *dulce*) is hardly grown in Germany, because this chemocultivar does not meet the requirements of the German Pharmacopoeia concerning the fenchone content of its essential oil. Sweet fennel is cultivated in France and southern European countries for the production of anethole, which is used as flavoring for spirituous liquors.

11.2.3 Market Trends

Analysis of the worldwide market of the fruits of *P. anisum, I. verum*, and *F. vulgare* is very difficult because the data available are incomplete, fragmentary, and sometimes conflicting. The Food and Agriculture Organization of the United Nations is the only official source of data relative to the principal economic parameters concerning the production and the worldwide trade of anise, badian, and fennel, though these data are cumulative for the three species. In fact, in this case, it is very hard to quantify the amount of every single one of these three spices that is produced, imported and exported in the world, because often many Institutes for Foreign Commerce give the same generic denomination to different products. In India, for example, there is some confusion regarding the usage of the term aniseed, as the terms "saunf" and "badyan" are concurrently used for *P. anisum, F. vulgare*, and *I. verum*; furthermore, the imports into India from Europe are that of *P. anisum*, whereas those from southeast Asia are that of *I. verum* and the exports from India refer to *F. vulgare*. The confusion apparently is caused by the similar odors of the three spices, a result of the presence in all three essential oils of a common component. Furthermore, anise, badian, and fennel are important industrial sources of essential oil, principally used today for the production of anethole, and the formulation of anise-flavored compound oils widely employed in pharmaceutical and foodstuff industries. The essential oils obtained from the above-mentioned three species are reciprocally employed.

Table 11.1 (source FAOSTAT) shows the worldwide situation, by country, of the crop production in metric tons (mt), including the quantities of the products in the market (marketed production) and the quantities consumed or used by the producers (auto-consumption); of the yield recorded in hectograms (100 grams) per hectare (hg/ha), which represents the harvested production per unit of harvested area; and of the area from which the crop is gathered (area harvested) recorded in hectares (ha). These data demonstrate that India has in the last 30 years been the leader in the production of these medicinal plants. The same source (FAOSTAT) supplied the data (Table 11.2 and Table 11.3) relative to the foreign trade (imports and exports) in quantitative forms, expressed in weight (mt) and in value terms in thousands of U.S. dollars. For this latter parameter, national currencies used as legal tender in international transaction by the countries are converted by using the average annual exchange rate (RH series) provided by the International Monetary Fund.

The global production (Figure 11.1), analyzed from 1961 to 1999, shows a positive trend with an annual average increase close to 2%. A proportional increasing trend has the global area harvested (Figure 11.2) and the harvested production per unit of harvested area (Figure 11.3).

In the last four decades, the worldwide trade of anise, badian, and fennel in value terms has always continuously and steadily increased both for the exports (Figure 11.4) and for the imports (Figure 11.5), which is not a result of a rise in prices, as the same positive trend is present when the imports and exports are recorded in quantitative forms (Figure 11.6 and Figure 11.7).

According to data of the Food and Agriculture Organization, India, Mexico, China, Turkey, and Egypt are the greatest producing countries, whereas the United States and the 15 European Union member states are the greatest worldwide importers.

The trade of anise, badian, and fennel follows, as mentioned above, different paths. The fruits of *P. anisum* found on the American or European markets come from Turkey, Spain, and Southeast

Asia (Ilyas, 1980), even if they are grown on small scale in different parts of the world. *Illicium verum* fruits are obtained from the markets of China, Hong Kong, Malaysia, Singapore, and Indonesia under the title "badyan" or "badian" seed or "star anise" (Ilyas, 1980). A mountain area (Bi Se) in the province of Guangxi is the biggest star anise-producing area in China. The star anise oil and star anise fruits from Bi Se have good quality and are being sold to many European countries and to the United States. *Foeniculum vulgare* fruits and their essential oil come from Egypt and India, which are the greatest exporter countries of this product in the world.

The most recent statistics, from 1997, demonstrate that anise, badian, and fennel represent 2.93% of the U.S. import market for spices (0.76% for anise or badian, and 2.17% for fennel) (Adkins and Hecker, 1997). It must be taken into consideration that the United States is the largest import market for spices in the world. Spices are generally considered to be any of the flavored or aromatic substances of vegetable origin that are obtained from tropical or other plants. They are commonly used as condiments or are employed for other purposes on account of their fragrance, preservative, or medicinal qualities. Spices are mostly grown and exported by developing and least-developed countries. India is the world's largest producer and exporter of spices, producing around 50 different varieties of spices, and is also a major consumer, accounting for approximately 20% of world spice consumption.

Anise import volumes fell by 10% in 1997, although import value increased by 6%. Turkey and China are the major suppliers (Figure 11.8). In 1997, India was also successful in the U.S. import market with fennel. In the three previous years, Egypt had consistently supplied 70 to 80% of import demand. In 1997, India (Figure 11.9) became the leading supplier, increasing their fennel exports to the United States fivefold. Overall, total fennel import volume increased by 20% in 1997, whereas import value increased by 15%.

New York City is the largest U.S. port of entry for spices, accounting for slightly more than 50% of total import value. The United States relies on imports for approximately two-thirds of its seasoning needs. An estimated 80 to 85% of spices are marketed in their whole, unground state but processed spices are increasing their market share.

The importation and distribution of spices and, thus, of anise, badian, and fennel, in the United States is carried out through established brokers, importers, and processors. There has recently been a trend toward direct buying, in which the processors or supermarket chains buy directly from sources in producing countries, bypassing the agent and importer. In addition, some U.S. spice processors have set up long-term alliances with farmers and governments in spice-producing countries. The U.S. retail sector is serviced by a small number of large processing and marketing companies that grind imported or domestically produced spices and pack them in a variety of containers for sale at retail outlets. Commercial practices for spices are similar to those of other commodities, but the uniqueness of the quality and availability of the spices from their traditional origins makes for a highly specialized market. Nearly two thirds of U.S. spice use takes place in the food-manufacturing industry, with retail food store outlets making up most of the remainder.

A trend toward less salt in foods has stimulated the use of more condiments to compensate for flavor loss. Despite these changes, retail sales of spices appear to be diminishing as a result of increasing sales of ready-to-eat foods, fast foods, and restaurant food. Recent bacteriological problems have put greater significance on essential oils. The advantages are consistency of quality, freedom from microorganisms, uniform dispersion in the product, and easy handling and storage.

The 15 European Union member states imported 158,978 metric tons of spices in 1996, a 2% increase from the previous year and a 53% rise from 1988 (Adkins and Hecker, 1997). Imports of anise, badian, and fennel represent 1.18%, 0.24%, and 3.85%, respectively, of the total European spice imports. The data relative to the European anise, badian, and fennel import trend are reported in Figure 11.10 in quantitative forms expressed in weight (mt) and in Figure 11.11 in value terms in thousands of U.S. dollars. Imports of anise rose more than 7% from 1995 and 97% from 1988, whereas imports of badian jumped by 13% from 1995 and 27% from 1988. Fennel imports were up by 25% compared with 1995 and by 44% compared with 1988.

TABLE 11.1
Crop Production in Metric Tons (mt), Yield in Hectograms per Hectare (hg/ha) and Area Harvested in Hectares (ha)

	1999			1989			1979			1969		
Anise, Badian, and Fennel	**Production (mt)**	**Yield (hg/ha)**	**Area Harv. (ha)**	**Production (mt)**	**Yield (hg/ha)**	**Area Harv. (ha)**	**Production (mt)**	**Yield (hg/ha)**	**Area Harv. (ha)**	**Production (mt)**	**Yield (hg/ha)**	**Area Harv. (ha)**
Afghanistan	200	6.667	300	800	7.273	1.1						
Argentina	2.8	8	3.5	2.2	7.097	3.1	1.180			698		
Bangladesh							600			300		
Bulgaria							261	503	5.187	1.400	3.795	3.689
China	26.75	9.386	28.5	17	7.391	23	9			4.5		
Cyprus				3	15	2	10	20	5	38	29.231	13
Denmark	129	5.609	230	1.618	8.09	2	1.475	13.446	1.097	310	8.986	345
Ecuador	10	4	25	141	28.2	50	5	1.923	26			
Egypt	22	8.462	26	23.8	8.655	27.5	11.5			1		
Ethiopia	100	6.25	160									
Ethiopia PDR				100	7.143	140	40	4.000	100	50	7.143	70
Georgia	30											
Greece	470	9.792	480	460	9.787	470	695	8.797	790	410	4.556	900
Guatemala	600	7.5	800	540	6.835	790	450			50		
Hungary	3	9.091	3.3	3	9.091	3.3	200			100		
India	110	3.667	300	67.593	3.96	170.693	92.414	5.084	181.786	43	4.526	95
Iran, Islamic Rep of	22	8.8	25	22.5	8.654	26	3.5	5.833	60	13	7.429	17.5
Kenya	100	5.263	190	100	5	200	200	6.667	300	400	8	500

Lebanon	95	19	50	28	12.174	23						
Madagascar	10	5	20	10	5	20	10	5	20			
Malawi	50	5	100	50	5	100						
Mexico	32	91.429	3.5	20.223	94.324	2.144	600	40	150	100	40	25
Morocco	23	10.455	22	23	10.455	22	11	5.5	20	11.31	5.655	20
Netherlands	200	10	200	1.126	17.35	649	3	12.019	2.496	4.226	15.751	2.683
Poland										3.4		
Romania	5	5.556	9	3	7.5	4						
Russian Federation	4	8	6									
Spain	400	2.5	1.6	1.768	5.051	3.5	1.398			1.9		
Syrian Arab Republic	9	3.6	25	7.1	7.553	9.4	1.5	4.167	3.6	2.887	4.224	6.834
Tunisia	9.5	7.6	12.5	9.5	7.6	12.5	8.6			7.5		
Turkey	22	6.286	35	8.1	4.451	18.2	2.5	7.143	3.5	2.3	7.419	3.1
Viet Nam	1.5	7.5	2	1.2	8	1.5	330	6.600	500	600	6	1
West Bank							13	4.062	32			
Yugoslavia SFR				1.5	7.143	2.1	2			1		
Zimbabwe	100	5	200	100	5.556	180						

TABLE 11.2
Foreign Trade (Exports) Expressed in Weight (metric tons) and in US$K

Anise, Badian, and Fennel Exports	1999		1989		1979		1969	
	Val	Qty (mt)	Val	Qty (mt)	Val	Qty (mt)	Val	Qty (mt)
Afghanistan	50	140	790	770				
Argentina	207	360	319	874	1.228	744	51	233
Australia	1.140	1.407	1.382	3.112				
Austria	1.476	842	198	75	47	27	3	8
Bangladesh	8	6	53	54	101	100	3	19
Barbados	59	29	7	5				
Belgium-Luxembourg	603	285	211	231	143	102	7	28
Bolivia	18	38					6	25
Bosnia and Herzegovina	1.000	600						
Brazil	1		1		8	6		
Bulgaria	3.400	24.000					410	1.261
Canada	5.304	7.673	81	72				
Chile	10	10	17	7				
China	4.819	5.159	7.245	7.4	2.601	4.892	42	151
Colombia								
Costa Rica	43	19						
Croatia	14	4						
Cyprus	1		25	35	10	5	3	5
Czech Republic	1.212	1.523						
Denmark	336	321	162	182	1.666	1.009	76	259
Ecuador	4	2			8	18		
Egypt	4.545	6.698	7.410	20.386	5.871	9.529	58	135
Estonia	109	134						
Ethiopia	28	20						
Ethiopia PDR			226	81				
Finland	339	368			5	4		
France	2.131	492	1.093	576	514	256	53	91
Germany	4.599	1.820	2.533	1.448	1.651	963	87	198
Greece	18	45	8	6	3	4	3	10
Guadeloupe								
Guatemala	10	70	1	2	189	165	13	39
Hungary	1.018	1.504	1.415	3.022	69	30		
India	12.000	21.000	7.623	9.186	8.412	10.508	530	1.302
Indonesia	129	275	71	15	1	6		
Iran, Islamic Rep of	13.109	11.547	18.195	22.026	4.332	3.047	2.880	12.538
Iraq							1	5
Israel			20	20	242	230		
Italy	1.053	313	744	245	841	363	195	347
Jamaica	2							
Japan					1		8	7
Jordan	6	1	28	41	24	45	8	48
Kenya			2	1	36	168	70	479
Korea, Republic of			40	5			2	3
Kuwait	19	16	40	18				
Latvia	7	3						
Lebanon	20	40	80	40	450	500	396	1.712
Lithuania	72	102						
Macedonia (FYR)	2.500	1.000						

TABLE 11.2 (continued)
Foreign Trade (Exports) Expressed in Weight (metric tons) and in US$K

Anise, Badian, and Fennel Exports	1999		1989		1979		1969	
	Val	Qty (mt)	Val	Qty (mt)	Val	Qty (mt)	Val	Qty (mt)
Madagascar	1	27						
Malawi	72	523	11	3				
Malaysia	249	4	144	256	45	150	14	71
Mauritius	6							
Mexico	46	67	190	78	144	186	21	54
Morocco	4.113	3.175	4.265	8.724	2.036	5.231	1.925	11.647
Myanmar	200	580						
Nepal			5	6				
Netherlands	5.774	5.446	4.091	5.497	6.357	4.275	1.247	3.664
New Zealand	22	14	46	244				
Nicaragua		1						
Norway	8	10						
Oman	178	101	7	9				
Pakistan	1.583	1.999	793	718	25	37		
Panama	3	1						
Peru	82	35	29	17	182	119		
Poland	999	1.108					1.116	3.374
Portugal	35	13	7	2	13	2	2	4
Romania	1.690	4.87	1.018	2.482				
Russian Federation	199	948						
Réunion			2					
Saint Lucia	1							
Saudi Arabia	14	15	77	63	68	155		
Singapore	9.485	7.954	6.707	9.128	7.372	8.081	719	3.160
Slovakia	14	51						
Slovenia	29	4						
South Africa	34	29	24	8	11	20	4	10
Spain	1.642	701	1.298	730	2.068	1.236	714	1.633
Sri Lanka	38	12	5	2	11	5		
Swaziland	1	1						
Sweden	133	26	18	1	26	18	3	4
Switzerland	36	11	18	6	13	4	2	3
Syrian Arab Republic	12.423	7.473	1.148	1.309	696	687	732	2.895
Tanzania, United Rep of	40	60						
Thailand	53	37	30	15				
Togo	7	3	7	2				
Trinidad and Tobago	152	174	2	1				
Tunisia	13	14	1.068	904	947	719	176	743
Turkey	18.158	12.449	16.341	21.726	596	512	39	105
Ukraine	750	296						
United Kingdom	1.355	496	392	136	367	175	37	76
United States	1.162	564	529	204				
Venezuela	17	25						
Viet Nam	2.800	1.400	900	500	30	104		
Yemen	77	27					17	57
Yugoslavia SFR	870	1.100	1.331	1.062	2.565	1.710		
Zambia	37	10						
Zimbabwe	23	35	15	20				

TABLE 11.3
Foreign Trade (Imports) Expressed in Weight (metric tons) and in US$K

Anise, Badian, and Fennel Imports	1999		1989		1979		1969	
	Val	Qty (mt)	Val	Qty (mt)	Val	Qty (mt)	Val	Qty (mt)
Albania		10						
Algeria	420	820	719	929	936	801	185	636
Angola							3	2
Argentina	588	348	317	284	722	351	61	91
Armenia	4	10						
Australia	840	527	413	277	308	276	69	217
Austria	2.270	2.005	1.334	1.661	1.697	1.285	293	913
Azerbaijan	7	3						
Bahamas	15	4	32	9				
Bahrain	702	632	24	19	2.641	19		
Bangladesh	3.409	2.600	5.265	4.743	1.800	1.124		
Barbados	38	12	11	7				
Belarus	360	261						
Belgium-Luxembourg	1.000	605	493	554	317	272	67	187
Belize	9	4						
Bhutan	34	3						
Bolivia	62	29	164	156	293	398	69	149
Botswana	24	7	3	3				
Brazil	8.626	6.927	7.649	5.536	3.928	1.699	504	1.089
Brunei Darussalam	130	120	80	70	45	40	5	16
Bulgaria	170	210						2
Cambodia							3	12
Cameroon	10	10	8	3	5	5	1	1
Canada	1.517	971	991	773				
Cape Verde	18	6	2	1			1	1
Chile	540	350	541	533	395	200	26	49
China	2.880	2.564	2.796	2.657	1.244	1.003	225	717
Colombia	2.250	1.734	1.930	2.795	3.883	1.831	465	1.085
Congo, Dem Republic of	1	3					13	1
Costa Rica	250	110	156	121				
Croatia	76	53						
Cyprus	115	96	44	55				
Czech Republic	538	394						
Denmark	396	245	135	153	95	74	16	53
Djibouti	100	70	68	59				
Dominican Republic	140	40	90	100	28	56	25	120
Ecuador	1.123	876	9	5	451	344	115	250
Egypt	569	485	775	1.723				
El Salvador	57	71	90	77				
Estonia	71	33						
Ethiopia	4	12						
Ethiopia PDR					9	8		
Faeroe Islands	1							
Fiji Islands	290	410	266	319	311	360	53	168
Finland	218	93	170	134	112	87	41	105
France	4.107	3.029	2.485	2.866	2.266	1.902	354	1.122
French Guiana			9	2				
French Polynesia	13	1	10		3		1	1

TABLE 11.3 (continued)
Foreign Trade (Imports) Expressed in Weight (metric tons) and in US$K

Anise, Badian, and Fennel Imports	1999		1989		1979		1969	
	Val	Qty (mt)	Val	Qty (mt)	Val	Qty (mt)	Val	Qty (mt)
Gabon	1	1						
Germany	10.909	10.330	7.741	9.282	7.529	7.067	1.208	4.271
Ghana	1							
Greece	640	743	439	527	129	92	29	57
Grenada			5	3			5	8
Guadeloupe			28	21	3	1		
Guatemala	70	70	37	64	402	143	37	47
Honduras	170	209	104	186				
Hungary	111	99	40	32			55	155
Iceland	23	9	15	10	15	7	3	7
India	1.100	1.400	1.413	905	781	631	529	570
Indonesia	3.666	12.522	3.988	7.137	1.885	3.773		
Iran, Islamic Rep of								2
Israel	564	320	552	700	226	126	42	103
Italy	3.43	1.655	1.052	1.024	877	867	151	514
Jamaica	20	20	66	30	62	30		
Japan	9.095	6.005	4.228	5.065	3.785	3.681	604	2.197
Jordan	578	789	212	343	260	249	55	171
Kazakhstan	16	42						
Kenya	244	184	39	39	241	142	60	168
Korea, Republic of	647	400	526	408	91	82	2	3
Kuwait	945	722	352	435				
Laos							1	6
Latvia	267	179						
Lebanon	980	430	400	500	1.300	1.500	501	1.719
Libyan Arab Jamahiriya	10	70	553	337	1.223	727	154	556
Lithuania	180	283						
Macedonia (FYR)	110	80						
Madagascar							3	5
Malawi			7	3				
Malaysia	9.067	12.464	4.851	8.763	2.711	5.803	694	3.734
Maldives	30	40						
Malta	70	21	13	5	3			
Martinique			1					
Mauritius	755	868						
Mexico	3.070	2.288	1.195	1.401	77	56	32	55
Moldova, Republic of	8	23						
Morocco	3.244	3.015	1.186	1.484	2.782	1.710	4	7
Mozambique							6	12
Myanmar	499	220					10	22
Nepal	200	190	3.523	3				
Netherlands	5.922	6.585	4.966	8.566	3.043	2.769	225	829
New Caledonia	14							
New Zealand	282	160	204	82	145	67	28	70
Nicaragua	5	6					52	127
Norway	389	131	207	129	176	85		
Oman	453	292	597	693				
Pakistan	2.073	10.209	408	1.626	1.169	1.475		

TABLE 11.3 (continued)
Foreign Trade (Imports) Expressed in Weight (metric tons) and in US$K

Anise, Badian, and Fennel Imports	1999		1989		1979		1969	
	Val	Qty (mt)	Val	Qty (mt)	Val	Qty (mt)	Val	Qty (mt)
Panama	53	22	34	11				
Papua New Guinea	5	1	3	1				
Paraguay	614	248	10	7			8	16
Peru	1.946	1.573	284	229	287	168	131	289
Philippines	120	69	31	44	31	29	6	10
Poland	265	268						
Portugal	486	263	196	174	304	146	37	84
Réunion			92	94	67	45	12	28
Romania	81	107						
Russian Federation	165	214						
Saint Lucia	30	10	12	8				
Sao Tome and Principe							2	2
Saudi Arabia	5.308	7.756	4.523	6.519	2.978	3.020		
Senegal	6	3	3	1				
Seychelles	27	10	13	9	21	16		
Singapore	10.088	12.040	7.938	12.342	10.152	12.885	1.147	4.791
Slovakia	464	462						
Slovenia	1.662	872						
South Africa	2.876	3.795	1.368	1.315	899	562	138	425
Spain	2.163	1.519	832	620	1.117	587	461	626
Sri Lanka	5.885	12.941	4.613	9.556	4.836	8.605	2.543	11.834
Swaziland	4	2						
Sweden	782	428	469	356	524	359	104	280
Switzerland	2.195	871	912	759	959	813	265	859
Syrian Arab Republic	392	193						
Tanzania, United Rep of					210	151	37	105
Tunisia	930	2.725	1.844	2.321	258	133		
Turkey	2.509	2.060	3	6				
Uganda							31	119
Ukraine	145	63						
United Arab Emirates					149	78		
United Kingdom	8.836	7.484	4.235	5.084	3.764	3.591	531	1.976
United States of America	21.668	18.942	14.683	14.487	19.234	14.899	2.335	7.345
Uruguay	83	56	34	28	35	14	2	2
USSR			767	373				
Venezuela	1.616	937	586	484	2.963	1.472	194	532
Yemen	536	917			400	330	165	577
Yugoslavia SFR	14	6	171	131	255	137	7	22
Zambia	4	29						
Zimbabwe	65	33	40	10				

The data in terms of volume and in value terms relative to the foreign trade (imports) of anise (Figure 11.12 and Figure 11.13), badian (Figure 11.14 and Figure 11.15) and fennel (Figure 11.16 and Figure 11.17) by country showed that Germany remains the largest spice importer of anise, badian, and fennel in the European Union, followed by the Netherlands for import of anise, by France for import of badian, and by Greece for import of fennel.

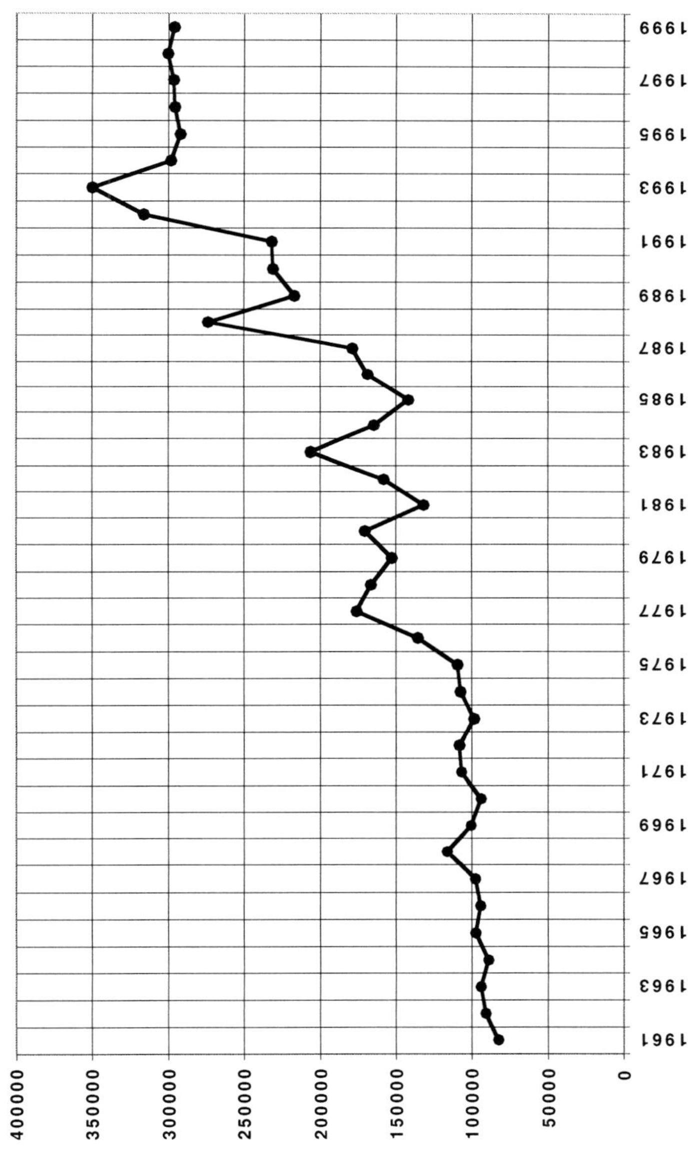

FIGURE 11.1 Anise, badian, fennel: Worldwide production trend

FIGURE 11.2 Anise, badian, fennel: Worldwide area harvested trend

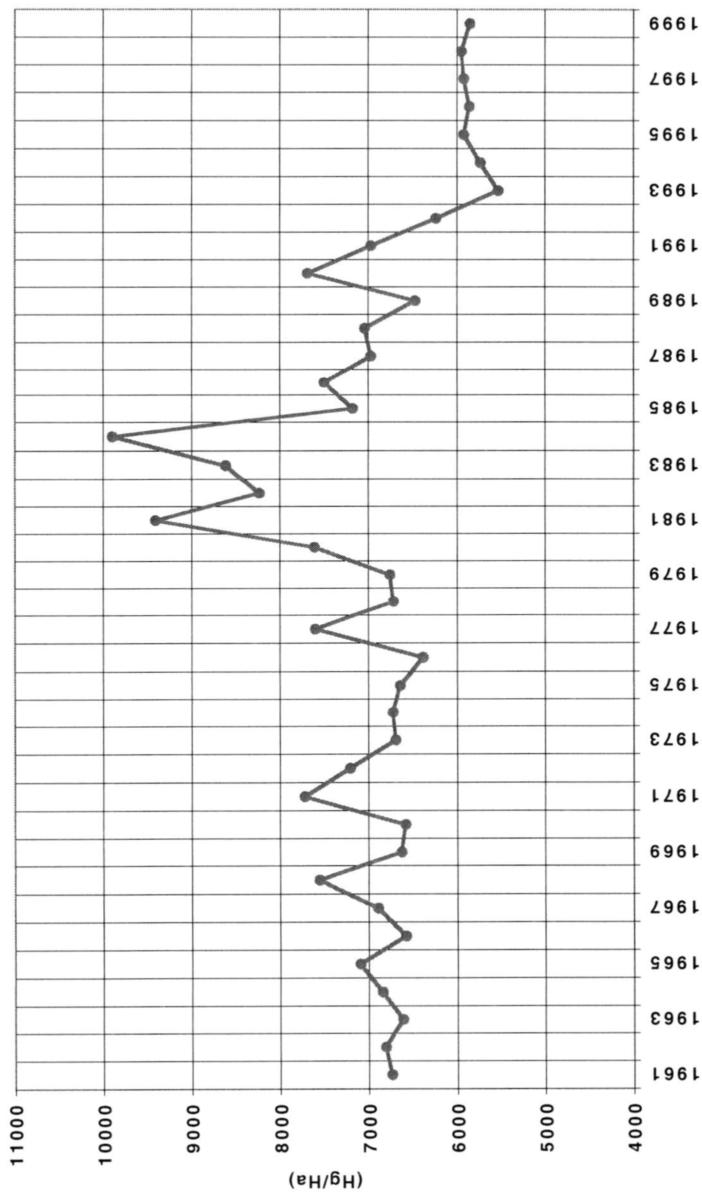

FIGURE 11.3 Anise, badian, fennel: Worldwide yield trend

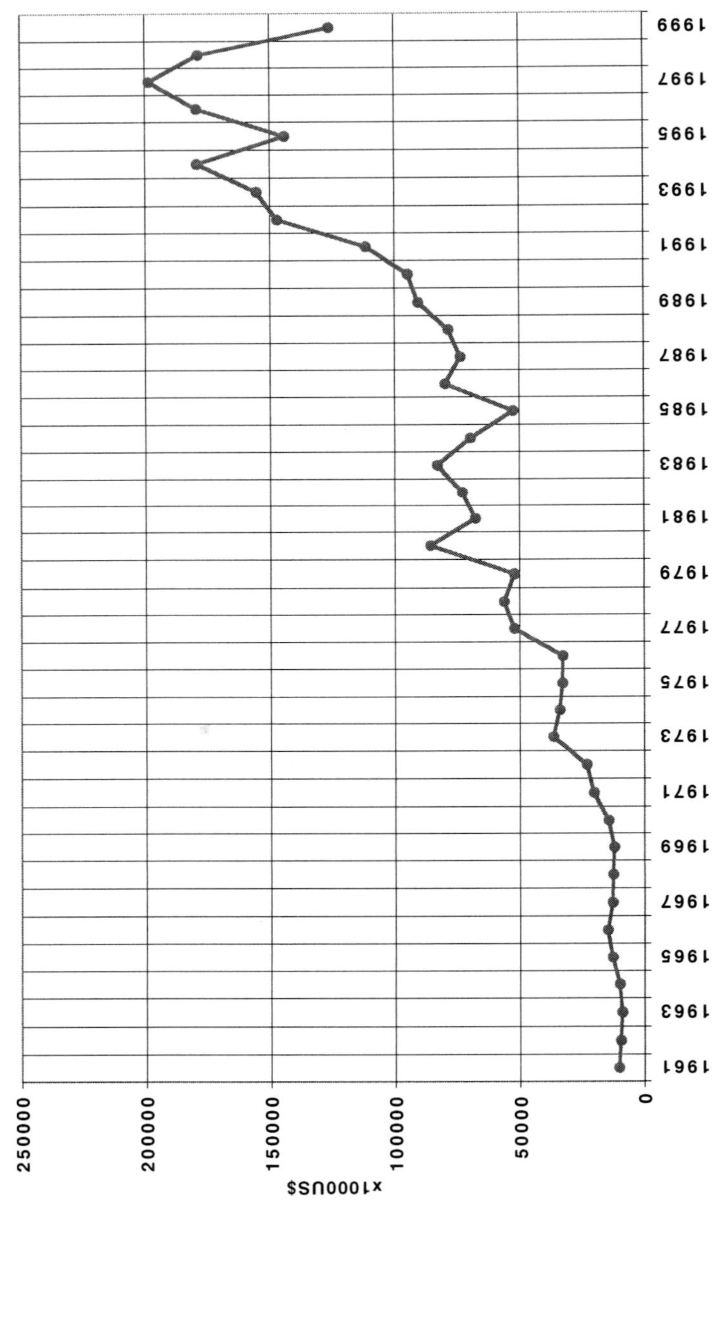

FIGURE 11.4 Anise, badian, fennel: Worldwide exports trend

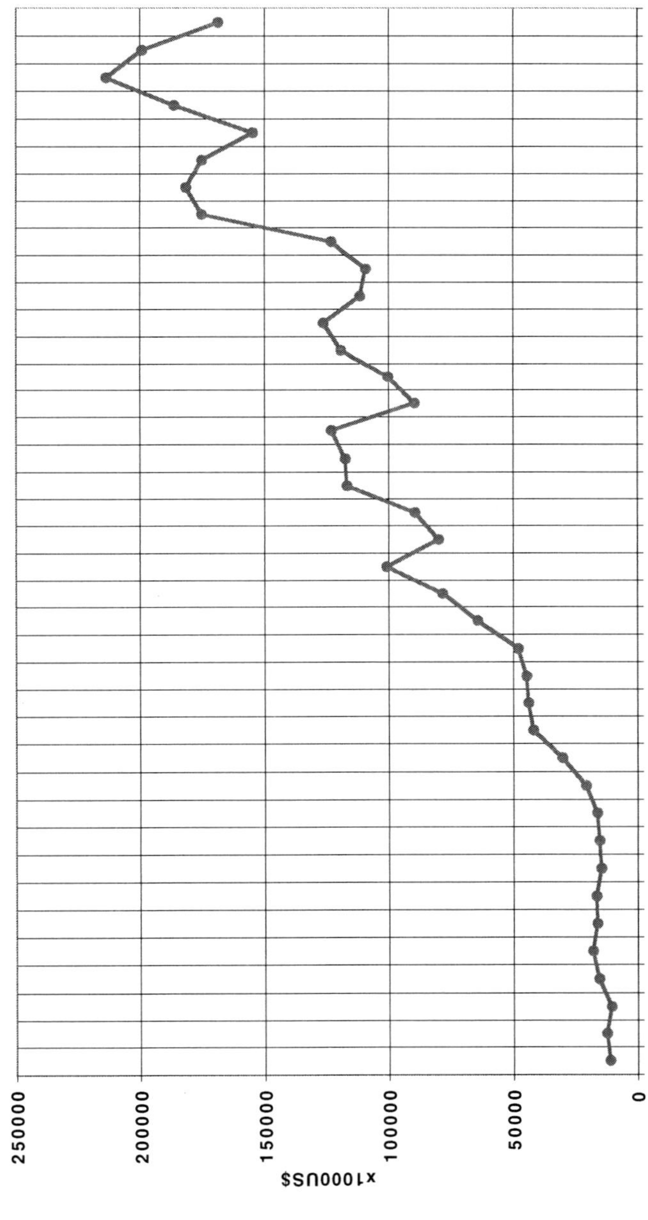

FIGURE 11.5 Anise, badian, fennel: Worldwide imports trend

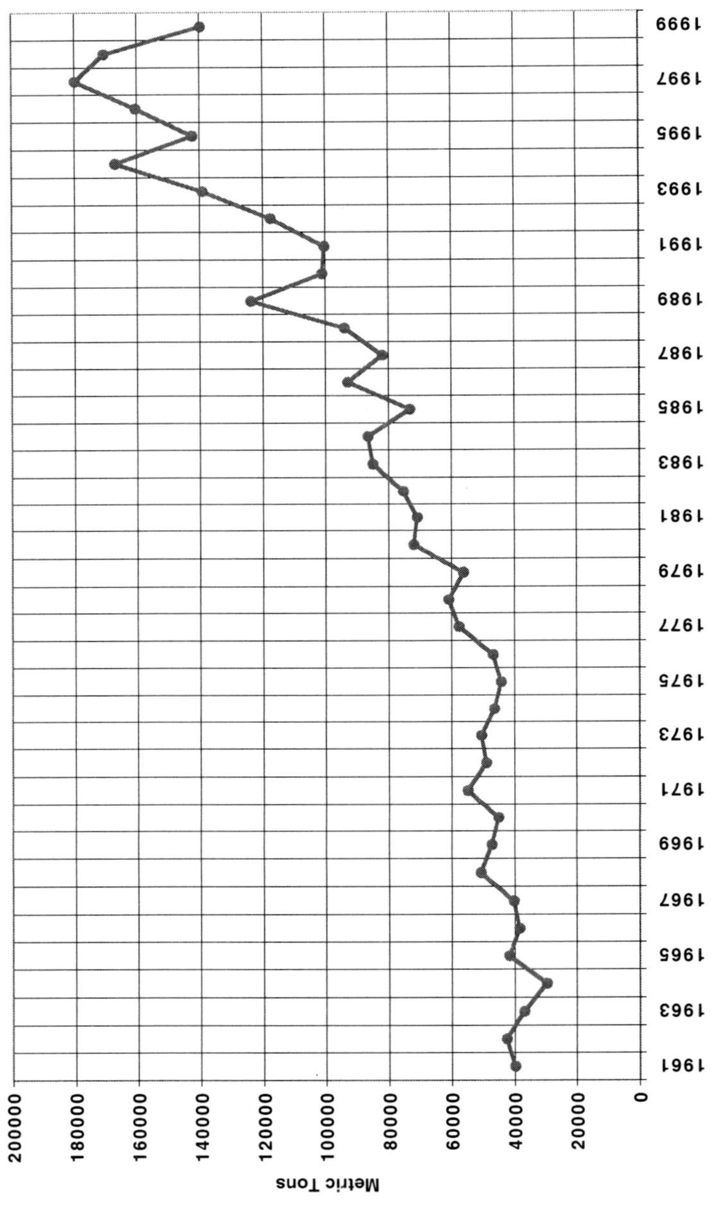

FIGURE 11.6 Anise, badian, fennel: Worldwide exports trend

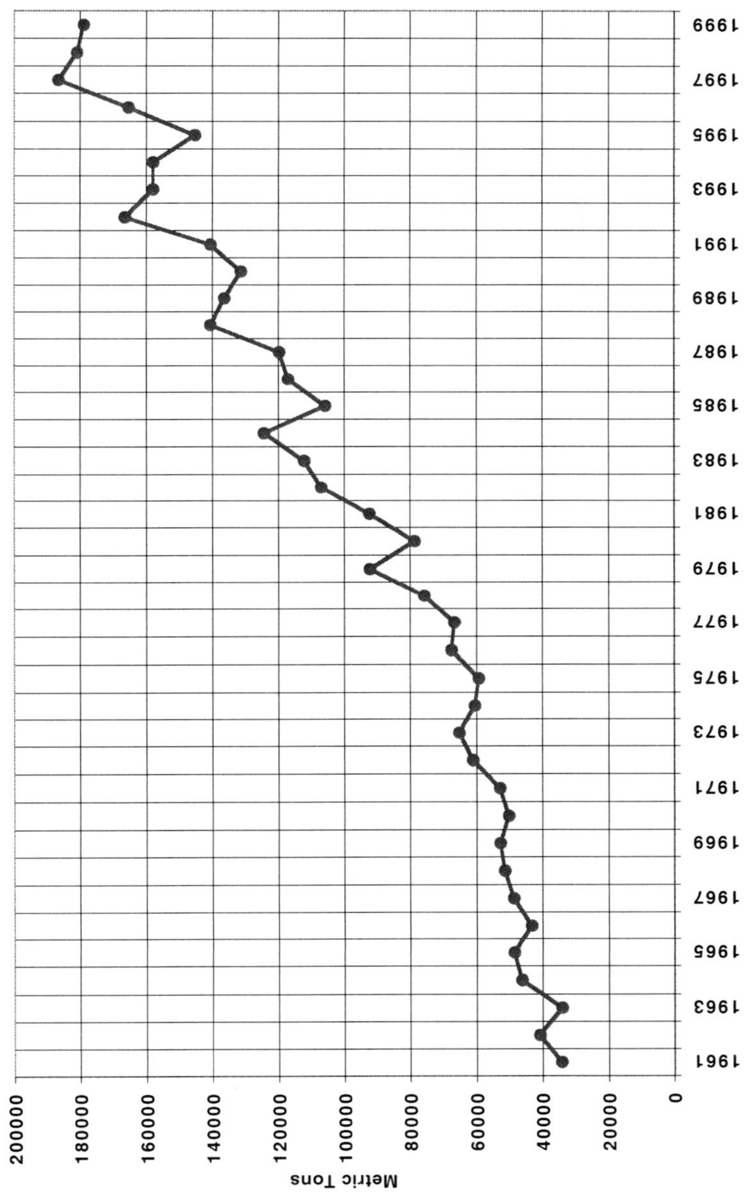

FIGURE 11.7 Anise, badian, fennel: Worldwide imports trend

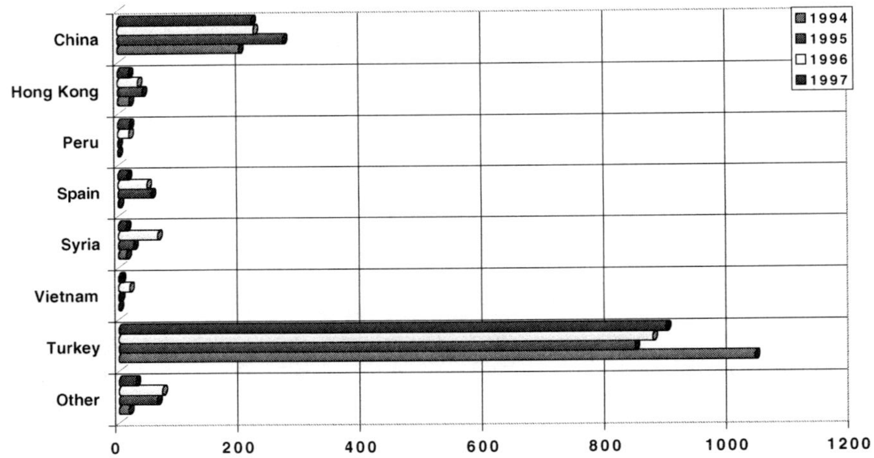

FIGURE 11.8 U.S. imports of anise or badian (metric tons), 1994–1997

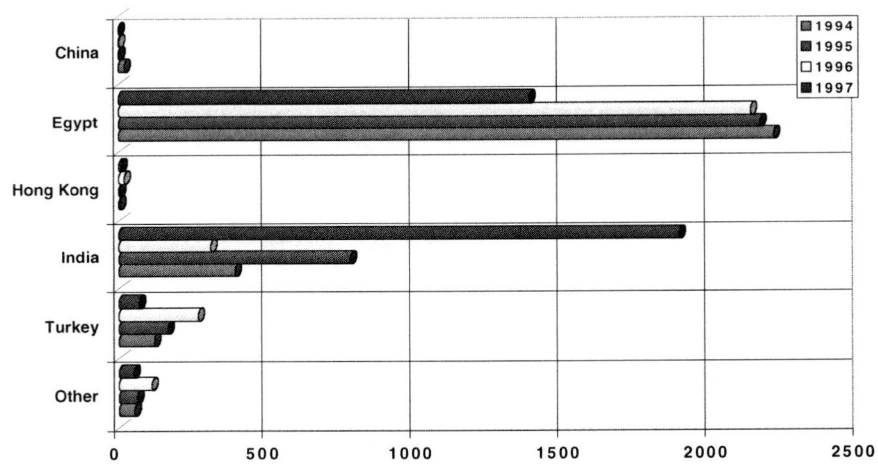

FIGURE 11.9 U.S. imports of fennel (metric tons), 1994–1997

In terms of value, the order of the five major importers remains the same as for volume. Imports of spice seeds rose without exception in 1996, with shipments of fennel and anise reaching record highs.

Finally, it needs to be pointed out that the data concerning the market of the essential oils obtained from anise, badian, and fennel are few. The most recent statistic concerns the imports in the United States of anise essential oil in the years 1996 and 1997 (Figure 11.18). These data provide, furthermore, a reliable valuation of the prices per kilogram of anise essential oil, which on the American market have been US$9.95 in 1996 and US$8.65 in 1997.

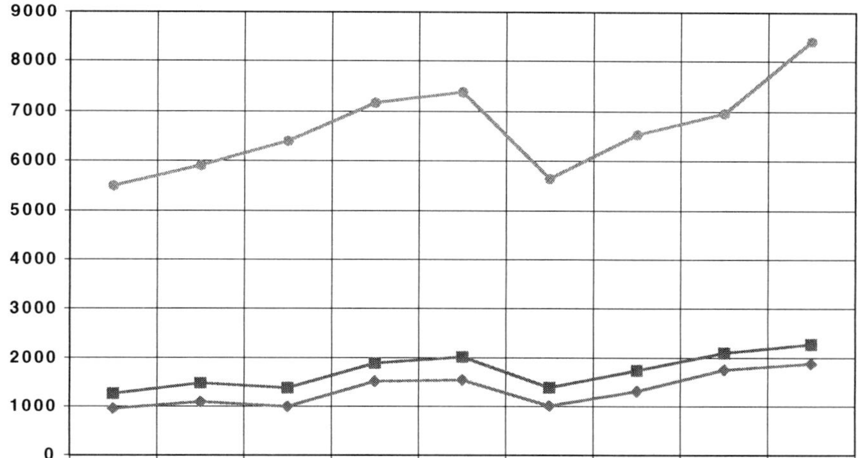

	1988	1989	1990	1991	1992	1993	1994	1995	1996
Fennel	4230	4419	5019	5280	5365	4256	4794	4862	6121
Badian	311	390	389	378	471	378	431	350	397
Anise	950	1091	988	1513	1544	1005	1309	1753	1879

FIGURE 11.10 European imports of anise, badian, and fennel, 1988–1996, volume (metric tons)

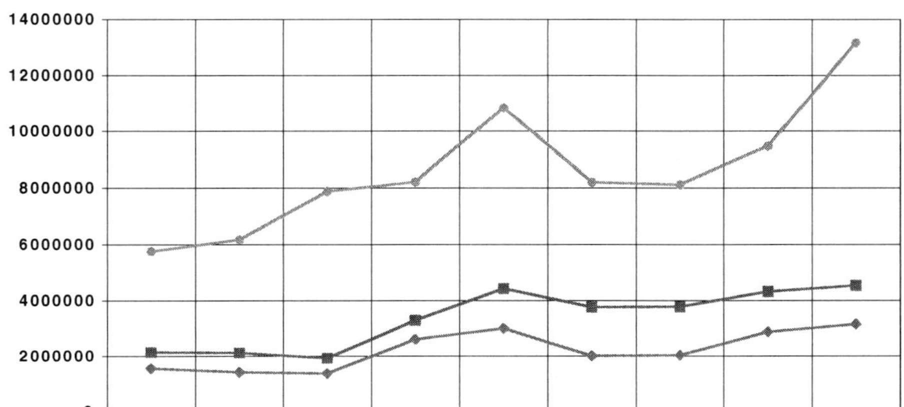

	1988	1989	1990	1991	1992	1993	1994	1995	1996
Fennel	3603000	4040000	5959000	4909000	6422000	4437000	4335000	5149000	8622000
Badian	596000	714000	563000	683000	1410000	1757000	1751000	1442000	1362000
Anise	1540000	1406000	1356000	2609000	3003000	2003000	2027000	2869000	3161000

FIGURE 11.11 European imports of anise, badian, and fennel, 1988–1996, value (US$)

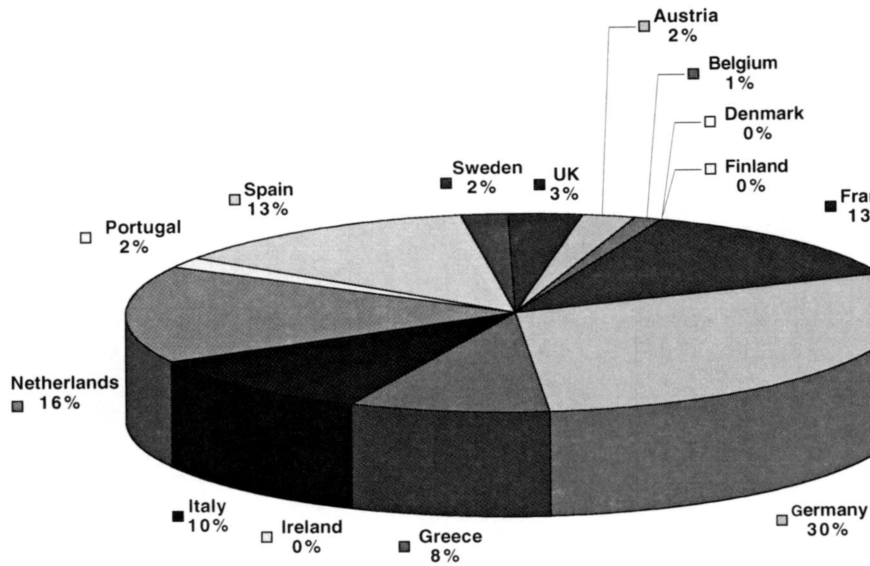

FIGURE 11.12 European anise imports, by country, 1996, volume (metric tons)

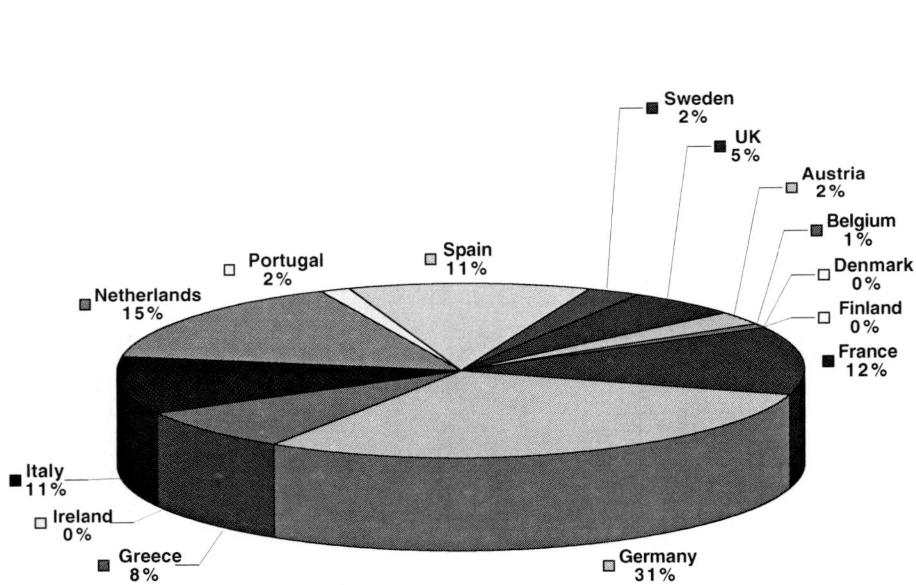

FIGURE 11.13 European anise imports, by country, 1996, value (US$)

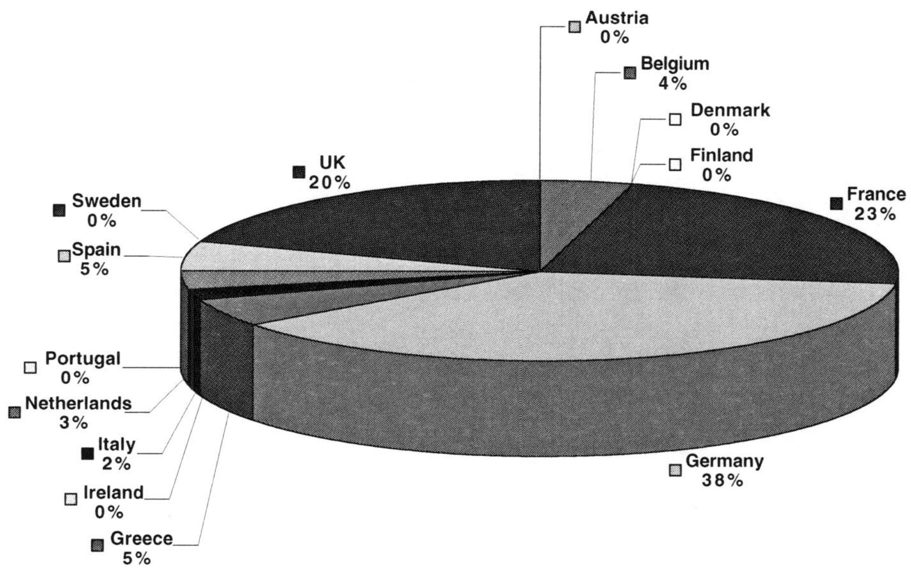

FIGURE 11.14 European badian imports, by country, 1996, volume (metric tons)

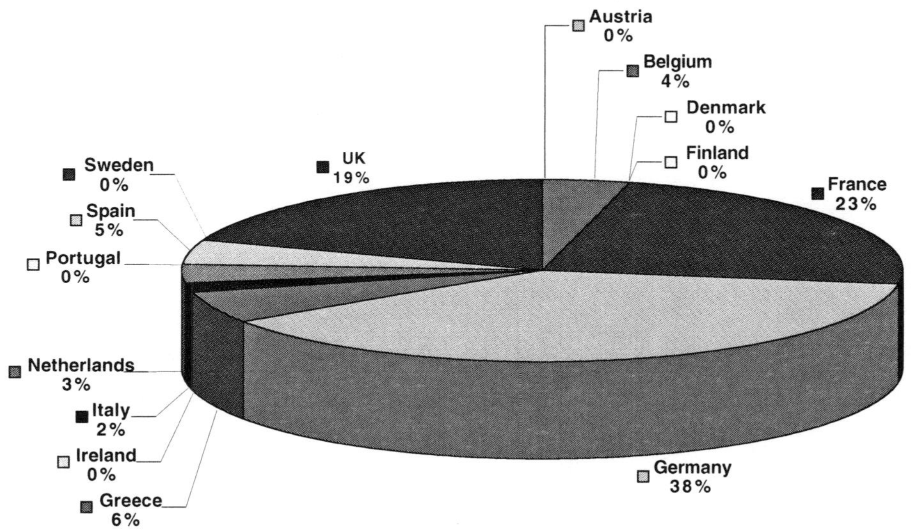

FIGURE 11.15 European badian imports, by country, 1996, value (US$)

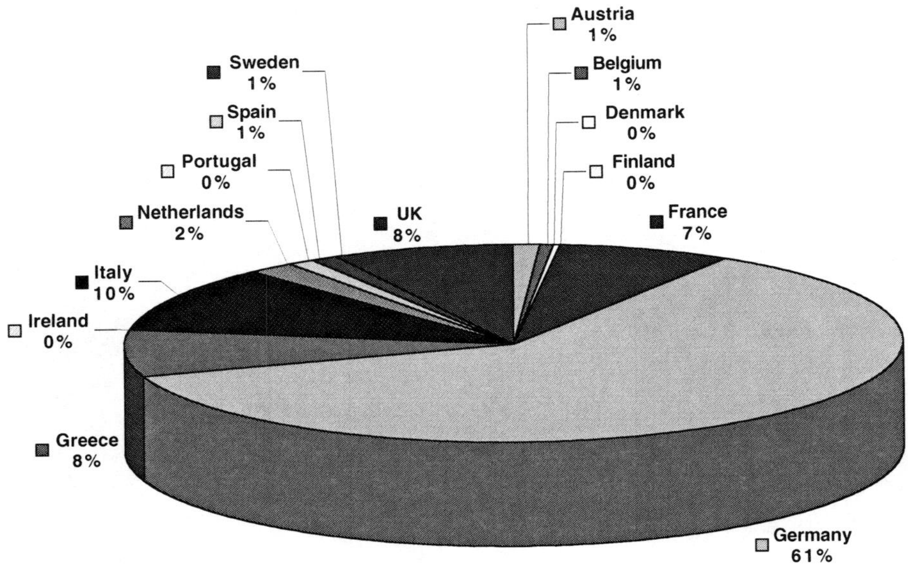

FIGURE 11.16 European fennel imports, by country, 1996, volume (metric tons)

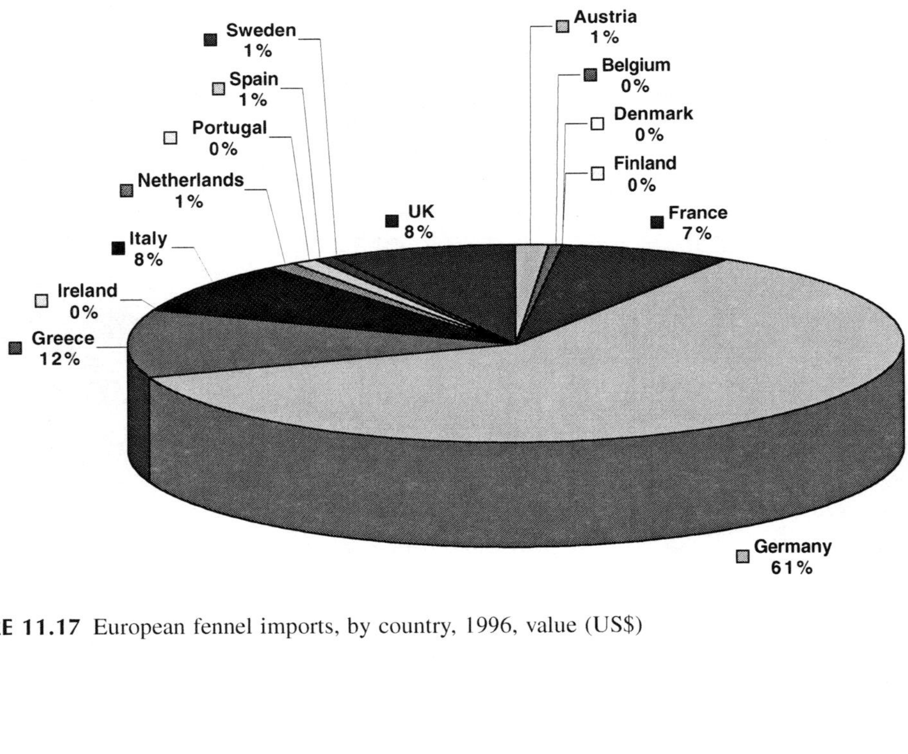

FIGURE 11.17 European fennel imports, by country, 1996, value (US$)

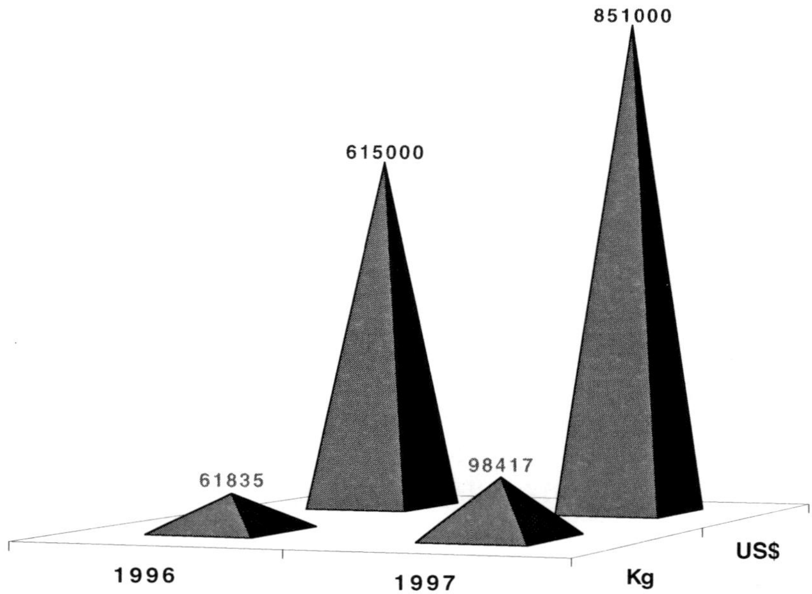

FIGURE 11.18 U.S. imports of anise essential oils, 1996–1997

References

Adkins, J. and Hecker, S., U.S. Import Market for Spices Holds Steady Publication produced by Fintrac Inc. under subcontract with Development Alternatives Inc. for the USAID-funded Asia Regional Agribusiness (RAP) Project. *Site designed by Fintrac Inc., under the auspices of the Asia Regional Agribusiness Project and through the sponsorship of the U.S. Agency for International Development and the U.S. Department of Agriculture*, 1997.

Adkins, J. and Hecker, S., An overview of Europe's increasing market for spices. *Market Asia,* January-February 1998 issue Publication produced by Fintrac Inc. under subcontract with Development Alternatives Inc. for the USAID-funded Asia Regional Agribusiness (RAP) Project. *Site designed by Fintrac Inc., under the auspices of the Asia Regional Agribusiness Project and through the sponsorship of the U.S. Agency for International Development and the U.S. Department of Agriculture.*

Bailey, L.H. and Bailey, E.Z., *Hothus Third.* MacMillan, New York, 1976.

Bianchini, F. and Corbetta, F., *Health Plants of the World. Atlas of Medicinal Plants.* Newsweek Books, 1977.

British Herbal Pharmacopoeia, British Herbal Medicine Association, Keighley, 1983.

Claus, E.P., and Tyler, V., *Pharmacognosy*, 5th ed., Lea & Febiger, Philadelphia, 1965.

Council of Europe, *Flavouring Substances and Natural Sources of Flavourings,* 3rd ed., Maisonneuve, Strasbourg, 1981.

Ghazanfar, S.A., *Handbook of Arabian Medicinal Plants*, CRC Press, Boca Raton, FL, 1994, p. 209.

Hornok, L., *Cultivation and Processing of Medicinal Plants*, Wiley, New York, 1992, p. 143–147.

Ilyas, M., Spices in India: III, *Econ. Bot.,* 34(3), 236–259, 1980.

Leung, A.Y., *Encyclopedia of Common Natural Ingredients Used in Food, Drugs and Cosmetics,* Wiley, New York, 1980.

Leung, A.Y. and Foster, S., *Enciclopedia delle Piante Medicinali Utilizzate Negli Alimenti, nei Farmaci e nei Cosmetici*, Planta medica, Aporie, Rome, 1999, p. 48–50.

Mitchell, J. and Rock, A., *Botanical Dermatology*, Greengrass, Vancouver, 1979.

Morton, J.F., *Atlas of Medicinal Plants of Middle America*, C. Thomas Publisher, Springfield, IL, 1981, p. 650–651.

Newall, C.A., Anderson, L.A., and Philipson, J.D., *Herbal Medicines. A Guide for Health-Care Professionals*, Pharmaceutical Press, London, 1996, p. 30–31.

Small, E., Confusion of common names for toxic and edible "star anise" (*Illicium*) species, *Econ. Bot.*, 50, 337–339, 1996.

Smith, A.C., The families *Illiciaceae* and *Schisandraceae*, *Sargentia,* 7, 1–224, 1947.

Zeven, A.C. and de Wet, J.M.J., *Dictionary of Cultivated Plants and Their Regions of Diversity*, Pudoc, Wageningen, 1982.

12 The Genus *Foeniculum*

Amit Krishna De

CONTENTS

12.1 INTRODUCTION

Fennel, an ancient seasoning herb, was well known to the ancient Egyptians, Romans, Indians, and Chinese. It is now cultivated in Mediterranean countries, in Romania, and in India. The use of fennel goes back to ancient India, when it was used as a condiment and culinary spice. In Greece, it was a symbol of success. In Rome, the young fennel shoots were used as food. Emperor Charlemagne is known to have encouraged its cultivation in Central Europe. Even today, it is indispensable in modern French and Italian cooking. India exports a substantial quantity of fennel to the United States, Singapore, the United Kingdom, the United Arab Emirates, Sri Lanka, Malaysia, Saudi Arabia, and Japan in a variety of forms including seeds, powder, and volatile oil. Throughout the world, fennel is called by different names in different languages (Table 12.1).

12.2 PLANT SOURCES

Fennel (*Foeniculim vulgare* Mil), a biennial or perennial aromatic stout herb, belongs to the family Umbelliferae (Figure 12.1). The dried, ripe fruit (seed) of cultivated varieties are small, oblong, ellipsoidal or cylindrical, straight or slightly curved, and greenish yellow or yellowish brown in colour. The plant is a biennial or perennial aromatic, stout, glabrous herb and grows to a height of 1.5 to 1.8 m.

Varieties and races differ in the size, odor, and taste of the fruits exist among wild and cultivated fennels, but they are hardly distinguishable from one another; they are regarded as races, varieties, or subspecies of *F. vulgare.* The varieties that yield commercially important volatile oils are referred generally to the subspecies capillaceum and placed under two distinct varieties: var. *vulgare* (Mill) and var. *dulce (*Mill.) Thellung (cultivated, yielding sweet, or Roman or Florence fennel oil). Var. *vulgare* is cultivated chiefly in Russia, Hungary, Germany, France, Italy, India, Japan, Argentina, and the United States. The cultivation of var. *dulce* is confined to France, Italy, and Macedonia in southern Europe. A different variety of fennel (var. *panmorium*) is also available in India.

12.3 CULTIVATION AND PLANT RAW MATERIAL

Fennel is a cold-weather crop and grows well at altitudes of 2000 m. It can also be grown as a summer-season crop in temperate regions. Dry and cold weather favors increased seed production.

0-415-32246-4/04/$0.00+$1.50

TABLE 12.1
Names of Fennel in Different Languages

Language	Name
Latin	*Foeniculum vulgare* Mill
English	Fennel
Spanish	Hinojo
French	Fenonil
German	Fenchel
Swedish	Fankal
Arabic	Shamar
Dutch	Venkel

FIGURE 12.1 *Foeniculum vulgare*

The seeds may be sown in any good soil in shallow drills 45 cm apart. Fennel thrives well on well-drained loamy to clay soils, which are rich in minerals and lime. However, heavy soils are more desirable than light soil for higher yield. When the plants are 7.5 to 10 cm high, they are thinned out to 30 cm apart. Other than occasional weeding and irrigation once a week, no further culture is required. In the plains sowing is done from October to November, and in the hills it is done from March to May. Fennel grows well in the mild climates of India and thrives in the sunny, slimy, well-drained loams of Western India (Gujrat, Rajasthan, and Uttar Pradesh).

Fennel requires frequent watering, with the frequency dependent on soil type and prevailing water conditions. Because all the fruits do not mature together, harvesting of the umbels has to be done four to five times at a 10- to 15-day interval. The umbels are dried, separated, and cleaned by winnowing. On average, a grain yield of 500 to 900 kg per hectare can be obtained. Improved varieties can yield up to 1500 kg per hectare.

Results from research undertaken since 1988 to underpin Italian fennel-seed production has been reported on (Landi, 1996). Considerable variation in plant size, vegetative cycle, and yields were observed because of variety differences and environmental variability. Late-autumn sowing in northern and central Italy can lead to crop failure caused by frost. Lower sowing rates and narrower row spacing have proved advantageous. As regards essential oil production, biomass

TABLE 12.2
Nutritive Value of Fennel Seeds (per 100 g)

Moisture	6.3
Protein	9.5
Fat	10
Minerals	13.4
Fiber	18.5
Carbohydrates	42.3
Calcium	1.3
Phosphorous	0.48
Iron	0.01
Sodium	0.09
Potassium	1.7
Vitamin A	1040 U
Vitamin B_1	0.41
Vitamin B_2	0.36
Niacin	6
Vitamin C	12
Volatile oil	1 to 1.5%

should be distilled at the wax ripe stage. Mechanization of harvesting has proved problematic, as there is variation in crop ripening, leaving the yields generally low and of poor quality.

An embryogenic suspension cell line of fennel, directly induced from hypocotyls, was developed (Umetsu et al., 1995). The suspension cells were cold-preserved at 4°C for up to 12 weeks and then cultured at 25° for 2 weeks. The packed cell volume of the cultured cells decreased as the cold-preservation period became longer. The number of cells with a size of 32 to 82 μm and having embryogenic potential diminished during the cold-preservation period. Cells that were cold-preserved for 2 to 6 weeks were capable of forming normal somatic embryos that were identical to those obtained from the cells without cold preservation. The resulting somatic embryos regenerated to develop into normal plantlets when cultured in hormone-free MS medium under illumination. Esterase isoenzyme analysis on polyacrylamide gel electrophoresis revealed the appearance of bands in the embryos treated with 4- and 6-week cold-preservation that differed from those in the embryos from the cells not treated with cold-preservation. Anethole was detected in the methanol extracts of embryos induced from cold-preserved cells at the same concentration as in extracts of embryos induced from cells without cold-preservation (2 to 7 mg/kg DW).

12.4 CHEMICAL CONSTITUENTS

Fennel seeds are used as spices and condiments. The nutritive analysis of fennel is given in Table 12.2 (Pruthi, 1979). Its caloric value is 370.

Flowers (of a local cultivar) were tagged at anthesis, and seeds were collected at weekly intervals after 40 days from anthesis for analysis of their biochemical constituents (Gupta et al., 1995). Moisture and protein contents decreased significantly with the advancement of seed development. Oil content ranged from 68 to 135.7 g/kg (DW basis), increasing with seed development. The oil content was greatest in mature seeds. The presence of phytate (11.35 to 13.10 mg/g), which affects the availability of Zn and Fe, was also observed. Fennel seeds were found to be a rich source of micro- and macro-elements. Neutral detergent fiber, acid detergent fiber, cellulose, and lignin contents increased significantly with the advancement of seed development. Oleic and linoleic fatty acids were found to be major fatty acids, and variable proportions of different fatty acids were observed.

The composition of the oil varies widely according to variety and origin. The volatile oil in fennel seeds ranged from 1.0 to 1.5%. Eastern European seeds have been reported to contain 4 to 6% oil. Fennel contains substantial amounts of fenchone and anethole. It also has a sweet taste. The fatty acids of the oil are palmitic acid and petroselinic acid, d-alpha pinene, camphene, d-alpha phellandrene, dipentene, d-fenchone, methyl chavicol, and foeniculinarin also present as seed oils. Water extracts of the seed show the presence of a thermostable growth-retarding substance. The diuretic properties of fennel are attributed to the presence of glycolic acid.

Studies of different samples of dry, ripe fruits of *F. vulgare* of different origins were performed in Tuscany, Italy. The essential oils isolated by hydrodistillation were analyzed by GC-MS; the 16 main constituents of each sample were identified, with *trans*-anethole, estragole, limonene, and fenchone being the most abundant. The amounts of *trans*-anethole and estragole were found to be inversely proportional (Miraldi, 1999). Twelve constituents have been identified that were isolated for the first time from fennel, including piperitenone and piperitenone oxide. The main constituents of the essential oils were limonene (52.4% in the Cannes essential oil and 56.9% in the Meve essential oil) and piperitenone oxide (22.1% in the Cannes essential oil and 14.2% in the Meve essential oil) (Badoc et al., 1994).

The morphological and chemical characteristics of 13 *F. vulgare* var. *vulgare* populations of different origins (including Hungarian, Italian, French, Korean, and Belgian cultivars) were compared over 2 years by J. Bernath and his group (1996). High stability was shown for morphological aspects such as relative leaf mass, seed size, and 1000-seed mass. The presence of (E)-anethole and methylchavicol in essential oils were stable characteristics. Three distinct chemotypes were observed based on the cluster analysis of the seed essential oil: fenchone-rich (31 to 42% fenchone), methylchavicol-rich (30 to 43% methylchavicol), and (E)-anethole-rich (60 to 85% anethole) chemotypes. Within the anethole group, a further four chemical subvariants of a lower rank (chemoforms) were observed based on discriminant analysis. The researchers concluded that morphological characters could not be used to support any intraspecific chemical classification.

Essential oil accumulation in developing fruits of *F. vulgare* sub sp. *vulgare* var. grown in Hungary during 1996 was studied (Bernath et al., 1999). Nine different stages of development of generative organs were distinguished for analysis. Histological investigations indicated that oil ducts (vittae) were present in the early stages of development of generative organs, even at the time of bud formation; their relative area (calculated to the cross section) was highest at the time of seed set (53.6%). Essential oil yield was highest (11.6%) during fruit development, when the relative ratio of oil ducts was high and the accumulation of assimilates (development of endosperm) had not started yet. Yield calculated from one schizocarp increased up to the final ripening stages (0.787 μL/fruit). Analysis of essential oil composition indicated that the accumulation ratio of myrcene continuously decreased throughout development, whereas the fenchone content showed an opposite trend: its ratio was 9.1% in buds, and it increased with development to 23.2%.

Some scientists have isolated a new chromanone glycoside and a new phenylethanoid glycoside from the water-soluble portion of the methanol extract of *F. vulgare*. An optical isomeric mixture of threoepoxyanethole was obtained from the ether soluble portion, and it was considered to be an auto-oxidation product of *trans*-anethole (Kitajima et al., 1999). In a study carried out in 1991 at Senigallia, Ancona, Italy, three varieties of *F. vulgare* subsp. capillaceum, namely sweet (var. dulce), bitter (var. vulgare), and Florence (var. azoricum), were assessed for agronomic and chemical parameters at full bloom, waxy seed, and seed-ripening stages (Marotti et al., 1992). Seed (fruit) yield and essential oil concentrations were highest in sweet fennel when the plants were harvested at the waxy seed stage. Florence and bitter fennel produced the highest biomass at full bloom. The main constituent of the essential oil was *trans*-anethole, which ranged from 81 to 88% in the oils from the seeds and from 49 to 75% in the oils from whole plants. The highest oil yields were obtained by harvesting at the waxy seed stage for all three varieties, and for sweet fennel in particular (62.0 kg/ha). Sowing at a rate of 5 kg/ha gave increased oil yields compared with a sowing rate of 10 kg/ha (22 and 16.8 kg/ha, respectively), but the 1000-seed weight was greater with the higher rate.

12.5 DISEASES AND PESTS

Fennel crop is attacked by a number of fungi, bacteria, and viruses, but powdery mildew, blight, and aphids are the most important pests (Singh and Singh, 1996). The crop is commonly attacked by powdery mildew (*Erysiphe polygoni*) at flowering stage in cloudy weather during February to March. This can be controlled by dusting sulfur at the rate of 15 to 20 kg per hectare or by spraying karathane 0.1% at 15 to 20-day intervals.

In contrast, blight (*Ramularia foeniculi*) appears in the month of January on lower and older leaves as minute, angular, brown, necrotic areas that later become large and are covered with greyish-white crumpet growth. In the later stages of blight, linear and rectangular spots cover the entire stem, peduncles, and fruits. Severely affected leaves shrivel and dry up. This process can be halted by spraying 0.2% difolatan or dithane M 45.

Infected soil is the most common source of inoculum. Earthing encourages contact of bacterial soft rot (*Erwinia carotovora*) with growth cracks, which occur naturally. Use of healthy seed material and removal of diseased plant debris is recommended. Crop rotation with nonhost plants reduces the inoculum potential in the soil.

Some other diseases reported in India include Umbel Blight, a fungal disease caused by *Colletatrichum capsici*. The disease is characterized by water-soaked lesions at the tips of the umbel, which later turns brownish black. Phyllod, a new virus disease, is characterized by stunting plants and by the malformation of leaves and chlorosis in the initial stages. In the later stages, umbelets transform into green, leaflike structures followed by abundant vegetative growth with shortened internodes and profuse branching from the axils, giving a bushy appearance to the plants.

Aphids attack at the flowering stage and can be checked by spraying endosulfan, dimethoate, monocrotophas and phosphamidon. The attack by *Thrips flavus* can be checked by any potent insecticide.

In 1993, commercial fields of fennel (*F. vulgare*) in the Salinas Valley, California, showed areas of declining plants. *Sclerotinia minor* was isolated from diseased fennel, and its pathogenicity was confirmed. This was the first report of *S. minor* causing a disease of fennel (Koike, 1994).

The effects of different concentrations of four fungicides, Bavistin (carbendazim), Dithane M-45 (mancozeb), Agrosan G N (phenylmercury acetate), and PCNB (quintozene), and of two antibiotics, griseofulvin and mycostatin, on the seed mycoflora and seed germination of *F. vulgare* were studied *in vitro* (Bharati et al., 1997). All the fungicides and antibiotics were found to be effective in reducing the seed mycoflora both qualitatively and quantitatively. The percentage of germination for all the fungicide-treated seeds was higher than that for the untreated ones except in the case of Agrosan G N, whereas both the antibiotics exhibited a phytotoxic effect on seed germination.

Callus induction and morphogenic response of several fennel populations from European companies were determined by genotype and hormonal treatment. Some 100% callus formation occurred only in the Francia Pernod population under the action of 2,4-D or NAA and kinetin. Only this latter hormonal treatment induced shoot regeneration. Plant regeneration was observed, especially in the Francia Pernod population. Calluses grown in the presence of 2,4-D or NAA plus kinetin showed considerable differences at cytological and histological levels, which were correlated to their different morphogenic capability (Anzidei et al., 1996).

12.6 THERAPEUTIC AND PHARMACOLOGIC PROPERTIES

Fennel is one of the most common culinary spices of the world. It is also a well-recognized medicinal plant and has been entered into the pharmacopoeia or the official list of the medicinal plants in all countries because of its volatile oil, which is a stimulant, an aromatic, and a carminative. Fennel is mostly used as a flavoring agent in food. Young leaves of the tree are also used raw or cooked with curry and are sometimes used in salad or as a vegetable. The seeds are generally used for flavoring and are distilled for the oil. The fruit is commonly used as a masticatory. Fennel is widely used for flavoring soups, sauces, pastries, confectionaries, bread rolls, liquors, meat dishes, and seasoning of prickles. The seed and the oil have a variety of medicinal properties.

The seeds are sweet, laxative, and aphrodisiac, and they arrest bleeding. They also relieve flatulence and promote the removal of catarrhal matter and phlegm from the bronchial tubes. Oil of fennel, distilled from the dry seeds, is aromatic, carminative, and antispasmodic. It is used in various carminative preparations.

The leaves of fennel are digestive, appetizing, and stimulating. They increase the secretion and discharge of urine. The aqueous ethanol extracts of *F. vulgare* var. *dulce* roots were tested for their diuretic activity in rats (Beaux, 1997). Pharmacologic evaluation revealed that the extracts increased urine flow and urinary sodium excretion. Similar diuretic effects were observed with doses of 50, 100, and 200 mg extract per kilogram.

The fennel is well known as a digestive aid. It may be given in small quantities to help young children digest carbohydrates. An infusion prepared by boiling a tablespoon of fennel seeds in 100 mL of water for half an hour is highly beneficial in indigestion, biliousness, flatulence, constipation, and atonic dyspepsia. Chewing of fennel seeds after meals prevents foul breath, indigestion, constipation, and vomiting.

Fennel is one of the safest herbs for colic, for helping the baby to release gas, and to relieve the stomach. Fennel may be used in combination with other herbs like peppermint and crushed caraway seeds or alone. A teaspoon of the herbs is boiled in a cup of water and allowed to steep in the water for about 20 minutes. This mixture is then strained and allowed to cool. This tea, given to the baby in his feed bottle, helps cure colic. Not more than a teaspoon or two should be given at a time.

Leaves of fennel are useful in respiratory disorders like asthma and bronchitis. The juice may be given in the treatment of such conditions. Eating fennel seeds with figs is also a good medicine for cough, bronchitis, and lung abscesses. Fennel seeds promote menstruation and regulate monthly periods. An infusion of the seeds can be given in painful menstruation and other menstrual irregularities. It is also believed that fennel benefits the eyes. Herbalists today recommend bathing for thickened, sore, or inflamed eyes with fennel tea. Regular application of the leaf-juice boiled with honey is said to cure conjunctivitis. The residue left after distillation of essential oil from the fruit is used as a rich cattle feed, as it contains 14 to 22% protein and 12 to 18.5% fat.

The pharmacological activity of an ethanol extract (500 mg/kg) of dried, ripe fruits of *F. vulgare* was examined in rats and mice. In rats, the extract showed diuretic activity and enhanced bile secretion. In mice, the extract exhibited analgesic, antipyretic, and timodepressive activities. The extract inhibited the growth of *Staphylococcus aureus* and *Bacillus subtilis*. The acute toxicity of the extract (0.5, 1, and 3 g/kg, p.o.) was investigated in mice; no deaths occurred, but reduced locomotor activity and piloerection were observed at the highest dose (Tanira et al., 1996).

It has also been reported that the LD_{50} values of the volatile oils of leaves *F. vulgare* was found to be 1075 mg/kg (intraperitoneally) in mice (Essway et al., 1995). The oil decreased the blood glucose concentration of normal rats at 1 hour. The high blood glucose concentrations of alloxan-induced diabetic rats were significantly decreased at 1 and 3 hours after injection of the volatile oil.

12.7 RECENT FINDINGS

F. vulgare has been found to be effective as an estrogenic agent in the treatment of idiopathic hirsutism, which is defined as the occurrence of excessive male pattern hair growth in women who have a normal ovulatory menstrual cycle and normal levels of serum androgens. Recently, the clinical response of idiopathic hirsutism to topical fennel extract was studied. The mean values of hair diameter reduction were found in patients receiving the creams containing 2% fennel (Javidnia et al., 2003).

Fennel seed oil has been shown to reduce intestinal spasms and increase motility of the small intestine, thus reducing infantile colic with no side effects. Dicyclomine hydrochloride, the only pharmacological treatment for infantile colic, has been found to cause serious side effects, including death (Alexandrovich et al., 2003).

A phenylpropanoid derivative, dillapional, was found to be an antimicrobial principle of the stems of *F. vulgare.* A coumarin derivative, scopoletin, was also isolated as a marginally antimicrobial agent along with inactive compounds from this plant (dillapiol, bergapten, imperatorin and psolaren) (Kwon et al., 2002).

References

Alexandrovich, I., Rakovitskaya, O., Kolmo, E., Sidorova, T., and Shushunov, S., The effect of fennel (*Foeniculum vulgare*) seed oil emulsion in infantile colic: A randomized, placebo-controlled study, *Altern. Ther. Health Med.,* 9(4), 58–61, 2003.

Anzidei, M., Vivona, L., Schiff, S., and Bennici, A, *In vitro* culture of *Foeniculum vulgare*: callus characteristics in relation to morphogenesis, *Plant Cell Tissue Organ Cult.*, 45(3), 263–268, 1996.

Badoc, A., Deffieux, G., Lamarti, A., Bourgeois, G., and Carde, J.P., Essential oil of *Foeniculum vulgare* Mill. (fennel) subsp. piperitum (Ucria) Cout. fruit, *J. Essential Oil Res.*, 6(4), 333–336, 1994.

Beaux, D., Fleurentin, J., and Mortier, F., Diuretic action of hydroalcohol extracts of *Foeniculum vulgare* var. dulce (D.C.) roots in rats, *Phytother. Res.*, 11(4), 320–322, 1997.

Bernath, J., Nemeth, E., Petheo, F., Mihalik, E., Kalman, K., and Franke, R., Regularities of the essential oil accumulation in developing fruits of fennel (*Foeniculum vulgare* Mill.) and its histological background, *J. Essential Oil Res.*, 11(4), 431–438, 1999.

Bernath, J., Nemeth, E., Kattaa, A., and Hethelyi, E., Morphological and chemical evaluation of fennel (*Foeniculum vulgare* Mill.) populations of different origin, *J. Essential Oil Res.*, 8(3), 247–253, 1996.

Bharati, Rai, Jariwala, S., and Kanak, Manjari, Efficacy of fungicides and antibiotics on seed mycoflora and seed germination of some spices, *Indian Phytopathol.*, 50(2), 261–265, 1997.

Gupta, Kaushalya, Thakral, K.K., Gupta, V.K., and Arora, S.K., Metabolic changes of biochemical constituents in developing fennel seeds (*Foeniculum vulgare*), *J. Sci. Food Agric.*, 68(1), 73–76, 1995.

Essway, G.S., Sobbhy, H.M., and El-Banna, H.A., The hypoglycemic effect of volatile oil of some Egyptian plants, *Vet. Med. J. Giza*, 43(2), 167–172, 1995.

Javidnia, K., Dastgheib, L., Mohammadi Samani, S., and Nasiri, A., Antihirsutism activity of Fennel (fruits of *Foeniculum vulgare*) extract. A double-blind placebo controlled study, *Phytomedicine,* 10(6–7), 455–458, 2003.

Kitajima, J., Ishikawa, T., Tanawa, Y., and Ida, Y., Constituents of fennel X. New chromanone and phenylethanoid glycosides and theo-epoxyanethole, *Chem. Pharmaceut. Bull.*, 47(10), 1448–1450, 1999.

Koike, S.T., First report of stem rot of fennel in the United States caused by *Sclerotinia minor*, *Plant Dis.* 78(7), 754, 1994.

Kwon, Y.S., Choi, W.G., Kim, W.J., Kim, W.K., Kim, M.J., Kang, W.H., and Kim, C.M., Antimicrobial constituents of *Foeniculum vulgare*, *Arch. Pharm. Res.,* 25(2), 154–157, 2002.

Landi, R., Cultural problems with seed fennel (*Foeniculum vulgare* Mill.) in Italy and possibilities for cultural diffusion. Atti convegno internazionale: Coltivazione e miglioramento di piante officinali, Trento, Italy, 2–3 *Giugno* 1994, 165–188, 1996.

Marotti, M., Dellacecca, V., Piccaglia, R., and Giovanelli, E., Agronomic and chemical evaluation of three "varieties" of *Foeniculum vulgare* Mill, in First World Congress on Medicinal and Aromatic Plants for Human Welfare (WOCMAP), Maastricht, Netherlands, 19–25 July 1992, Palevitch, D., Simon, J.E., and Mathe, A., Eds., *Acta Horticult.* 331, 63–69, 1993.

Miraldi, E., Comparison of the essential oils from ten *Foeniculum vulgare* Miller samples of fruits of different origin, *Flavour Fragrance J.*, 14(6), 379–382, 1999.

Pruthi, J.S., Spices and condiments, National Book Trust, New Delhi, 1979, 1999.

Singh, V.B. and Singh, K., *Spices*, New Age International, Delhi, 1996, p. 123–130.

Tanira, M.O.M., Shah, A.H., Mohsin, A., Ageel, A.M., and Qureshi, S. Pharmacological and toxicological investigations on *Foeniculum vulgare* dried fruit extract in experimental animals, *Phytother. Res.*, 10(1), 33–36, 1996.

Umetsu, H., Wake, H., Saitoh, M., Yamaguchi, H., and Shimomura, K., Characteristics of cold-preserved embryogenic suspension cells in fennel, *Foeniculum vulgare* Miller, *J. Plant Physiol.*, 146(3), 337–342, 1995.

Index

D

E

F

L

M

N

O

P

Q

R

S

T

U

V

W

Y